Medical Intelligence Unit

Glycoproteins and Human Disease

Inka Brockhausen, Ph.D.
William Kuhns, M.S., M.D.
Department of Biochemistry
Research Institute
Hospital for Sick Children
University of Toronto
Toronto, Ontario, Canada

Springer-Verlag
Berlin Heidelberg GmbH

R.G. Landes Company
Austin

MEDICAL INTELLIGENCE UNIT
GLYCOPROTEINS AND HUMAN DISEASE

R.G. LANDES COMPANY
Austin, Texas, U.S.A.

Springer

ISBN 978-3-662-21962-1

While the authors, editors and publisher believe that drug selection and dosage and the specifications and usage of equipment and devices, as set forth in this book, are in accord with current recommendations and practice at the time of publication, they make no warranty, expressed or implied, with respect to material described in this book. In view of the ongoing research, equipment development, changes in governmental regulations and the rapid accumulation of information relating to the biomedical sciences, the reader is urged to carefully review and evaluate the information provided herein.

Library of Congress Cataloging-in-Publication Data

Brockhausen, Inka, 1944-
Glycoproteins and human disease / Inka Brockhausen, William Kuhns.
p. cm. — (Medical intelligence unit)
Includes bibliographical references and index.
ISBN 978-3-662-21962-1 ISBN 978-3-662-21960-7 (eBook)
DOI 10.1007/978-3-662-21960-7

1. Glycoproteins—Metabolism—Disorders. I. Kuhns, William, M.D. II. Title. III. Series.
[DNLM: 1. Glycoproteins. 2. Disease QU 55 B863g 1996]
RC632.G57B76 1996
616.3'995—dc20
DNLM/DLC
for Library of Congress

96-43740
CIP

Publisher's Note

R.G. Landes Company publishes six book series: *Medical Intelligence Unit, Molecular Biology Intelligence Unit, Neuroscience Intelligence Unit, Tissue Engineering Intelligence Unit, Biotechnology Intelligence Unit* and *Environmental Intelligence Unit.* The authors of our books are acknowledged leaders in their fields and the topics are unique. Almost without exception, no other similar books exist on these topics.

Our goal is to publish books in important and rapidly changing areas of bioscience and environment for sophisticated researchers and clinicians. To achieve this goal, we have accelerated our publishing program to conform to the fast pace in which information grows in bioscience. Most of our books are published within 90 to 120 days of receipt of the manuscript. We would like to thank our readers for their continuing interest and welcome any comments or suggestions they may have for future books.

Shyamali Ghosh
Publications Director
R.G. Landes Company

ABBREVIATIONS

AFP	alpha feto-protein
AIDS	acquired immunodeficiency syndrome
ALL	acute lymphocytic leukemia
AML	acute myeloid leukemia
βAPP	β-amyloid precursor protein
BHK cells	baby hamster kidney cells
CAM	cell adhesion molecule
CDGS	carbohydrate deficient glycoprotein syndrome
CEA	carcino-embryonic antigen
CF	cystic fibrosis
CFTR	cystic fibrosis transmembrane conductance regulator
CHO cells	Chinese hamster ovary cells
CLL	chronic lymphocytic leukemia
CML	chronic myelogenous leukemia
CRD	carbohydrate recognition domain
CSF	colony stimulating factor
Dol-P	dolichol-phosphate
Dol-PP	dolichol-pyrophosphate
DTT	dithiothreitol
EBV	Epstein-Barr virus
EC	endothelial cells
ECM	extracellular matrix

EGF	epidermal growth factor
ELISA	enzyme linked immunosorbent assay
EPO	erythropoietin
ER	endoplasmic reticulum
FSH	follicle stimulating hormone
Fuc	fucose
Gal	galactose
GalNAc	N-acetylgalactosamine
G-CSF	granulocyte colony stimulating factor
Glc	glucose
GlcNAc	N-acetylglucosamine
GM-CSF	granulocyte macrophage colony stimulating factor
GP	glycoprotein
GPI	glycosylphosphatidylinositol
GVHD	graft versus host disease
HBV	hepatitis B virus
hCG	human chorionic gonadotropin
HEMPAS	hereditary erythroblastic multinuclearity with acidified serum lysis test
HIV	human immunodeficiency virus
HPLC	high performance liquid chromatography
HPAEC	high pressure anion exchange chromatography
HSV	Herpes Simplex virus

HUVEC	human umbilical cord vein endothelial cells
IGF	insulin-like growth factor
LAD	leukocyte adhesion deficiency
lamp	lysosomal associated membrane protein
LH	luteinizing hormone
Man	mannose
Man-6-P	mannose-6-phosphate
MHC	major histocompatibility complex
NCAM	neural cell adhesion molecule
Neu5Ac	5-N-acetylneuraminic acid
Neu5Gc	5-N-glycolylneuraminic acid
NK	natural killer
NMR	nuclear magnetic resonance
OA	osteoarthritis
PAPS	3-phospho-adenosine-5'-phosphosulphate
PGM	phosphoglucomutase
PHA	phytohemagglutinin
PMN	polymorphonuclear leukocytes
PSA	polysialic acid
SA	sialic acid
TNF	tumor necrosis factor
tPA	tissue plasminogen activator
TRH	thyrotropin-releasing hormone

TSH	thyroid stimulating hormone
UEA	Ulex europaeus agglutinin
uv	ultraviolet light
vWF	von Willebrand factor
WAS	Wiscott-Aldrich syndrome
WGA	wheat germ agglutinin
Xyl	xylose

CONTENTS

Acknowledgments

This work was supported by the Medical Research Council of Canada and the Canadian Cystic Fibrosis Foundation. The authors greatly appreciate the advice and critical reading of the manuscript by Drs. Steven Itzkowitz, Bill Jourdian and John Schutzbach.

CHAPTER 1

ABSTRACT

Glycoproteins are widely distributed among different species, cell types and tissues. The sugar moieties of mammalian glycoproteins show significant changes in their structures and relative occurrences during growth, development and differentiation and have been implicated in many biological functions. Diverse alterations of these carbohydrate chains occur in diseases such as cancer, metastasis, leukemia, inflammatory and other diseases. The glycosylation abnormalities found in disease are the result of complex alterations of oligosaccharide assembly by glycosyltransferases, mainly in the Golgi apparatus. These alterations affect early, intermediate and mainly late glycoprotein processing steps, depending on the disease type. Knowledge of the biological functions of glycoproteins and control mechanisms in the biosynthesis of glycans is instrumental for the development of better diagnosis and treatment and the understanding of the glycobiology of human disease.

CHAPTER 2

INTRODUCTION

Mammalian glycoproteins carry thousands of different glycan structures which comprise mainly the N-linked (β-N-acetyl-D-glucosamine-Asn) and O-linked (α-N-acetyl-D-galactosamine-Ser or -Thr) types of sugar chains (N-glycans and O-glycans, respectively). Minor structures are also found as a single O-GlcNAc (β-N-acetyl-D-glucosamine-Ser/Thr-linked) or as O-Fuc-linked (α-L-fucose-Thr/Ser), O-Glc-linked (β-D-glucose-Thr/Ser) and O-Man-linked (α-D-mannose-Ser) glycan chains. Other major classes of glycosylated mammalian proteins include the collagens and proteoglycans; the latter may also carry N- and O-glycans typically found on glycoproteins. O- or N-glycans are known to change during growth, development, oncogenic transformation and differentiation of cells. The different sizes, chemical, physical and antigenic properties of glycans determine the variety of biological functions which are now slowly being unraveled. Since glycoproteins are involved in many aspects of the immune system, in fertilization, cell adhesion, hormone action and receptor functions as well as in bacterial and viral binding and infectivity and other functions (see chapter 9), it is not surprising that altered carbohydrate structures lead to abnormal functions and altered behavior of diseased cells.

There is a wealth of literature on the structural and biosynthetic changes of glycosylation in disease. However, reports are often conflicting and reveal our poor understanding of the complex regulatory mechanisms in the biosynthesis of glycoproteins. In most cases we can only establish associations and correlations and not causative relationships between glycoprotein glycan structures, their biosynthesis and the phenotype of a diseased cell. Abnormal growth and differentiation as a consequence of disease, for example, due to inflammation, may be responsible for glycan changes. The changed phenotype of diseased cells may aggravate the clinical picture but may also be protective in nature or may be exploited in therapeutic interventions. There are very few genetic defects of enzymes involved in glycoprotein assembly as many of these defects may be incompatible with life or a developing organism.

In this book, we will describe glycoprotein structural abnormalities in disease as related to their biosynthesis, and the current knowledge of glycan structures and functions, as well as their biosynthesis in a broader context. Extensive reviews of glycoprotein structure and biosynthesis have been written.[1-12] We will not focus on glycolipids, proteoglycans and degradative pathways of glycolipids and glycoproteins where many inborn errors of metabolism have been described.

References

1. Beyer TA, Sadler JE, Rearick JI et al. Glycosyltransferases and their use in assessing oligosaccharide structure and structure-function relationships. In:Mesiter ed. Adv Enzym Vol 52. New York:Wiley and Sons, 1981; 23-175.
2. Kornfeld R, Kornfeld S. Assembly of asparagine-linked oligosaccharides. Ann Rev Biochem 1985; 54:631-664.
3. Brockhausen I. Clinical aspects of glycoprotein synthesis. Crit Rev Clin Lab Sci 1993; 301; 65-151.
4. Brockhausen I. Biosynthesis of O-glycans of the N-acetylgalactosamine-α-Ser/Thr linkage type. In:Montreuil J, Vliegenthart JFG, Schachter H eds. Glycoproteins. New York:Elsevier Pub 1995; 201-260.
5. Schachter H, Brockhausen I. The biosynthesis of serine/threonine N-acetylgalactosamine linked carbohydrate moieties. In:Allen H, Kisailus E eds. Glycoconjugates, Composition, Structure and Function. New York:Marcel Dekker, 1992; 263-332.
6. Schachter H. Biosynthetic controls that determine the branching and microheterogeneity of protein-bound oligosaccharides. Biochem Cell Biol 1986; 64:163-181.
7. Schachter H. Enzymes associated with glycosylation. Curr Opin Struct Biol 1991; 1:755-765.
8. Schachter H. Biosynthesis. 2c. Glycosyltransferases involved in the synthesis of N-glycan antennae. In:Montreuil J, Schachter H, Vliegenthart JFG eds. Glycoproteins. The Netherlands:Elsevier Science B.V. 1995; 153-199.
9. Schauer R. Chemistry, metabolism and biological functions of sialic acids. Adv Carbohydr Chem Biochem 1982; 40:131-234.
10. Field MC, Wainwright LJ. Molecular cloning of eukaryotic glycoprotein and glycolipid glycosyltransferases:a survey. Glycobiology 1995; 5:463-472.
11. Verbert A. Biosynthesis. 2b. From $Glc_3Man_9GlcNAc_2$-protein to $Man_5GlcNAc_2$-protein:transfer 'en bloc' and processing. In:Montreuil H, Schachter H, Vliegenthart JFG eds. Glycoproteins. The Netherlands:Elsevier Science B.V. 1995; 145-152.
12. Hounsell EF, Davies MJ, Renouf DV. O-linked protein glycosylation structure and function. Glycoconj J 1996; 13:19-26.

CHAPTER 3

Structures of α-N-Acetylgalactosamine-Ser/Thr-Linked Oligosaccharides (O-glycans)

O-glycans have been described on many mammalian and non-mammalian secreted and membrane-bound glycoproteins and may contain GalNAc, Gal, Fuc, GlcNAc and sialic acids, and may be sulfated. Sialic acids may be found as O-acetyl-, N-acetyl-, N-glycolyl- and other derivatives. Mucins comprise the main class of O-glycan and contain glycoproteins which may consist of approximately 50-80% by weight of O-glycans. Mucins are the major glycoprotein components of mucous secretions and also occur on cell membranes.

3.A. STRUCTURAL ANALYSIS

O-glycans of mucins exhibit extreme oligosaccharide microheterogeneity; they may be released from glycoproteins in the reduced form by β-elimination in alkaline borohydride solution. Alternative chemical methods for O-glycan, release such as the use of hydrazine, have been suggested. Galβ1-3GalNAc-, a common O-glycan that can often be exposed by sialidase treatment, can be released by a specific endo-N-acetylgalactosaminidase (O-glycanase). Another type of O-glycanase has been described that cleaves more complex O-glycans.[1] O-glycans may be degraded by the sequential release of sugars by exo-glycosidases, a method that may also be useful for establishing O-glycan structures. Trifluoromethylsulfonic acid removes but also degrades O-glycans as well as the peptide. A mild version of this treatment has been developed where GalNAc remains attached to the peptide and only complex chains are cleaved.[2,3]

The heterogeneous mixture of O-glycans released from glycoproteins has to be separated for detailed analysis of O-glycan structures. Purification of oligosaccharides can be achieved by electrophoresis, high pressure liquid chromatography (HPLC), paper or thin layer chromatography, gel filtration, ion exchange or affinity chromatography and the use of lectins or antibodies. The sugar composition of hydrolysates may be determined by chemical methods which require large amounts of material, by gas chromatography using volatile derivatives, amino sugar analysis using an amino acid analyzer, by high pressure liquid chromatography, high pressure anion exchange chromatography with pulsed amperometric detection (HPAEC), or other chromatographic methods requiring the availability of sugar standards. Anomeric linkages between sugars or the presence of certain glycan groups may be determined with lectins, by immunological methods using specific mono- or polyclonal antibodies, or by the use of glycosidases that cleave terminal or internal sugars in specific linkages.

Proton nuclear magnetic resonance of O-glycans in D_2O solution, which exchanges hydroxyl and amide hydrogens for deuterium, has been extremely useful in establishing sugar composition and complex glycan structures. This method is nondestructive and can often identify the entire O-glycan structures, including the anomeric linkages based on characteristic chemical shifts and coupling constants of the hydrogens involved. This requires the previous analysis of standard compounds of the same structure, but new structures have also been determined using the total assignment of all carbon-linked hydrogen signals, mainly by two-dimensional NMR techniques.

Mass spectrometry can establish the molecular weight of glycans, and alone or in combination with partial chain degradation will also give information of structural elements. Fast atom bombardment, electrospray as well as laser desorption mass spectrometry techniques have been successful in elucidating both O- and N-glycan structures. Often glycopeptides may be obtained by partial protease digestion of O-glycoproteins either using mild protease digestion or an O-sialoglycoprotease which cleaves peptide near sialic acid rich O-glycosylation sites.[4] Glycopeptides have been successfully analyzed by mass spectrometry. However, it appears difficult to assign the various O-glycan structures to their respective Thr/Ser-linked glycosylation sites since carbohydrates are usually cleaved during the process of mass spectrometry before the peptide bonds break.

Methylation analysis, i.e. permethylation, hydrolysis, peracetylation and reduction to obtain partially methylated and acetylated sugar derivatives, followed by separation by gas chromatography and identification of derivatives by electron-impact mass spectrometry to yield characteristic mass/charge patterns is the method of choice to unequivocally determine the linkage patterns between sugar residues.

3.B. COMMON O-GLYCAN CORE STRUCTURES

GalNAc α-O-glycosidically linked to Ser or Thr is the only structure common to all O-glycans and is often referred to as the Tn antigen (Table 1). Both the Tn and the sialylated Tn antigen (sialylα2-6GalNAcα-Thr/Ser) have been reported as cancer-associated antigens although they are also found in normal colonic and submaxillary mucins and other glycoproteins. At least eight different core structures have been reported to occur in O-glycans with core 1 and 2 structures being the more prevalent ones in non-mucin glycoproteins and core 1 to 4 being most common in mucins:[5,6]

core 1	Galβ1-3GalNAcα-
core 2	GlcNAcβ1-6(Galβ1-3)GalNAcα-
core 3	GlcNAcβ1-3GalNAcα-
core 4	GlcNAcβ1-6(GlcNAcβ1-3) GalNAcα-
core 5	GalNAcα1-3GalNAcα-
core 6	GlcNAcβ1-6GalNAcα-
core 7	GalNAcα1-6GalNAcα-
core 8	Galα1-3GalNAcα-

The various O-glycan core structures are expressed in a tissue-specific fashion.

Mono- or di-sialylated core 1 structures, +/-sialylα2-6(+/-sialylα2-3Galβ1-3)GalNAcα-, are often components of serum or cell surface glycoproteins but also occur in mucins. Core 1 may be substituted by a number of sugar residues; the unusual unsubstituted core 1 occurs as a cancer-associated antigen (T- or Thomsen-Friedenreich antigen, Table 1).[7] Core 3, 4, 5, 7, and 8 structures have only been found in mucins. Although core 5 occurs in mucins from several species, in humans it appears to be an oncofetal structure appearing in meconium[8] and in human adenocarcinoma.[9] Oligosaccharides with core 6 have only been reported on human glycoproteins to date.[10] However, the origin of core 6 may be through cleavage of the Galβ1-3 residue of core 2 by a specific β-galactosidase in vivo or during oligosaccharide purification. Core 7 was reported in bovine submaxillary mucin[11] and core 8 was recently described in human bronchial mucin.[12] Core 1, 3, 5, 7 and 8 may be substituted at the GalNAc-Ser/Thr moiety by a terminal α2-6-linked sialic acid. In addition, core 1 and 3 may be branched with β1-6-linked GlcNAc linked to GalNAc to form core 2 and 4, respectively, and may be further elongated. With the increased sensitivity of structural analysis, it is possible that other minor core structures will be found.

3.C. COMPLEX O-GLYCAN STRUCTURES

O-glycan chains may contain 20 or more sugars and various branch points. O-glycan core structures in human glycoproteins may be elongated by GlcNAcβ1-3 and Galβ1-3/4 residues (N-acetyl-lactosamine

Table 1. Blood group and tissue antigens

Antigen	Structure
Tn	GalNAcα-Thr/Ser
Sialyl-Tn	SAα2-6GalNAcα-Thr/Ser
T	Galβ1-3GalNAcα-Thr/Ser
Sialyl-T	SAα2-3Galβ1-3GalNAcα-Thr/Ser, SAα2-6(Galβ1-3)GalNAcα-Thr/Ser
A	GalNAcα1-3Galβ- | α1-2 Fuc
B	Galα1-3Galβ- | α1-2 Fuc
H or O	Galβ- | α1-2 Fuc
i	Gal1-4GlcNAcβ1-3Galβ-
I	Galβ1-4GlcNAc | β1-6 Galβ1-4GlcNAcβ1-3Galβ-
Forssman	GalNAcα1-3GalNAcβ1-3Gal-
S^d (Cad)	GalNAcβ1-4(SAα2-3)Galβ-
P^k	Galα1-4Galβ1-4Glcβ-Cer
P_1	Galα1-4Galβ1-4GlcNAc
P	GalNAcβ1-3Galα1-4Galβ1-4Glcβ-Cer
M* glycophorin antigen	H_2N-Ser-Ser*-Thr*-Thr*-Gly-
N* glycophorin antigen	H_2N-Leu-Ser*-Thr*-Thr*-Glu-

Table 1 continues on next page

Table 1. (continued)

	Type 1 chains	Type 2 chains
Le^a	Galβ1-3GlcNAcβ1-3Gal- | α1-4 Fuc	
Le^b	Galβ1-3GlcNAcβ1-3Gal- | α1-2 | α1-4 Fuc Fuc	
Le^x (SSEA-1)		Galβ1-4GlcNAcβ1-3Gal- | α1-3 Fuc
Le^y		Galβ1-4GlcNAcβ1-3Gal- | α1-2 | α1-3 Fuc Fuc
Sialy-Le^a		SAα2-3Galβ1-3GlcNAcβ1-3Gal- | α1- 4 Fuc
Sialyl-Le^x		SAα2-3Galβ1-4GlcNAcβ1-3Gal- | α1-3 Fuc
Sialyl-dimeric Le^x		SAα2-3Galβ1-4GlcNAcβ1-3Galβ1-4GlcNAc- | α1-3 | α1-3 Fuc Fuc SA | α6 SAα2-3Galβ1-3GalNAc-Ser/Thr

attachment positions of the O-glycans * Variations in amino acid sequences surrounding three O-glycosylation sites give rise to MN blood groups. Complete antigenicity requires intact unmodified oligosaccharides.

structures), they may be fucosylated, sialylated, sulfated, and may contain blood group or tissue antigens (Table 1). Poly-N-acetyl-lactosamine structures may be branched to form GlcNAcβ1-6 (GlcNAcβ1-3)Gal- structures. This is important for the conversion of i to I antigenicity (Table 1). Linear poly-N-acetyl-lactosamines may be present with unusual GlcNAcβ1-6 residues in human glycoproteins.[13] Unusual sugars at the nonreducing termini include GlcNAcα1-4 or GalNAcβ1-3/4. The only internal α-linked sugars in humans are GalNAcα-linked to Ser/Thr and sialic acid within the polysialic acid chains while a number of α-linked sugars may be found internally in other species.[14,15] Certain non-human structures may be recognized by preformed antibodies (for example against Galα1-3Gal- epitopes) or may trigger the formation of human antibodies directed against xenoantigens.[16]

Glycosaminoglycan chains of proteoglycans may be highly sulfated with both N- and O-sulfation along repeating disaccharides.[17] N-and O-glycans of proteoglycans and glycoproteins may also be sulfated. Sulfated O-glycans may have important roles in diseases such as cystic fibrosis, cancer and inflammation. The endothelial ligand for L-selectin was found to contain sulfated O-glycans.[18] The i antigenicity of keratan sulfate can be masked by sulfation of poly-N-acetyl-lactosamine.[19]

Glycosylation of mucins generates a great diversity of O-glycans which may be neutral, sialylated or sulfated.[20,21] Sulfate most frequently occurs linked to Gal and

GlcNAc residues of N-acetyl-lactosamine sequences.[21-27] Most commonly, sulfation is found at the 3-position of Gal and the 6-position of GlcNAc of core structures 1 to 4,[28] although 6-sulfated Gal has also been described.[22-24] In rat gastric and salivary mucins, linear core 1 structures have been identified with sulfate attached to the 6-position of GlcNAc[29-31] while sialomucin from rat mammary adenocarcinoma cells exhibits core 2 structures with 6-sulfated-GlcNAc.[32] In hog gastric mucin, 6-sulfated-GlcNAc is apparently the only sulfated sugar.[33,34] The major O-glycans isolated from human meconium glycoproteins include sialylated, partially α3-fucosylated and 3-sulfated-Gal structures with O-glycan core 3.[35] The same Gal or GlcNAc sugar residues which carry sulfate may also bear fucose or sialic acid.[36-38] Thus there is great glycan heterogeneity which imposes a problem when considering the biological functions of individual chains.

References

1. Iwase H, Ishii I, Ishihara K et al. Release of oligosaccharides possessing reducing end N-acetylgalactosamine from mucus glycoprotein in Streptomyces SP. OH-11242 culture medium through action of endo-type glycosidase Biochem Biophys Res Com 1988; 151:422-428.
2. Gerken TA, Butenhof KJ, Shogren R. Effects of glycosylation on the conformation and dynamics of O-linked glycoproteins: carbon-13 NMR studies of ovine submaxillary mucin. Biochem 1989; 28:5536-5543.
3. Gerken TA, Gupta R, Jentoft N. A novel approach for chemically deglycosylating O-linked glycoproteins. The deglycosylation of submaxillary and respiratory mucins. Biochem 1992; 31:639-648.
4. Otulakowski GL, Shewan PE, Udoh AE et al. Proteolysis of sialoglycoproteins by Pasteurella haemolytica cytotoxic culture supernatant. Infect Immun 1991; 42:54.
5. Schachter H, Brockhausen I. The biosynthesis of Serine/Threonine)-N-acetylgalactosamine-linked carbohydrate moieties. In: Allen H, Kisailus EC eds. Glycoconjugates, Composition, Structure and Function, Marcel Dekker, New York. 1992:263-332.
6. Hounsell EF, Davies MJ, Renouf DV. O-linked protein glycosylation structure and function. Glycoconj J 1996; 13:19-26.
7. Springer G. T and Tn, general carcinoma antigens. Science 1984; 224:1198-1206.
8. Hounsell EF, Lawson AM, Feeney J et al. Structural analysis of the O-glycosidically linked core-region oligosaccharides of human meconium glycoproteins which express oncofoetal antigens. Eur J Biochem 1985; 148:367-377.
9. Kurosaka A, Nakajima H, Funakoshi I et al. Structures of the major oligosaccharides from human rectal adenocarcinoma glycoprotein. J Biol Chem 1983; 258:11594-11598.
10. Brockhausen I. Biosynthesis of O-glycans of the N-acetylgalactosamine-α-Ser/Thr linkage type. In: Montreuil J, Vliegenthart JFG, Schachter H eds. Glycoproteins. New York: Elsevier Pub 1995; 201-260.
11. Chai WG, Hounsell EF, Cashmore GC et al. Neutral oligosaccharides of bovine submaxillary mucin. A combined mass spectrometry and 1H-NMR study. Eur J Biochem 1992; 203:257-268.
12. van Halbeek H, Strang A-M, Lhermitte M et al. Structures of monosialyl oligosaccharides isolated from the respiratory mucins of a non-secretor (O, Le^{a+b-}) patient suffering from chronic bronchitis. Characterization of a novel type of mucin carbohydrate core structure. Glycobiol 1994; 4:203-219.
13. Hanisch FG, Uhlenbruck G, Peter-Katalinic J et al. Structures of neutral O-linked polylactosaminoglycans on human skim milk mucins. A novel type of linearly extended poly-N-acetyllactosamine backbones with Galβ(1-4) GlcNAcβ(1-6) repeating units. J Biol Chem 1989; 264:872-883.
14. Strecker G, Wieruszeski J-M, Michalski J-C et al. Primary structure of neutral and acidic oligosaccharide-alditols derived from the jelly coat of the Mexican axolotl. Occurrence of oligosaccharides with fucosyl(α1-3)fucosyl(α1-4)-3-deoxy-D-glycero-D-galacto-nonulosonic acid and galactosyl(α1-4)[fucosyl(α1-2)]galactosyl(β1-4)-N-acetylglucosamine sequences. Eur J Biochem 1992; 207:995-1002.

15. Strecker G, Wieruszeski JM, Plancke Y et al Primary structure of 12 neutral oligosaccharide-alditols released from the jelly coats of the anuran Xenopus laevis by reductive β-elimination. Glycobiology 1995; 5:137-146.
16. Galili U, Macher B, Buehler J et al. Human natural anti-α-galactosyl IgG, II The specific recognition of α(1-3) linked galactose residues. J Exp Med 1985; 162:573.
17. Kusche M, Oscarsson L, Reynertson R et al. Biosynthesis of heparin. J Biol Chem 1991; 266:7400-7409.
18. Imai Y, Rosen S. Direct demonstration of heterogenous sulfated O-linked carbohydrate chains on an endothelial ligand for L-selectin. Glycoconj J 1993; 10:34-39.
19. Tang P, Scudder P, Mehmet H et al. Sulfate groups are involved in the antigenicity of keratin sulfate and mask i antigen expression on their poly-N-acetyllactosamine backbones. Eur J Biochem 1986; 160:537-545.
20. Lamblin G, Boersma A, Lhermitte M et al. Further characterization, by a combined HPLC ^{1}H-NMR approach, of the heterogeneity displayed by the neutral carbohydrate chains of human bronchial mucins. Eur J Biochem 1984; 140:227-236.
21. Lamblin G, Rahmoune H, Wieruszeski J-M et al. Structure of two sulfated oligosaccharides from respiratory mucins of a patient suffering from cystic fibrosis. Biochem J 1991; 275:199-206.
22. Mawhinney T, Adelstein E, Gayer D et al. Structural analysis of monosulfated side-chain oligosaccharides isolated from human tracheobronchial mucous glycoproteins. Carbohydr Res 1992; 223:187-207.
23. Mawhinney T, Adelstein E, Morris D et al. Structure determination of five sulfated oligosaccharides derived from tracheobronchial mucus glycoproteins. J Biol Chem 1987; 262:2994-3001.
24. Mawhinney T, Landrum D, Gayer D et al. Sulfated sialyl-oligosaccharides derived from tracheobronchial mucous glycoproteins of a patient suffering from cystic fibrosis. Carbohydr Res 1992; 235:179-197.
25. Lo-Guidice J, Wieruszeski J, Lemóine J et al. Sialylation and sulfation of the carbohydrate chains in respiratory mucins from a patient with cystic fibrosis. J Biol Chem 1994; 269:18794-18813.
26. Hokke C, Damm J, Kamerling J et al. Structure of three acidic O-linked carbohydrate chains of porcine zona pellucida glycoproteins. FEBS Lett 1993; 329:29-34.
27. Hokke C, Damm J, Penninkhof B et al. Structure of the O-linked carbohydrate chains of porcine zona pellucida glycoproteins. Eur J Biochem 1994; 221:491-512.
28. Strecker G, Wieruszeski J, Martel C et al. Determination of the structure of sulfated tetra- and pentasaccharides obtained by alkaline borohydride degradation of hen ovomucin. Glycoconj J 1987; 4:329-337.
29. Carter SR, Slomiany A, Gwozdzinski K et al. Enzymatic sulfation of mucus glycoprotein in gastric mucosa. J Biol Chem 1988; 263:11977-11984.
30. Goso Y, Hotta K. Chromatographic mapping of sulfated oligosaccharides from rat gastrointestinal mucins. Glycoconj J 1991; 8:210.
31. Goso Y, Hotta K. Types of oligosaccharide sulfation, depending on mucus glycoprotein source, corpus or antral, in rat stomach. Biochem J 1989; 264:805-812.
32. Hull S, Carraway K. Sulfation of the tumor cell surface sialomucin of the 13762 rat mammary adenocarcinoma. J Cell Biochem 1989; 40:67-81.
33. Slomiany B, Meyer K. Isolation and structural studies of sulfated glycoproteins of hog gastric mucosa. J Biol Chem 1972; 247:5062-5070.
34. Slomiany B, Meyer K. Oligosaccharides produced by acetolysis of blood group active (A+H) sulfated glycoproteins from hog gastric mucin. J Biol Chem 1973; 248:2290-2295.
35. Capon C, Leroy Y, Wieruszeski J et al. Structures of O-glycosidically linked oligosaccharides isolated from human meconium glycoproteins. Eur J Biochem 1989; 182:139-152.
36. Lo-Guidice JM, Wieruszeski JM, Lemoine J et al. Acidic oligosaccharides isolated from respiratory mucins of a patient suffering from cystic fibrosis. Glycoconj J 1993; 10:313.

37. Gowda D, Bhavanandan V, Davidson E. Structures of O-linked oligosaccharides present in the proteoglycans secreted by human mammary epithelial cells. J Biol Chem 1986; 261:4935-4939.

38. Capon C, Laboisse CL, Wieruszeski J-M et al. Oligosaccharide structures of mucins secreted by the human colonic cancer cell line CL.16E. J Biol Chem 1992; 267:19248-19257.

CHAPTER 4

Biosynthesis of O-Glycans

The structures and relative amounts of glycan chains found in a glycoprotein are the result of a sensitive regulation of biosynthesis. In the O-linked biosynthetic pathways, sugars are added individually from nucleotide-sugar donors predominantly in the Golgi apparatus by the sequential action of glycosyltransferases (Table 2). The ordered sequence of glycosylation reactions is guided by the relative activity levels and specificities of glycosyltransferases and the intracellular localization of enzymes and substrates. The glycosyltransferases that synthesize N- and O-glycans are apparently arranged in the various compartments of the Golgi as an assembly line where they are placed in the order of their action. With certain exceptions, glycosyltransferases can synthesize only one type of linkage between defined sugar residues.

The Golgi-localized glycosyltransferases studied to date are all type 2 membrane bound proteins or glycoproteins that have a short amino terminal end directed towards the cytoplasm, a hydrophobic uncleaved membrane anchor region flanked by charged amino acids, a stem region that is typically rich in Pro and also Thr and Ser, and a catalytically active region at the carboxy terminal that is appropriately folded and directed towards the lumen of the Golgi. Proteolytic cleavage at the stem region can occur during enzyme purification or in the acute phase response. This will produce soluble active enzyme that may be secreted; some of the serum enzymes may originate from this type of proteolytic cleavage. The membrane anchor region and the peptide adjacent to it contribute to the localization of the enzyme within an assembly line in the Golgi apparatus (see section 8.B).

The general reactions of Golgi glycosyltransferases acting on glycoproteins in vivo are:

Nucleotide-glycosyl donor + peptide or glycosyl-peptide acceptor substrate → nucleotide + glycosyl-peptide product

This is also the general reaction for glycosyltransferases acting in other cellular compartments. However, in the endoplasmic reticulum (ER), the glycosyl-donor for some reactions is dolichol-monophosphate- or dolichol-pyrophosphate-sugars. Glycosyltransferase reactions are often stimulated in vitro by detergent and divalent metal ions, and may be

Table 2.

1. Glycosyltransferases assembling N-glycans:

Oligosaccharyl-transferase, Dolichol-diphospho-oligosaccharide protein oligosaccharyltransferase E.C.2.4.1.119;
GlcNAc-1-phospho-transferase, UDP-GlcNAc: dolichyl-phosphate GlcNAc-transferase, E.C.2.4.1.15;
β4-Man-transferase, Chitobiosyldiphosphodolichol β-mannosyltransferase, E.C. 2.4.1.142;
Dol-P Man synthase, Dolichyl-phosphate β-mannosyltransferase, E.C. 2.4.1.83;
Dol-P-Glc synthase, Dolichyl-phosphate β-glucosyltransferase, E.C. 2.4.1.117;
lysosomal enzyme GlcNAc-1-phosphotransferase, UDP-GlcNAc: lysosomal enzyme GlcNAc-1-phosphotransferase, E.C. 2.7.8.17
GlcNAc-transferase I, UDP-GlcNAc: Manα1-3-R (GlcNAc to Manα1-3) β1,2-GlcNAc-transferase I, E.C. 2.4.1.101;
GlcNAc-transferase II, UDP-GlcNAc: Manα1-6-R (GlcNAc to Manα1-6) β1,2-GlcNAc-transferase II, E.C. 2.4.1.143;
GlcNAc-transferase III, UDP-GlcNAc: R_1-Manβ1-4-R_2 (GlcNAc to Manβ1-4) β1,4-GlcNAc-transferase III, E.C.2.4.1.144;
GlcNAc-transferase IV, UDP-GlcNAc: R_1-Manα1-3-R_2 (GlcNAc to Manα1-3) β1,4-GlcNAc-transferase IV, E.C.2.4.1.145;
GlcNAc-transferase V, UDP-GlcNAc: R_1-Manα1-6-R_2 (GlcNAc to Manα1-6) β1,6-GlcNAc-transferase V;
GlcNAc-transferase VI, UDP-GlcNAc: R_1-Manα1-6-R_2 (GlcNAc to Manα1-6) β1,4-GlcNAc-transferase VI;
α3-sialyltransferase (N), CMP-sialic acid: Galβ1-3/4GlcNAc-R (sialic acid to Gal) α2,3-sialyltransferase, E.C.2.4.99.6;
α6-sialyltransferase (N), CMP-sialic acid: Galβ1-4GlcNAc-R (sialic acid to Gal) α2,6-sialyltransferase, E.C.2.4.99.1;
α6-Fuc-transferase, GDP-Fuc: R_1–GlcNAcβ-Asn-R_2 (Fuc to GlcNAc) α1,6-Fuc-transferase, E.C.2.4.1.68.

2. Glycosyltransferases assembling O-glycans:

polypeptide α-GalNAc-transferase, UDP-GalNAc: polypeptide α-GalNAc-transferase, E.C.2.4.1.41;
core 1 *β3-Gal-transferase,* UDP-Gal: GalNAc-R β1,3-Gal-transferase, E.C.2.4.1.122;
core 2 *β6-GlcNAc-transferase,* UDP-GlcNAc: Galβ1-3GalNAc-R (GlcNAc to GalNAc) β1,6-GlcNAc-transferase, E.C.2.4.1.102;
core 3 *β3-GlcNAc-transferase,* UDP-GlcNAc: GalNAc-R β1,3-GlcNAc-transferase, E.C.2.4.1.147;
core 4 *β6-GlcNAc-transferase,* UDP-GlcNAc: GlcNAcβ1-3GalNAc-R (GlcNAc to GalNAc) β1,6-GlcNAc-transferase, E.C.2.4.1.148;
core 5 *α3-GalNAc-transferase,* UDP-GalNAc: GalNAc-R α1,3-GalNAc-transferase;
core 6 *β6-GlcNAc-transferase,* UDP-GlcNAc: GalNAc-R β1,6-GlcNAc-transferase;
elongation *β3-GlcNAc-transferase,* UDP-GlcNAc: Galβ1-3(R_1)GalNAc-R_2 (GlcNAc to Gal) β1,3-GlcNAc-transferase, E.C.2.4.1.146;
α3-sialyltransferase (O), CMP-sialic acid: Galβ1-3GalNAc-R (sialic acid to Gal) α2,3-sialyltransferase, E.C.2.4.99.4;
α6-sialyltransferase (O) I, CMP-sialic acid: R_1-GalNAc-R_2 (sialic acid to GalNAc) α2,6-sialyltransferase, E.C.2.4.99.3;
α6-sialyltransferase (O) II, CMP-sialic acid: sialylα2-3Galβ1-3GalNAc-R (sialic acid to GalNAc) α2,6-sialyltransferase, E.C.2.4.99.7.

Table 2 continues on next page

Table 2. (continued)

3. Glycosyltransferases acting on both N- and O-glycans:

blood group i β3-GlcNAc-transferase, UDP-GlcNAc: Galβ1-4-R β1,3-GlcNAc-transferase, E.C.2.4.1.149;
blood group I β6-GlcNAc-transferase, UDP-GlcNAc: GlcNAcβ1-3Galβ1-4-R (GlcNAc to Gal) β1,6-GlcNAc-transferase, E.C.2.4.1.150;
blood group Cad or S^d β4-GalNAc-transferase, UDP-GalNAc: sialic acid α2-3Galβ-R (GalNAc to Gal) β1,4-GalNAc-transferase;
blood group A α3-GalNAc-transferase, UDP-GalNAc: Fucα1-2Galβ-R (GalNAc to Gal) α1,3-GalNAc-transferase, E.C.2.4.1.40;
blood group B α3-Gal-transferase, UDP-Gal: Fucα1-2Galβ-R (Gal to Gal) α1,3-Gal-transferase, E.C.2.4.1.37;
β4-Gal-transferase, UDP-Gal: GlcNAc-R β1,4-Gal-transferase, E.C.2.4.1.38;
β3-Gal-transferase, UDP-Gal: GlcNAc-R β1,3-Gal-transferase;
α3-Gal-transferase, UDP-Gal: Galβ1-4GlcNAc-R α1,3-Gal-transferase, E.C.2.4.1.124;
α2-Fuc-transferase, GDP-Fuc: Galβ-R α1,2-Fuc-transferase, E.C.2.4.1.69;
α3-Fuc-transferase, GDP-Fuc: Galβ1-4GlcNAcβ-R (Fuc to GlcNAc) α1,3-Fuc-transferase, E.C.2.4.1.152;
α3/4-Fuc-transferase, GDP-Fuc: Galβ1-3/4GlcNAcβ-R (Fuc to GlcNAc) α1,3/4-Fuc-transferase, E.C.2.4.1.65;
α6-sialyltransferase, CMP-sialic acid: sialylα2-3Galβ1-4 GlcNAc-R (sialic acid to GlcNAc) α2,6-sialyltransferase; E.C.2.4.99.1
α8-sialyltransferase, CMP-sialic acid: sialylα2-3/8Gal-R (sialic acid to sialic acid) α2,8-sialyltransferase, E.C.2.4.99.8.

modified by many different factors. In in vitro assays, the peptide portion of the acceptor can often be replaced by a hydrophobic or other aglycon group. Polypeptide α-GalNAc-transferase, the α6-sialyltransferase I (O) and possibly other enzymes (Table 2) form an exception and can only act on peptide-containing substrates. The addition of the first sugar, GalNAc, to peptide occurs mainly in the cis-Golgi.[1] A number of studies suggest that, in addition, some O-glycosylation may take place in the ER or throughout later Golgi compartments. Since the distribution of the enzyme catalyzing the first step of O-glycosylation, polypeptide α-GalNAc-transferase, may be broader under certain cellular conditions,[2] it is conceivable that a redistribution of O-glycan processing glycosyl transferases may occur in diseased cells. In addition, factors influencing the concentration and transport of substrates and products, and factors influencing enzyme activities may lead to a shift in the ratio of final O-glycan structures in disease.

Fuc-transferases and sialyltransferases do not require divalent metal ions, but most of the UDP-sugar utilizing enzymes do require Mn^{2+} or similar metal ions. For reasons that are yet unclear, all β6-GlcNAc-transferases form an exception since these enzymes can transfer GlcNAc from UDP-GlcNAc in the presence of EDTA.

The cDNA encoding the first enzyme in O-glycan biosynthesis (polypeptide α-GalNAc-transferase), the enzyme synthesizing core 2 (core 2 β6-GlcNAc-transferase) and various enzymes involved in the elongation and termination of O-glycans have been cloned (Table 2). The cDNA encoding transferases synthesizing O-glycan cores 1, and 3 to 8 have not yet been isolated.

4.A. NUCLEOTIDE SUGAR TRANSPORT

Sugar nucleotides required as donors for glycosylation reactions are synthesized in the cytosol (UDP-GlcNAc, UDP-GalNAc, UDP-Glc, UDP-Gal, GDP-Man,

GDP-Fuc) and in the nucleus (CMP-sialic acid) from the respective sugar precursors by extensive interconversions of sugar-derivatives and sugar-nucleotides.[3] Sugar nucleotides are transported into the ER and into the Golgi apparatus by specific antiport systems which remove the free nucleotides that may act as inhibitors for glycosylation reactions. The transport into the ER appears to be restricted to UDP-GlcNAc and UDP-Glc. The transport and metabolism of sugar nucleotides may be regulated during differentiation. This is expected to influence their concentration at the site of glycoprotein synthesis and ultimately affect glycosylation reactions.

4.B. INITIATION OF O-GLYCAN SYNTHESIS

Studies using antibodies to deglycosylated mucin or to the polypeptide α-GalNAc-transferase, or using lectins and radioactive sugar precursors or subcellular fractionation suggest that O-glycosylation is initiated in the cis-Golgi. The first enzyme of the O-glycan biosynthetic pathways, polypeptide α-GalNAc-transferase, which transfers GalNAc to peptide from UDP-GalNAc, has been purified and cloned from several species including human.[4-6] It appears that a family of enzymes exists that can transfer GalNAc with high efficiency to Thr. Other members of this family remain to be characterized and transfer to Ser has yet to be demonstrated in vitro. It is possible that another enzyme exists that is responsible for GalNAc transfer to Ser. However, the low levels of GalNAc transfer to Ser by purified polypeptide α-GalNAc-transferases or crude homogenates in vitro may be an artifact of the enzyme assays, or due to lack of the appropriate glycoprotein substrate, or cofactors.

Site-directed recombination of the gene encoding a polypeptide α-GalNAc-transferase resulted in specific deletion of the catalytic region of the enzyme in thymocytes of mice. The specific defect in thymocytes and peripheral T-cells does not appear to delete O-glycan biosynthesis. Therefore, multiple, distinct polypeptide α-GalNAc-transferase genes may exist.[7]

The peptide structure of the acceptor substrate plays an important role in the activity of the enzyme. Non-mucin glycoproteins often carry O-glycans on very specific Thr or Ser residues, suggesting that there is site-directed glycosylation. Pro is usually found near an O-glycosylation site, and it may serve to expose Ser/Thr residues or to confer a favorable three-dimensional structure to the peptide that allows Ser or Thr to be glycosylated. Not only the peptide portion of substrates but also existing glycosylation influences O-glycosylation.[8] In vitro, subsequent O-glycosylation is more difficult than primary O-glycosylation. Thus the O-glycosylation is site-directed and may be due to selective accessibility of the site due to protein folding and glycosylation, and to specific structural factors within the peptide chain.

4.C. SYNTHESIS OF O-GLYCAN CORES

The synthesis and elongation of O-glycan core structures proceeds via many possible reactions, as illustrated in Figure 1. The enzymes assembling and elongating O-glycan cores may compete for a common substrate if they are co-localized in the Golgi. The addition of certain sugar residues may facilitate or may prevent further processing. For example, in branching reactions of O-glycans, a three-linked residue is usually added before a six-linked sugar branch can be introduced; this led us to propose the 3 before 6 rule.[9] Sialylation often terminates chains by inhibiting the addition of neutral sugars. For example, the formation of sialylα2-6GalNAc prevents O-glycan core formation and extension. The synthesis of core structures appears to be significantly influenced by the peptide structure and glycosylation of substrates.[10]

4.C.1. Core 1 Synthesis

O-glycan core 1 is synthesized by the ubiquitous core 1 β3-Gal-transferase (Fig. 1, Table 2), which adds a β1-3-linked

Gal residue to GalNAc-R where R may be one of many membrane-bound or secreted glycoproteins or mucins. The enzyme is responsible for the synthesis of the T-antigen. The activity is inversely correlated with the presence of Tn-antigen-reactive *Helix pomatia* lectin binding sites in human breast.[11]

In vitro the enzyme can also act on synthetic substrates where R is Thr, Ser, peptide, benzyl, phenyl or another synthetic aglycon group. The enzyme has been

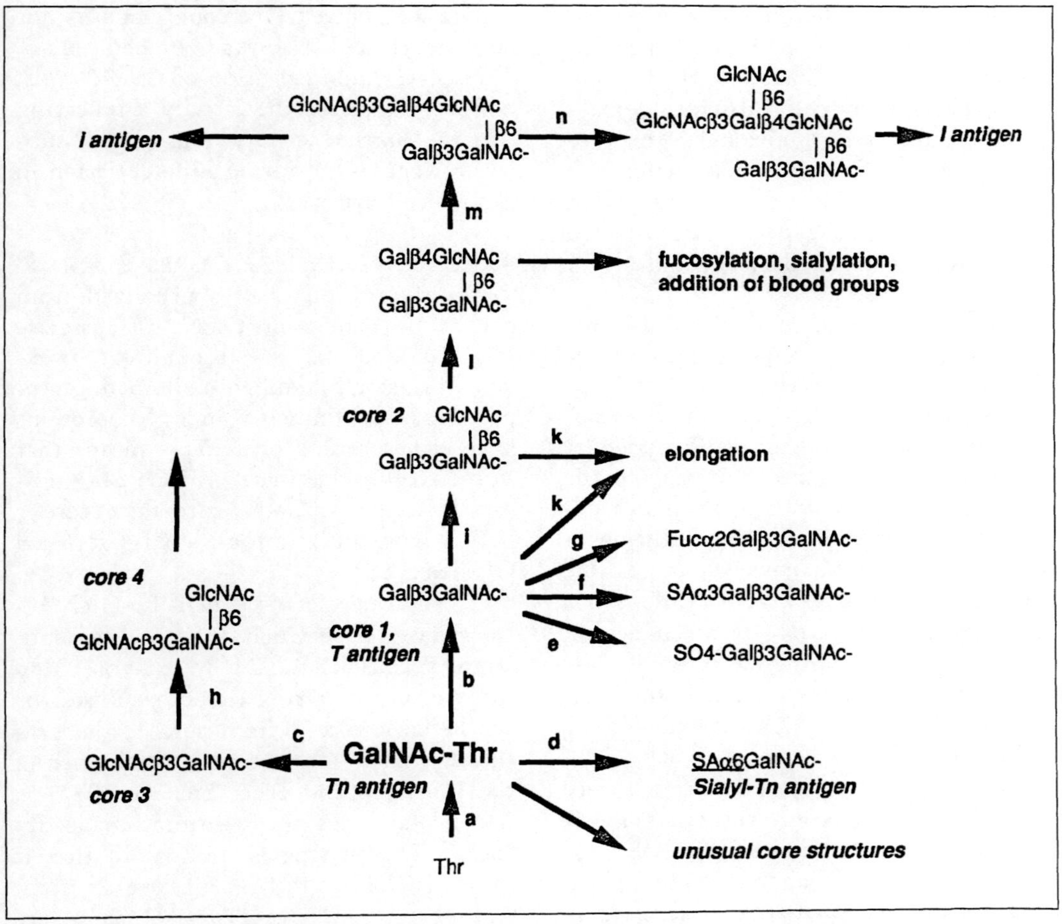

Fig. 1. Biosynthesis of O-glycans. O-glycans are assembled in the Golgi apparatus. The initial step is the transfer of GalNAc to Thr or Ser of the peptide chain by polypeptide α-GalNAc-transferase (step a) in the cis-Golgi compartment; this is followed by the sequential addition of sugars to the growing O-glycan chain. GalNAc linked to Thr/Ser is the only sugar common to all O-glycans. The drawing shows the core structures 1 to 4 commonly found in mucins, there are also less common core structures (cores 5 to 8). Cancer cells often express short O-glycans such as GalNAc-Thr/Ser (Tn antigen), sialic acidα2-6GalNAc-Thr/Ser (sialyl-Tn antigen) which cannot be converted further, and the unprocessed core 1 structure (T antigen). Core structures may be elongated, sialylated, fucosylated and sulfated. Mucin O-glycans also tend to carry blood group antigens, depending on the secretor status. SA, sialic acid. The conversions of O-glycan structures are catalyzed by enzymes as indicated.
Step a, polypeptide α-GalNAc-transferase; b, core 1 β3-Gal-transferase; c, core 3 β3-GlcNAc-transferase; d, α6-sialyltransferase I (O); e, core 1 sulfo-transferase; f, α3-sialyltransferase (O); g, α2-Fuc-transferase; h, core 4 β6-GlcNAc-transferase; i, core 2 β6-GlcNAc-transferase; k, elongation β3-GlcNAc-transferase; l, β4-Gal-transferase; m, i β3-GlcNAc-transferase; n, I β6-GlcNAc-transferase.

partially purified from several tissues[5] but its gene has not been cloned. Partially purified core 1 β3-Gal-transferase from rat liver is exceptionally stable at 4°C. The activity is very sensitive to modifications in the amino acid sequences of the peptide moieties of substrates and the presence of additional sugar residues in the substrate.[10]

This suggests that not only the first step of O-glycan processing but also the following steps are controlled by the peptide backbone of the substrate, either through a direct recognition by the processing enzymes or by secondary effects on peptide structure and peptide and glycan conformation. This type of site-directed processing has also been proposed for N-glycan processing.[12]

The specificity of core 1 β3-Gal-transferase requires the 2-N-acetyl group, the 3- and 4- but not the 6-hydroxyl of GalNAc substrate.[13] The enzyme does not act on sialylα2-6GalNAcα-R, the product of sialyltransferase I, probably due to the negative charge of sialic acid that is blocking the transfer of Gal.[14] The addition of α6-linked sialic acid, however, is possible after the formation of core 1. Core 1 β3-Gal-transferase from rat liver can act on core 6 (Fig. 1) to synthesize core 2.[13] This may be significant for certain human tissues such as ovaries which have the ability to form core 6.[15]

Although core 1 formation is a very early event in O-glycan synthesis, subcellular fractionation studies suggest the presence of core 1 β3-Gal-transferase in the same fraction as β4-Gal-transferase, a trans-Golgi enzyme.[16] Core 1 β3-Gal-transferase is active in most cell types and may therefore be used as a positive control for the assays of glycosyltransferases. There are a few examples of cell types lacking the transferase activity. Human T-lymphoblastoid cell line Jurkat as well as Tn-erythrocytes from patients with permanent mixed field agglutinability, the Tn syndrome,[17,18] form an exception and are deficient in core 1 β3-Gal-transferase[19] and therefore useful to study the biological role of core 1. Recently, the human colon cancer cell line LSC has been shown to lack core 1 β3-Gal-transferase activity.[20] These cells therefore exhibit a high expression of Tn and sialyl-Tn antigens.[21] Induction of differentiation of T cells from Tn syndrome patients by Na-butyrate[22] reactivated the enzyme, which indicated that it was downregulated rather than deleted due to a mutation. The enzyme could not be similarly reactivated in Jurkat[22] or LSC cells.[20] Enterocytic differentiation of Caco-2 cells, however, increased the activity[23] indicating that the enzyme activity level may be subject to regulation during differentiation in certain cell types.

4.C.2. Synthesis of Cores 2 and 4

O-glycan core 2 is synthesized from core 1 by core 2 β6-GlcNAc-transferase (Fig. 1, Table 2).[24-26] Although core 6 exists in a small number of human glycoproteins, most mucins and glycoproteins have either core 1 or 2. This means that in most cells and tissues, the Galβ1-3 residue is added to GalNAc to form core 1 before the addition of a β1-6 linked GlcNAc.

Several species of core 2 β6-GlcNAc-transferases exist which differ by their substrate specificities. The M enzyme from many mucin-secreting tissues catalyzes the synthesis of core 2 from core 1, the synthesis of core 4 from core 3, the branch of the I antigen and other branches (Fig. 1, Table 1) and was purified from bovine trachea.[27] The enzyme is unusual in that it can add GlcNAc in β1-6 linkage to either Gal or to GalNAc, provided these are 3-substituted by Gal or GlcNAc.

The L enzyme from many non-mucin-secreting tissues such as leukocytes and certain cell lines[28] catalyzes only the conversion of core 1 to core 2. It is possible that other species exist with different specificities.[28,29] Core 2 β6-GlcNAc-transferase L in leukemia cells has a more restricted substrate recognition than the M enzyme and requires most of the substituents of the Galβ1-3GalNAcα- substrate sugar rings.[29] The enzyme does not require peptide in the substrate and can act on free reducing oligosaccharides of core 1 structure or on Galβ1-3 GalNAc-R derivatives where R can be a variety of synthetic groups.

Any type of substitution of core 1, i.e. sialylation, fucosylation (Fig. 2), sulfation and elongation, will prevent core 2 β6-GlcNAc-transferase action.[5,24,30,31] Thus the GlcNAcβ1-6 branch of core 2 structures has to be formed first before the Gal moiety can be modified to form various types of core 2 glycans. The cDNA encoding the L enzyme was recently isolated from HL60 cells by expression cloning[32] using an antibody recognizing a core 2 containing hexasaccharide of leukosialin on expressing cell surfaces. The gene has been localized to human chromosome 9q21.[33]

Core 2 β6-GlcNAc-transferase activity appears to be regulated during cellular differentiation,[23] T-cell activation[34] and is altered in leukemia and immunodeficiency disease.[35-38] The enzyme is not expressed in certain breast cancer cells (see chapter 11.K).

Nishio et al[39] utilized mRNA differential display to identify changes in gene expression induced by metabolic stimuli such as hyperglycemia or diabetes. A cDNA clone (DH1) was characterized which is differentially expressed in the ventricles of streptozotocin (STZ) diabetic rats. Analysis of a cDNA sequence-inferred polypeptide showed 85% identity with human core 2 β6-GlcNAc-transferase. Transient transfection of Cos-7 cells by this

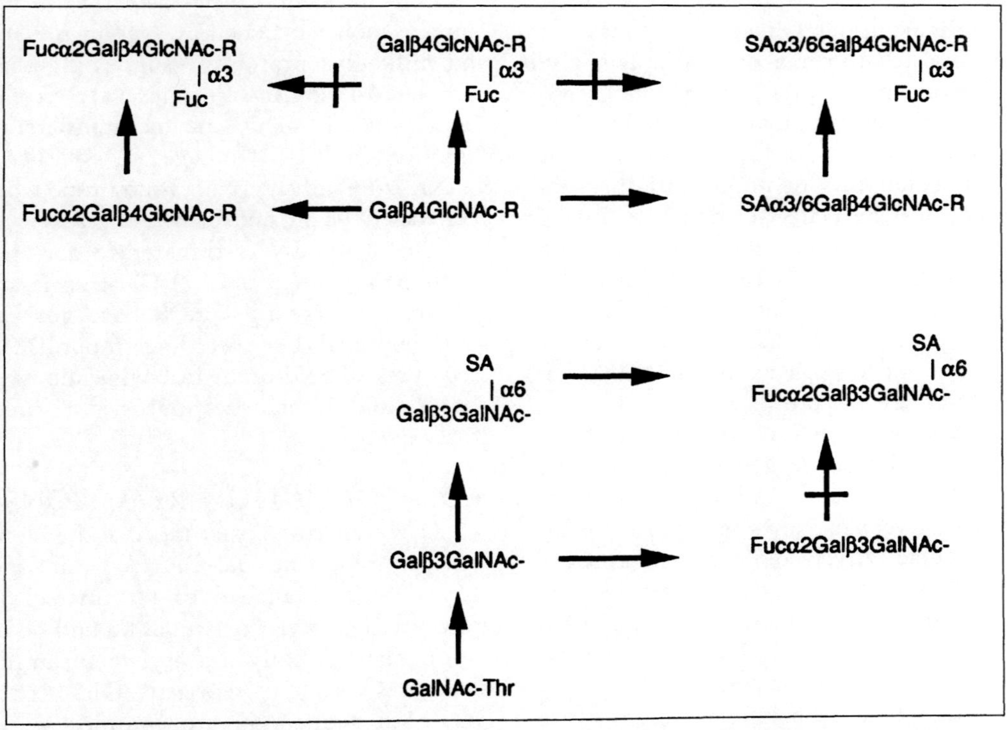

Fig. 2. Sialylation and fucosylation of N- and O-glycans. The terminal Galβ1-4GlcNAc structures of elongated N-glycans can be sialylated by α3-sialyltransferase (N) and α6-sialyltransferase (N). These reactions occur before fucosylation since several α3- and α4-fucosyltransferases can act on sialylated substrates but sialyltransferases generally do not act on fucosylated substrates. Similarly, the α3-Fuc-transferase acts on termini carrying the H-determinant (Fucα1-2 Gal) but the α2-Fuc-transferase cannot act on the Le^x determinant. O-glycans are similarly sialylated and fucosylated except that there is mainly α3-sialylation of Galβ1-4GlcNAc termini. A similar situation is seen in the early O-glycan pathways where α2-fucosylation of core 1 occurs after α6-sialylation of GalNAc but not vice versa.

cDNA produces core 2 β6-GlcNAc-transferase activity. It has been suggested that the DH1 gene may contain cardiac-specific core 2 β6-GlcNAc-transferase promoter elements which can be modulated by hyperglycemia or diabetes. This finding may be related to a specific form of cardiomyopathy observed in human diabetes.

4.C.3. Synthesis of Cores 3 to 8

O-glycan core 3, which is the biosynthetic intermediate of core 4, is synthesized by core 3 β3-GlcNAc-transferase (Fig. 1, Table 2). This enzyme is enriched in colonic tissue in several species,[40] but based on the occurrence of core 3 structures, should also be present in other cell types such as human salivary and respiratory tissues. Core 3 β3-GlcNAc-transferase acts on natural and synthetic GalNAc-R substrates where R may be peptide or a synthetic hydrophobic group. Since core 3 has not been reported to occur on non-mucin glycoproteins, the expression of core 3 β3-GlcNAc-transferase may be restricted to mucin-secreting cell types. The enzyme specificity appears to require all the substituents of the GalNAc moiety.[28] Core 3 β3-GlcNAc-transferase cannot act on sialylα2-6GalNAc-R but sialylation of core 3 structures is common in colonic mucins,[41,42] suggesting that the sialic acid is added to the 6-position of GalNAc after the formation of core 3.

Although GalNAcα-benzyl and related compounds are good substrates, the amino acid sequence and glycosylation of GalNAc-glycopeptides significantly influences the activity.[8] Colon cancer tissue exhibits decreased activity of this enzyme compared to normal tissue.[43,44] Cultured human colonic cancer cells do not express detectable levels of the activity[20,23,28,45] and this may result from the fact that these cells are transformed or because they are grown in culture. However, small amounts of core 3 and 4 structures have been found in mucins secreted from human colon cancer cell lines.[46]

The enzyme synthesizing core 4 is probably core 2 β6-GlcNAc-transferase M since core 4 β6-GlcNAc-transferase activity is always accompanied by the activity synthesizing core 2, even in a highly purified preparation of bovine tracheal tissue.[27] Although many tissues and cells exhibit the activity, the occurrence of core 4 structures is restricted due to the lack of core 3 β3-GlcNAc-transferase activities in most tissues. The core 4 activity relative to core 2 activity decreases in colon cancer[43] and during tumorigenic progression of polyposis coli cells;[45] this may indicate a specific loss of the M enzyme in colon cancer.

Core 5 is found in several mammalian and non-mammalian species. In humans, core 5 has been reported in adenocarcinoma and meconium.[47,48] The core 5 α3-GalNAc-transferase activity has been reported in one patient with adenocarcinoma but remains to be characterized.[49]

Core 6 has only been detected in human glycoproteins such as ovarian mucins and milk glycoproteins,[4,5] suggesting that core 6 β6-GlcNAc-transferase activity is human-specific. This enzyme transferring GlcNAc in β1-6 linkage to synthetic GalNAcα-R without prior 3-substitution has been described in human ovarian tissue.[50]

The α6-GalNAc-transferase activity synthesizing core 7 and α3-Gal-transferase activity synthesizing core 8 have not yet been described, but according to published structures, should occur in bovine submaxillary[51] and human respiratory[52] tissues, respectively.

4.D. ELONGATION REACTIONS

O-glycan core structures may be elongated by repeating Gal-GlcNAc units and they may be branched or terminated by sialylation, fucosylation, sulfation and other modifications. Poly-n-acetyl-lactosamine structures (single or repetitive GlcNAcβ1-3Galβ1-4 sequences) are commonly expressed in most cell types. These sequences are recognized by antibodies to the i blood group antigen (Table 1) which can detect the linear i antigen on erythrocytes in early fetal life. The i antigen is regulated during development with the appearance of the branched I blood group antigen at a later stage of development, probably due to the induction of a β6-GlcNAc-transferase synthesizing the I antigen which masks the i antigen.[53]

The addition of a GlcNAcβ1-3 residue to the Gal residues of core 1 and 2 structures is catalyzed by elongation β3-GlcNAc-transferase which is present in many tissues and cell types.[23,38,54-55] The enzyme is distinguished from i β3-GlcNAc-transferase adding the i antigen and core 3 β3-GlcNAc-transferase by its substrate specificity and tissue distribution.[5,40]

β4-Gal-transferase and i β3-GlcNAc-transferase act sequentially to build up poly-n-acetyl-lactosamine chains (Fig. 1, Table 2). These enzymes only require the terminal GlcNAc or Gal residues of substrates, respectively, and act on O-glycans, N-glycans and glycolipids. However, they are probably influenced by steric and conformational factors affecting substrate structures.

The β4-Gal-transferase is ubiquitous and can act on GlcNAc termini of glycoconjugates. The enzyme has been purified and characterized from many tissues.[9,56] In the mammary gland, the specificity of the enzyme is regulated by α-lactalbumin, which binds to the β4-Gal-transferase resulting in a change in the specificity of the enzyme to favor lactose synthesis. The cDNA encoding this enzyme has been cloned from various species including human[57] and the human gene is localized to chromosome 9 p13-p21.[58,59] Two forms of the enzyme exist, a short form and a 13 amino acid longer form both may be glycosylated and are expressed in a tissue-specific manner.[60] A cell surface bound form also exists.[61,62] The cell surface expression of Gal-transferase on sperm cells is thought to be important for the interaction with terminal GlcNAc residues on the zona pellucida of the egg[62] during fertilization.

Another enzyme from pig trachea involved in type 1 chain (Table 1) elongation adds Galβ1-3 residues to terminal GlcNAc residues of O-glycan core 3 and other structures.[63] β3-Gal-transferase activity has also been described in human colon cancer COLO 205 cells. This enzyme is probably involved in the synthesis of type 1 chains on glycolipids and the Galβ1-3GlcNAc structures in glycoproteins.[64] The i β3-GlcNAc-transferase has been described in many tissues. This enzyme exhibits relatively low activity in vitro but must be active in vivo based on the fact that poly-n-acetyl-lactosamine chains are very common and may consist of many repeating units. This suggests that the intracellular organization of glycosylation reactions within the Golgi may be extremely important for efficient synthesis, and this organization is disrupted in in vitro assay systems.

The I antigen branch is synthesized by I β6-GlcNAc-transferase which adds a GlcNAcβ1-6 branch to the subterminal Gal residue of (+/-Galβ1-4) GlcNAcβ1-3Gal structures.[65] Several activities have been described that are distinguished by their substrate specificities but all synthesize the GlcNAcβ1-6Gal branch. Some of these activities require substitution of the GlcNAcβ1-3 residue, while others only act on unsubstituted substrates or on internal Gal residues of Galβ1-4 GlcNAcβ1-3 Galβ1-4 R substrates.[66] In addition, β6-GlcNAc-transferases have been described that synthesize linear GlcNAc1-β6Gal linkages.[67] I β6-GlcNAc-transferase is restricted in its occurrence and must be under developmental regulation. The cDNA of one member of the I β6-GlcNAc-transferase family has been cloned by expression cloning from human teratocarcinoma cells[33] but the activity encoded by this cDNA remains to be characterized. The putative catalytic domain of the enzyme has significant homology to core 2 β6-GlcNAc-transferase L and both genes are found on human chromosome 9q21.[33]

4.E. TERMINAL REACTIONS

Reactions terminating O-glycans often involve the same enzymes as those terminating N-glycans. These include the addition of α-linked sugars such as Fuc, sialic acid and blood group determinants. O-glycans may also be terminated by unusual α-linked sugars such as GlcNAcα1-4, or by sulfate. Reactions specific for O-glycans include the synthesis of sialylα2-3 Galβ1-3 (R-)GalNAc- and sialylα2-6GalNAc-, to form sialylated T and Tn antigens (Table 1). These linkages are formed by families of sialyltransferases of which several members have recently been cloned (Table 2).[68]

4.E.1. Sialyltransferases Acting on O-Glycans

Sialic acids in glycoproteins are mostly in the terminal position but may also be found internally in sialylα2-8 sialic acid structures of polysialic acid chains, for example, in cell adhesion molecules.[68] The transfer of sialic acid residues may inhibit chain elongation and branching reactions and may terminate chain growth.

Sialyltransferases (Table 2, Fig. 2) have regions of homology and similarity within the catalytic domain named sialyl motif which has been useful in cloning sialyltransferases.[69-72] This region has been shown by site directed mutagenesis to be involved in CMP-sialic acid binding.[72]

The α3-sialyltransferase sialylating the Galβ1-3 residue of core 1 or core 2 oligosaccharides has been purified and characterized.[5,29,68] cDNA cloning from pig submaxillary glands and other species including human[68,73,74] revealed the existence of an α3-sialyltransferase family. The α3-sialyltransferase (O) activity is present in most tissues and cell types and probably acts on glycolipids and O-glycans with terminal Galβ1-3GalNAc- structure. The enzyme from human placenta and acute myeloid leukemia (AML) cells acts on glycoproteins and glycopeptides but does not require peptide in the substrate and accepts Galβ1-3 GalNAc-R derivatives where R may be nitrophenyl or benzyl or another synthetic group. The enzyme activity is typically high in acute and chronic myelogenous leukemia cells,[75,76] in colon cancer[43] and in breast cancer cells[77] compared to normal mammary epithelial cells. The increase is associated with higher sialylation and shorter chain length of O-glycans. Although the intracellular localization of α3-sialyltransferase (O) has not yet been determined, it is conceivable that the increase will prevent elongation, branching to form core 2, fucosylation and other terminal reactions. An exception is the addition of sialic acid to the 6-position of GalNAc which is possible after α3-sialylation of core 1. Thus in the synthesis of the common disialylated core 1 structure sialylα2-6 (sialylα2-3 Galβ1-3) GalNAc-, found in serum glycoproteins, mucins and cell surface glycoproteins, the α2-3-linked sialic acid is introduced before α2-6 linked sialic acid.

The various species of α3-sialyltransferases have overlapping but different substrate specificities. However, it is not known whether sialylα2-3Galβ1-4 linkages on N- and O-glycans are synthesized by the same α3-sialyltransferase. The expression of α3-sialyltransferase (O) is decreased upon Ha-ras oncogene transfection in rat fibroblasts[78] and is also regulated during the maturation of thymocytes.[79]

Several different types of α6-sialyltransferases may sialylate GalNAc-peptide. α6-Sialyltransferase (O) I from mucin-secreting tissues acts on GalNAc, Galβ1-3 GalNAc- and sialylα2-3 Galβ1-3 GalNAc- and requires peptide in the substrate. The enzyme has been purified from porcine submaxillary glands[80] and its cDNA was cloned from chicken embryos.[71] Another α6-sialyltransferase from fetal calf liver and other tissues requires sialylα2-3 Galβ1-3 GalNAcα-R as a substrate (previously named α6-sialyltransferase II[81,82]) where R can be peptide or a hydrophobic group. A third α6-sialyltransferase acts only on the disaccharide Galβ1-3 GalNAc-R if R is a peptide and does not act on GalNAc. The latter enzyme has been cloned from chicken testes.[83] The addition of blood group ABO determinants to the Gal moiety of core 1 is possible after the addition of sialic acid to form sialic acid α2-6(Galβ1-3) GalNAc.

The sialyl-Tn antigen, sialylα2-6GalNAc-Thr/Ser, is common in mucins secreted from submaxillary glands which exhibit low activity of core 1 β3-Gal-transferase compared to α6-sialyltransferase (O) I. Cancer cells often show a high expression of the sialyl Tn antigen[84] and this may be regulated by α6-sialyltransferase (O) I in rat colon cancer cells.[20]

A number of α8-sialyltransferases attach sialic acid in α2-8 linkage to either sialylα2-3 or sialylα2-8 linked residues to form polysialic acids. The enzymes are regulated in a tissue-specific and developmentally-specific fashion. α8-Sialyltransferases are highly expressed in developing rat brain[85] and human fetal but not adult brain.[86] The attachment of polysialic acids regulates the function of neural cell adhesion molecules (NCAM, discussed in chapter 9.A).[87]

4.E.2. Fucosyltransferases

At least two types of α2-Fuc-transferases (Table 2, Fig. 2) are involved in synthesizing the blood group H determinant, Fucα1-2Galβ-, on glycoproteins or glycolipids and are present in human serum (Table 1). The human blood group H-dependent enzyme is found in hematopoietic cells, and the secretor gene (Se-gene)-dependent α2-Fuc-transferase occurs in secretory tissues and many mucin-secreting cells. The human blood group H-dependent α2-Fuc-transferase has been purified from several tissues and human serum and has been shown to be a glycoprotein.[88] The gene was cloned and localized to human chromosome 19.[89] The "Bombay" type blood group lacks the H determinant, and individuals with this condition do not express the blood group H-dependent α2-Fuc-transferase. They may therefore show a serious transfusion reaction if given universal donor blood with blood group O(H) due to the presence of anti blood group H antibodies in their serum.[90] The "para Bombay" individuals also lack this enzyme but still contain the secretory gene-dependent α2-Fuc-transferase in their secretory tissues and therefore do not have anti H antibodies. The Se-gene-dependent α2-Fuc-transferase has also been purified and cloned and the human Se-gene is closely linked to that of the H-dependent enzyme.[90,91] The Se gene-dependent enzyme is expressed in a tissue-specific fashion along the intestines with high levels in the stomach and in the upper intestines and lower levels in the colon.[92]

Normal mouse milk is deficient in the H-antigen. To study the biosynthetic controls of the H-antigen, transgenic mice were produced[93] using a fusion gene containing cDNA encoding human α2-Fuc-transferase. The transgenic mice produced a high amount of α2-fucosylated glycoproteins and oligosaccharides in milk.

The expression of Lewis determinants (Table 1) is regulated during development and in many diseases, and varies between tissues and cell types. Le^x structures have been shown to be involved in important biological functions such as cell adhesion, and they change expression in many human tumors (discussed in chapter 11.A). At least five different α3-Fuc-transferases (III to VII, Table 2) distinguished by their specificities and expression in various cell types, can synthesize the X or Le^x determinants of type 2 chains (Table 1). As a result, the characteristics and specificities of these activities differ between tissues.

The cDNAs encoding these five α3-Fuc-transferases have been cloned.[94-98] The Le^a and Le^b blood group-dependent α4-Fuc-transferase (α3-Fuc-transferase III) is interesting since it can synthesize two different Fuc-GlcNAc linkages of type 1 and type 2 chains. The enzyme is found in secretions and many mucin-producing cell types and has been highly purified from human milk.[99] The human gene has been localized to chromosome 19. α3-Fuc-transferases III to VI (but not VII) act on neutral substrates such as Galβ1-4 GlcNAcβ-R to synthesize the Le^x determinant whereas α3-Fuc-transferases III, V, VI and VII use sialylated substrates sialylα2-3Galβ1-4GlcNAc to synthesize sialyl-Le^x. The α3-sialyltransferase involved in the synthesis of sialyl-Le^x may not act after the addition of the Le^x determinant. In addition, α6-sialylation of Gal prevents the synthesis of the sialyl-Le^x determinant.[100] Therefore, sialyl-Le^x is made by the action of α3-sialyltransferase followed by α3-Fuc-transferase. Le^x cannot be converted to Le^y since the α2-Fuc-transferase does not act after the α3-Fuc-transferase. Thus, Le^y is synthesized by the action of α2-Fuc-transferase followed by α3-Fuc-transferase on type 2 chains.

4.F. ADDITION OF TERMINAL BLOOD GROUP ANTIGENS

Blood group O determinants synthesized by α2-Fuc-transferase can be found on erythrocytes and on secreted mucins, as Fucα1-2Galβ1-3/4 GlcNAc-R or Fucα1-2 Galβ1-3 GalNAc- (core 1) terminal structures. Blood group O can be further converted to blood group A or B by A-dependent α3-GalNAc-transferase or B-dependent α3-Gal-transferase (Table 2).[101] Blood group O persons do not express A or B transferases. Both enzymes have been highly purified and cloned.[102-104] Blood group A and B are structurally very similar, yet the small difference is extremely important for assuring compatibility in blood transfusions since A persons have anti-B and B persons have anti-A antibodies in their serum. The two enzymes are highly homologous and differ only by a few amino acids, which determine the binding specificity of UDP-GalNAc or UDP-Gal resulting in GalNAc or Gal attachment, respectively.[102,104] Both genes are localized to human chromosome 9q34.

An α3-Gal-transferase that synthesizes the Galα1-3Galβ- linkage resembling blood group B, but without Fuc, exists in many mammalian species but not in humans and Old World monkeys.[105-107] Although a functional gene for the α3-Gal-transferase is absent from humans, a homologous pseudogene in humans has been localized to chromosome 9q33-34 and 12q14-15.[108-109] Antibodies against Galα1-3Gal- and possibly other foreign carbohydrate determinants have been found in human serum and these may be involved in xenograft rejection following transplantation.[105,110]

The Cad or S^d blood group antigen (Table 1) is synthesized by the β4-GalNAc-transferase that acts on sialylated N- and O-glycans and glycolipids with the sequences sialylα2-3Galβ1-4GlcNAc or sialylα2-3Galβ1-3GalNAc.[111-115] The enzyme may be secreted into the urine of S^{d+} individuals.[114] The expression of β4-GalNAc-transferase is regulated during the development of the mouse and is expressed in cytotoxic T-cells and in a tissue-specific fashion along the intestine.[115]

4.G. BIOSYNTHESIS OF SULFATED O-GLYCANS

The biosynthesis of sulfate esters requires a supply of inorganic sulfate derived from exogenous or endogenous sources and activation to a sulfate donor substrate 3'-phosphoadenosine 5'-phosphosulfate (PAPS) in the presence of specific sulfotransferases and acceptor substrates. Ultrastructural studies using high iron diamine suggest that the trans side of the Golgi apparatus and the trans-Golgi network are the compartments functionally involved in the sulfation of glycoconjugates.[116]

In the biosynthetic pathways of mucin-type oligosaccharides, a GlcNAc-6-sulfotransferase activity has been described in the Golgi fraction of rat stomach[117,118] acting on rat gastric mucin; the product was characterized as 6-sulfated-GlcNAcβ1-3Galβ1-3GalNAc. The enzyme is thought to be different from the sulfotransferase that synthesizes sulfatide (3-sulfo-Gal-ceramide), but similar to the GlcNAc-6-sulfotransferase acting on rat sublingual gland.[119] Goso and Hotta[120] found that GlcNAc-6-sulfo-transferase transfers sulfate only to terminal and not internal GlcNAcβ residues, and not to GlcNAcα–R. Galβ1-4 (6-sulfate) GlcNAc- sequences of glycoproteins are thus synthesized by sulfotransferase prior to Gal-transferase action; this is consistent with findings that β4-Gal-transferase can act on 6-sulfated-GlcNAc-R but not on 3- or 4- sulfated GlcNAc.[31] Sulfated oligosaccharide structures differ according to their positions in the gastrointestinal tract.[121]

Another mucin sulfotransferase from rat colon was characterized which synthesizes 3-sulfated-Galβ1-3GalNAc-R. Several partial O-glycan structures, including core 1, core 2 and N-acetyl-lactosamine, are good substrates for sulfotransferases from rat and human colon and other sources.[31] Processing of core 1 by core 2 β6-GlcNAc transferase is strongly inhibited by sulfation of the Gal residue, indicating that GlcNAcβ1-6 branching needs to precede sulfation in the O-glycan core 1 processing pathway.

Microsomal fractions from human airway mucosa contain a 3-sulfotransferase activity that acts on methyl galactosides or reduced carbohydrate chains released from human respiratory mucins with terminal Gal residues. Sulfate is added exclusively to the 3-position of the terminal Gal residues of N-acetyl-lactosamine sequences, consistent with the occurrence of these sulfated structures in human lung mucin. It is possible that the rat colonic 3-sulfotransferase described above is also present in human airways but does not act on Galβ1-3 GalNAc-ol sequences due to the fact that GalNAc is in the reduced form. The airway 3-sulfotransferase is different in properties from the enzyme involved in sulfatide synthesis.[122]

The precise role of sulfotransferases in sulfated O-glycan synthesis will be defined as pure enzyme and cDNA clones become available. Apparently sulfotransferases acting on various glycoconjugates have a concensus motif with the sequence GXXGXXK which is a putative binding sequence for PAPS.[123]

The availability of inorganic sulfate and of PAPS as a sulfate donor is essential for the esterification of substrates by sulfotransferases. This was shown in studies of a sulfate transport defect that could lead to decreased PAPS synthesis in genetically predisposed persons with osteochondrodysplasia. The defect is caused by mutations at the locus which encodes a sulfate transporter, a protein which is missing in patients' cells.[124] Its absence is associated with chondrodysplasia and perinatal death.

References

1. Roth J, Wang Y, Eckhardt AE et al. Subcellular localization of the UDP-N-acetyl-D-galactosamine: polypeptide N-acetylgalactosaminyltransferase-mediated O-glycosylation reaction in the submaxillary gland. Proc Nat Acad Sci USA 1994; 91:8935-8939.
2. Perez-Vilar J, Hidalgo J, Velasco A. Presence of terminal N-acetylgalactosamine residues in subregions of the endoplasmic reticulum is influenced by cell differentiation in culture. J Biol Chem 1991; 266: 23967-23976.
3. Abeijon C, Hirschberg CB. Topography of glycosylation reactions in the endoplasmic reticulum. Trends Biochem Sci 1992; 17:32-36.
4. Brockhausen I, Schachter H. Glycosyltransferases involved in N- and O-glycan biosynthesis. In: Gabius HJ, Gabius S eds. Glycoproteins: Status and Perspectives. Weinheim, Germany: Chapman & Hall In press, 1996.
5. Brockhausen I. Biosynthesis of O-glycans of the N-acetylgalactosamine-α-Ser/Thr linkage type. In: Montreuil J, Vliegenthart JFG, Schachter H eds. Glycoproteins. New York: Elsevier Pub 1995:201-260.
6. White T, Bennett EP, Takio K et al. Purification and cDNA cloning of a human UDP-N-acetyl-α-D-galactosamine: polypeptide N-acetylgalactosaminyltransferase. J Biol Chem 1995; 270:24156-24165.
7. Hennet T, Hagen F, Tabak LA et al. T-cell-specific deletion of a polypeptide N-acetylgalactosaminyltransferase gene by site-directed recombination. Proc Nat Acad Sci USA 1995; 92:12070-12074.
8. Brockhausen I, Möller G, Merz G et al. Control of mucin synthesis: the peptide portion of synthetic O-glycopeptide substrates influences the activity of O-glycan core 1 UDPgalactose: N-acetyl-α-galactosaminyl-R β3-galactosyltransferase. Biochemistry 1990; 29:10206-10212.
9. Schachter H, Brockhausen I. The biosynthesis of serine/threonine N-acetylgalactosamine linked carbohydrate moieties. In: Allen H, Kisailus E Eds. Glycoconjugates, Composition, Structure and Function. New York: Marcel Dekker, 1992:263-332.
10. Granovsky M, Bielfeldt T, Peters S, et al. O-glycan core 1 UDP-Gal: GalNAc β3-galactosyltransferase is controlled by the amino acid sequence and glycosylation of glycopeptide substrates. Eur J Biochem 1994; 221:1039-1046.

11. Zhuang D, Yousefi S, Dennis JW. Tn antigen and UDP-Gal:GalNAcα-R β1-3galactosyltransferase expression in human breast carcinoma. Cancer Biochem Biophys 1991; 12:185-198.
12. Savvidou G, Klein M, Grey A et al. Possible role for the peptide-oligosaccharide interactions in differential oligosaccharide processing at Asparagine-107 of the light chain and asparagine-297 of the heavy chain in a monclonal $IgG_1κ$. Biochemistry 1984; 23:3736-3740.
13. Brockhausen I, Möller G, Pollex-Krüger A et al. Control of O-glycan synthesis in rat liver: Specificity and inhibition of O-glycan core 1 UDP-Gal: GalNAca-R β3-galactosyl-transferase. Biochemistry 1992; Cell Biol 70:99-108.
14. Brockhausen I. unpublished.
15. Yazawa S, Abbas SA, Madiyalakan R et al. N-acetyl-β-D-glucosaminyltransferases related to the synthesis of mucin-type glycoproteins in human ovarian tissue. Carbohydr Res 1986; 149:241-252.
16. Elhammer A, Kornfeld S. Two enzymes involved in the synthesis of O-linked oligosaccharides are localized on membranes of different densities in mouse lymphoma BW5146 cells. J Cell Biol 1984; 99:327-331.
17. Cartron J, Andrev J, Cartron J et al. Demonstration of T-transferase deficiency in Tn-polyagglutinate blood samples. Eur J Biochem 1978; 92:111-119.
18. Thurnher M, Clausen H, Fierz W et al. T-cell clones with normal or defective O-galactosylation from a patient with permanent mixed-field polyagglutinability. Eur J Immunol 1992; 22:1835-1842.
19. Piller V, Piller F, Fukuda M. Biosynthesis of truncated O-glycans in the T cell line Jurkat. J Biol Chem 1990; 265: 9264-9271.
20. Brockhausen I, Dickinson N, Ogata S et al. Enzymatic basis for the high Sialyl-Tn expression in a colon cancer cell line. International Symp of Glycoconjugates, Seattle. Glycoconj J 1995; 12:566.
21. Ogata S, Chen A, Itzkowitz S. Use of model cell lines to study the biosynthesis and biological role of cancer-associated sialosyl-Tn antigen. Cancer Res 1994; 54:4036-4044.
22. Thurnher M, Rusconi S, Berger E. Persistant repression of a functional allele can be responsible for galactosyltransferase deficiency in Tn syndrome. J Clin Invest 1993; 91:2103-2110.
23. Brockhausen I, Romero PA, Herscovics A. Glycosyltransferase changes upon differentiation of CaCo-2 human colonic adenocarcinoma cells. Cancer Res 1991; 51:3136-3142.
24. Williams D, Longmore GD, Matta KL et al. Mucin synthesis. II. Substrate specificity and product identification studies on canine submaxillary gland UDP-GlcNAc:Galβ1-3GalNAc (GlcNAc to GalNAc) β6-N-acetylglucosaminyltransferase. J Biol Chem 1980; 255:11253-11261.
25. Williams D, Schachter H. Mucin synthesis. I. Detection in canine submaxillary glands of an N-acetylglucosaminyltransferase which acts on mucin substrates. J Biol Chem 1980; 255:11247-11252.
26. Wingert WE, Cheng P. Mucin biosynthesis: characterization of rabbit small intestinal UDP-N-acetylglucosamine: galactose β-3-N-acetylgalactosaminide (N-acetylglucosamine to N-acetylgalactosamine) β6-N-acetylglucosaminyltransferase. Biochemistry 1984; 23:690-697.
27. Ropp PA, Little MR, Cheng P-W. Mucin biosynthesis: purification and characterization of a mucin β6-N-acetylglucosaminyltransferase. J Biol Chem 1991; 266:23863-23871.
28. Vavasseur F, Yang J-M, Dole K et al. Synthesis of O-glycan core 3: characterization of UDP-GlcNAc: GalNAc β3-N-acetylglucosaminyltransferase activity from colonic tissues and lack of the activity in human cancer cell lines. Glycobiology 1995; 5:351-357.
29. Kuhns W, Rutz V, Paulsen H et al. Processing O-glycan core 1, Galβ1–3GalNAcα-R. Specificities of core 2, UDP-GlcNAc: Galβ1-3GalNAc-R(GlcNAc to GalNAc) β-6-N-acetylglucosaminyltransferase and CMP-sialic acid: Galβ1-3GalNAc-R α3-sialyltransferase. Glycoconj J 1993; 10:381-394.
30. Brockhausen I, Matta KL, Orr J et al. Mucin synthesis. Conversion of R_1-β1-3Gal-R_2 to R_1-β1-3(GlcNAcβ1-6)Gal-R_2 and of R_1-β1-3GalNAc-R_2 to R_1-β1-3(GlcNAcβ1-6)GalNAc-R_2 by a β6-N-acetyl-gluco-

saminyltransferase in pig gastric mucosa. Eur J Biochem 1986; 157:463-474.

31. Kuhns W, Jain R, Matta K et al. Characterization of a novel mucin sulfotransferase activity synthesizing sulfated O-glycan core 1, 3-sulfate-Galβ1-3GalNAcα-R. Glycobiology 1995; 5:689-697.
32. Bierhuizen MFA, Fukuda M. Expression cloning of a cDNA encoding UDP-GlcNAc:Galbeta1-3-GalNAc-R (GlcNAc to GalNAc) beta1-6GlcNAc transferase by gene transfer into CHO cells expressing polyoma large tumor antigen. Proc Natl Acad Sci USA 1992; 89:9326-9330.
33. Bierhuizen M, Mattei MG, Fukuda M. Expression of the developmental I antigen by a cloned human cDNA encoding a member of a beta-1,6-N-acetylglucosaminyltransferase gene family. Genes Dev 1993; 7:468-478.
34. Piller F, Piller V, Fox R et al. Human T lymphocyte activation is associated with changes in O-glycan biosynthesis. J Biol Chem 1988; 263:15146-15150.
35. Saitoh O, Piller F, Fox R et al. T-lymphocytic leukemia expresses complex, branched O-linked oligosaccharides on a major sialoglycoprotein, leukosialin. Blood 1991; 77:1491-1499.
36. Piller F, Le Deist F, Weinberg KI et al. Altered O-glycan synthesis in lymphocytes from patients with Wiscott-Aldrich syndrome. J Exp Med 1991; 173:1501-1510.
37. Higgins E, Siminovitch K, Zhuang D et al. Aberrant O-linked oligosaccharide biosynthesis in lymphocytes and platelets from patients with the Wiskott-Aldrich syndrome. J Biol Chem 1991; 266:6280-6290.
38. Brockhausen I, Kuhns W, Schachter H et al. Biosynthesis of O-glycans in leukocytes from normal donors and from patients with leukemia: increase in O-glycan core 2 UDP-GlcNAc:Galβ1-3GalNAcα–R (GlcNAc to GalNAc) β(1-6)-N-acetylglucosaminyltransferase in leukemic cells. Cancer Res 1991; 51:1257-1263.
39. Nishio Y, Warren C, Buczek-Thomas J et al. Identification and characterization of a gene regulating enzymatic glycosylation which is induced by diabetes and hyperglycemia specifically in rat cardiac tissue. J Clin Invest 1995; 96:1759-1767.
40. Brockhausen I, Matta KL, Orr J et al. Mucin synthesis. VI. UDP-GlcNAc:GalNAc-R β3-N-acetylglucosaminyl-transferase and UDP-GlcNAc:GlcNAcβ1-3GalNAc-R (GlcNAc to GalNAc) β6-N-acetylglucosaminyltransferase from pig and rat colon mucosa. Biochemistry 1985; 24:1866-1874.
41. Podolsky DK. Oligosaccharide structures of human colonic mucin. J Biol Chem 1985; 260:8262-8271.
42. Podolsky DK. Oligosaccharide structures of isolated human colonic mucin species. J Biol Chem 1985; 260:15510-15515.
43. Yang J, Byrd JC, Siddiki B et al. Alterations of O-glycan biosynthesis in human colon cancer tissues. Glycobiology 1994; 4:873-884.
44. King M, Chan A, Roe R et al. Two different glycosyltransferase defects that result in GalNAcα-0-peptide (Tn) expression. Glycobiology 1994; 4:267-279.
45. Vavasseur F, Dole K, Yang J et al. O-glycan biosynthesis in human colorectal adenoma cells during progression to cancer. Eur J Biochem 1994; 222:415-424.
46. Capon C, Laboisse CL, Wieruszeski J-M et al. Oligosaccharide structures of mucins secreted by the human colonic cancer cell line CL.16E. J Biol Chem 1992; 267: 19248-19257.
47. Kurosaka A, Nakajima H, Funakoshi I et al. Structures of the major oligosaccharides from human rectal adenocarcinoma glycoprotein. J Biol Chem 1983; 258:11594-11598.
48. Hounsell EF, Lawson AM, Feeney J et al. Structural analysis of the O-glycosidically linked core-region oligosaccharides of human meconium glycoproteins which express oncofoetal antigens. Eur J Biochem 1985; 148: 367-377.
49. Kurosaka A, Funakoshi I, Matsuyama M et al. UDP-GalNAc:GalNAc-mucin α-N-acetylgalactosamine transferase activity in human intestinal cancerous tissues. FEBS Lett 1985; 190:259-262.
50. Yazawa S, Abbas SA, Madiyalakan R et al. N-acetyl-β-D-glucosaminyltransferases related to the synthesis of mucin-type glycoproteins in human ovarian tissue. Carbohydr

Res 1986; 149:241-252.
51. Chai WG, Hounsell EF, Cashmore GC et al. Neutral oligosaccharides of bovine submaxillary mucin. A combined mass spectrometry and 1H-NMR study. Eur J Biochem 1992; 203:257-268.
52. van Halbeek H, Strang A-M, Lhermitte M et al. Structures of monosialyl oligosaccharides isolated from the respiratory mucins of a non-secretor (O, Le^{a+b-}) patient suffering from chronic bronchitis. Characterization of a novel type of mucin carbohydrate core structure. Glycobiol 1994; 2:203-219.
53. Feizi T, Kapadia A, Gooi HC et al. Human monoclonal auto-antibodies detect changes in expression and polarization of the Ii antigens during cell differentiation in early mouse embryos and teratocarcinomas. In: Muramatsu T, Gachelin G, Moscona AA, Ikawa Y, eds. Teratocarcinoma and Embryonic Cell Interactions. NY: Academic Press 1982; 201-215.
54. Brockhausen I, Williams D, Matta KL et al. Mucin synthesis. III. UDP-GlcNAc:Galβ1-3(GlcNAcβ1-6)GalNAc-R (GlcNAc to Gal) β3-N-acetylglucosaminyltransferase, an enzyme in porcine gastric mucosa involved in the elongation of mucin-type oligosaccharides. Can J Biochem Cell Biol 1983; 61:1322-1333.
55. Brockhausen I, Orr J, Schachter H. Mucin synthesis. The action of pig gastric mucosal UDP-GlcNAc:Galβ1-3(R_1)GalNAc-R_2 (GlcNAc to Gal) β3-N-acetylglucosaminyltransferase on high molecular weight substrates. Can J Biochem Cell Biol 1984; 62:1081-1090.
56. Beyer TA, Sadler JE, Rearick JI et al. Glycosyltransferases and their use in assessing oligosaccharide structure and structure-function relationships. In: Adv Enzym Vol 52. Mesiter Ed. New York:Wiley and Sons, 1981; 23-175.
57. Field MC, Wainwright LJ. Molecular cloning of eukaryotic glycoprotein and glycolipid glycosyltransferases: a survey. Glycobiology 1995; 5:463-472.
58. Duncan AMV, McCorquodale MM, Morgan C et al. Chromosomal localization of the gene for a human galactosyltransferase (GT-1). Biochem Biophys Res Comm 1986; 141:1185-1188.
59. Shaper NL, Shaper JH, Bertness V et al. The human galactosyltransferase gene is on chromosome 9 at band p13. Somat Cell Mol Genet 1986; 12:633-636.
60. Harduin-Lepers A, Shaper JH, Shaper NL. Characterization of two cis-regulatory regions in the murine β1,4-galactosyltransferase gene: evidence for a negative regulatory element that controls initiation at the proximal site. J Biol Chem 1993; 268:14348-14359.
61. Shur BD. Expression and function of cell surface galactosyltransferase. Biochim Biophys Acta 1989; 988:389-409.
62. Shur BD. Cell surface β1,4-galactosyltransferase: twenty years later. Glycobiology 1991; 1:563-575.
63. Sheares BT, Carlson DM. Characterization of UDP-galactose:2-acetamido-2-deoxy-D-glucose 3β-galactosyltransferase from pig trachea. J Biol Chem 1983; 258:1081-1090.
64. Holmes EH, Levery SB. Biosynthesis of fucose containing lactoseries glycolipids in human colonic adenocarcinoma COLO 205 cells. Arch Biochem Biophys 1989; 274:633-647.
65. Piller F, Cartron JP, Maranduba A et al. Biosynthesis of blood group I antigens. Identification of a UDP-GlcNAc: GlcNAcβ1-3Gal(-R) β1-6(GlcNAc to Gal)N-acetylglucosaminyltransferase in hog gastric mucosa. J Biol Chem 1984; 259:13385-13390.
66. Gu J, Nishikawa A, Matsuura N et al. Deficiency of β(1-6) N-acetylglucosaminyltransferase involved in the biosynthesis of blood group I antigen in the liver of LEC rats. Jpn J Cancer Res 1992; 83:878-884.
67. Basu M, Basu S. Biosynthesis in vitro of Ii core glycosphingolipids from neolactotetrosylceramide by beta 1-3- and beta 1-6-N-acetylglucosaminyltransferases from mouse T-lymphoma. J Biol Chem 1984; 259:12557-12562.
68. Harduin-Lepers A, Recchi MA, Delannoy P. 1994, The year of sialyltransferases. Glycobiology 1995; 5:741-758.
69. Kitagawa H, Paulson JC. Cloning of a novel α2-3-sialyltransferase that sialylates glycoprotein and glycolipid carbohydrate groups.

J Biol Chem 1994; 269:1394-1401.

70. Drickamer K. A conserved disulphide bond in sialyltransferases. Glycobiology 1993; 3:2-3.
71. Kurosawa N, Hamamoto T, Lee Y-C et al. Molecular cloning and expression of GalNAc α2,6-sialyltransferase. J Biol Chem 1994; 269:1402-1409.
72. Datta AK, Paulson JC. The sialyltransferase "sialylmotif" participates in binding the donor substrate CMP-NeuAc. J Biol Chem 1995; 270:1497-1500.
73. Gillespie W, Kelm S, Paulson JC. Cloning and expression of the Gal beta 1, 3GalNAc alpha 2,3-sialyltransferase. J Biol Chem 1992; 267:21004-21010.
74. Chang ML, Eddy RL, Shows TB et al. Three genes that encode human beta-galactoside alpha 2,3-sialyltransferases. Structural analysis and chromosomal mapping studies. Glycobiology 1995; 5:319-325.
75. Baker MA, Kanani A, Brockhausen I et al. Presence of cytidine 5'-monophospho-N-acetylneuraminic acid: Galβ1-3GalNAc-R α(2-3)-sialyltransferase in normal human leukocytes and increased activity of this enzyme in granulocytes from chronic myelogenous leukemia petients. Cancer Res 1987; 47:2763-2766.
76. Kanani A, Sutherland DR, Fibach E et al. Human leukemic myeloblasts and myeloblastoid cells contain the enzyme cytidine 5'-monophosphate-N-acetylneuraminic acid:Galβ1-3GalNAc α(2-3)-sialyltransferase. Cancer Res 1990; 50:5003-5007.
77. Brockhausen I, Yang J-M, Burchell J et al. Mechanisms underlying aberrant glycosylation of MUC1 mucin in breast cancer cells. Eur J Biochem 1995; 233:607-617.
78. Delannoy P, Pelczar H, Vandamme V et al. Sialyltransferase activity in FR3T3 cells transformed with ras oncogene: decreased CMP-Neu5tc: Galβ1-3GalNAc α-2,3-sialyltransferase. Glycoconj J 1993; 10:91-98.
79. Gillespie W, Paulson J, Kelm S et al. Regulation of α2,3-sialyltransferase expression correlates with conversion of peanut agglutinin $(PNA)^+$ to PNA^- phenotype in developing thymocytes. J Biol Chem 1993; 268:3801-3804.
80. Sadler JE, Rearick JI, Hill RL. Purification to homogeneity and enzymatic characterization of an α-N-acetylgalactosaminide α2-6-sialyltransferase from porcine submaxillary glands. J Biol Chem 1979; 254:5934-5941.
81. Bergh MLE, van den Eijnden DH. Aglycon specificity of fetal calf liver and ovine and porcine submaxillary gland α-N-acetylgalactosaminide α2-6 sialyltransferase. Eur J Biochem 1983; 136:113-118.
82. Bergh MLE, Hooghwinkel GJM, van den Eijnden DH. Biosynthesis of the O-glycosidically linked oligosaccharide chains of fetuin. J Biol Chem 1983; 258:7430-7436.
83. Kurosawa N, Kojima N, Inoue M et al. Cloning and expression of Galβ1,3GalNAc-specific GalNAc α2,6-sialyltransferase. J Biol Chem 1994; 269:19048-19053.
84. Itzkowitz SH, Yuan M, Montgomery CK et al. Expression of Tn, sialosyl-Tn, and T antigens in human colon cancer. Cancer Res 1989; 49:197-204.
85. Livingston BD, Paulson JC. Polymerase chain reaction cloning of a developmentally regulated member of the sialyltransferase gene family. J Biol Chem 1993; 268:11504-11507.
86. Nakayama J, Fukuda MN, Fredette B et al. Expression cloning of a human polysialyltransferase that forms the polysialylated neural cell adhesion molecule present in embryonic brain. In Proc Natl Acad Sci USA 1995; 92:7031-7035.
87. Troy F. Polysialylation: from bacteria to brains. Glycobiology 1992; 2:5-23.
88. Sarnesto A, Köhlin T, Thurin J et al. Purification of H-gene-encoded β-galactoside α1-2 fucosyltransferase from human serum. J Biol Chem 1990; 265:15067-15075.
89. Larsen RD, Ernst LK, Nair RP et al. Molecular cloning, sequence and expression of a human GDP-L-fucose:beta-D-galactoside 2-alpha-L-fucosyltransferase cDNA that can form the H blood group antigen. Proc Natl Acad Sci USA 1990; 87:6674-6678.
90. Kelly RJ, Ernst LK, Larsen RD et al. Molecular basis for H blood group deficiency in Bombay (O_h) and para-Bombay individuals. Proc Natl Acad Sci USA 1994; 91:5843-5847.
91. Kelly RJ, Rouquier S, Giorgi D. Sequence

and expression of a candidate for the human secretor blood group alpha(1,2)fucosyltransferase gene (FUT2)—Homozygosity for an enzyme-inactivating nonsense mutation commonly correlates with the nonsecretor phenotype. J Biol Chem 1995; 270:4640-4649.
92. Kuhns WJ, Schoentag R. Carcinoma-related alterations of glycosyltransferases in human tissues. Cancer Res 1981; 41:2767-2772.
93. Prieto PD, Mukerji P, Kelder B et al. Remodeling of mouse milk glycoconjugates by transgenic expression of a human glycosyltransferase. J Biol Chem 1995; 270:29515-29519.
94. Kukowska-Latallo JF, Larsen RD, Nair RP et al. A cloned human cDNA determines expression of a mouse stage-specific embryonic antigen and the Lewis blood group α(1,3/1,4)fucosyltransferase. Genes & Development 1990; 4:1288-1303.
95. Lowe JB, Kukowska-Latallo JF, Nair RP et al. Molecular cloning of a human fucosyltransferase gene that determines expression of the Lewis X and VIM-2 epitopes but not ELAM-1-dependent cell adhesion. J Biol Chem 1991; 266:17467-17477.
96. Weston BW, Nair RP, Larsen RD et al. Isolation of a novel human α(1,3) fucosyltransferase gene and molecular comparison to the human Lewis blood group α(1,3/1,4) fucosyltransferase gene. Syntenic, homologous, nonallelic genes encoding enzymes with distinct acceptor substrate specificities. J Biol Chem 1992; 267:4152-4160.
97. Weston BW, Smith PL, Kelly RJ. Molecular cloning of a fourth member of a human alpha(1,3)fucosyltransferase gene family. Multiple homologous sequences that determine expression of the Lewis x, sialyl Lewis x, and difucosyl sialyl Lewis x epitopes. J Biol Chem 1992; 267:24575-24584.
98. Sasaki K, Kurata K, Funayama K et al. Expression cloning of a novel alpha 1,3-fucosyltransferase that is involved in biosynthesis of the sialyl Lewis x carbohydrate determinants in leukocytes. J Biol Chem 1994; 269:14730-14737.
99. Prieels J, Monnom D, Dolmans M et al. Copurification of the Lewis blood group N-acetylglucosaminide α1-4 fucosyltransferase and N-acetylglucosaminide α1-3 fucosyltransferase from human milk. J Biol Chem 1981; 256:10456-10463.
100. Skacel PO, Edwards AJ, Harrison CT et al. Enzymic control of the expression of the X determinant (CD15) in human myeloid cells during maturation: the regulatory role of 6'-sialyltransferase. Blood 1991; 78:1452-1460.
101. Watkins WM. Biosynthesis. 5. Molecular basis of antigenic specificity in the ABO, H and Lewis blood-group systems. In: Glycoproteins. Montreuil H, Schachter H, Vliegenthart JFG eds. The Netherlands: Elsevier Science B.V. 1995; 313-390.
102. Yamamoto F, Hakomori S. Sugar-nucleotide donor specificity of histo-blood group A and B transferases is based on amino acid substitutions. J Biol Chem 1990; 265:19257-19262.
103. Yamamoto F, Marken J, Tsuji T et al. Cloning and characterization of DNA complementary to human UDP-GalNAc:Fucα1-2Galα1-3GalNAc transferase (Histo-blood group A transferase) mRNA. J Biol Chem 1990; 265:1146-1151.
104. Yamamoto F, Clausen H, White T et al. Molecular genetic basis of the histo-blood group ABO system. Nature 1990; 345:229-233.
105. Galili U, Shohet SB, Kobrin E et al. Man, apes and old world monkeys differ from other mammals in the expression of α-galactosyl epitopes on nucleated cells. J Biol Chem 1988; 263:17755-17762.
106. Joziasse DH, Shaper JH, Van den Eijnden DH et al. Bovine α1→3-galactosyltransferase: isolation and characterization of a cDNA clone. Identification of homologous sequences in human genomic DNA. J Biol Chem 1989; 264:14290-14297.
107. Larsen RD, Rajan VP, Ruff MM et al. Isolation of a cDNA encoding a murine UDPgalactose:β-D-galactosyl-1,4-N-acetyl-D-glucosaminide α-1,3-galactosyltransferase: expression cloning by gene transfer. Proc Natl Acad Sci USA 1989; 86:8227-8231.
108. Joziasse DH, Shaper JH, Jabs EW et al. Characterization of an α1→3-galactosyltransferase homologue on human chro-

mosome 12 that is organized as a processed pseudogene. J Biol Chem 1991; 266:6991-6998.
109. Shaper NL, Lin S-P, Joziasse DH et al. Assignment of two α-1,3-galactosyltransferase gene sequences (GGTA1 and GGTA1P) to chromosomes 9q33-q34 and 12q14-q15. Genomics 1992; 12:613-615.
110. Good AH, Cooper DKC, Malcolm AJ et al. Identification of carbohydrate structures that bind human antiporcine antibodies: Implications for discordant xenografting in humans. Transplant Proc 1992; 24:559-562.
111. Piller F, Blanchard D, Huet M et al. Identification of a α-NeuAc-(2-3)-β-D-galactopyranosyl N-acetyl-β-D-galactosaminyltransferase in human kidney. Carbohydr Res 1986; 149:171-184.
112. Serafini-Cessi F, Dall'Olio F, Malagolini N. Characterization of N-acetyl-β-D-galactosaminyltransferase from guinea-pig kidney involved in the biosynthesis of Sd[a] antigen associated with Tamm-Horsfall glycoprotein. Carbohydr Res 1986; 151:65-76.
113. Takeya A, Hosomi O, Kogure T. Identification and characterization of UDP-GalNAc:NeuAcα2-3Galβ1-4Glc(NAc) β1-4(GalNAc to Gal) N-acetylgalactosaminyltransferase in human blood plasma. J Biochem (Tokyo) 1987; 101:251-259.
114. Serafini-Cessi F, Malagolini N, Dall'Olio F. Characterization and partial purification of β-N-acetylgalactosaminyltransferase from urine of Sd(a+) individuals. Arch Biochem Biophys 1988; 266:573-582.
115. Dall'Olio F, Malagolini N, Di Stefano G et al. Postnatal development of rat colon epithelial cells is associated with changes in the expression of the β1-4 N-acetylgalactosaminyltransferase involved in the synthesis of Sd[a] antigen and of α2-6 sialyltransferase activity towards N-acetyllactosamine. Biochem J 1990; 270:519-524.
116. Albedi F, Cassano A, Ciaralli F et al. Ultrastructural identification of sulfated glycoconjugates in the Golgi apparatus in human colonic absorptive cells. Histochem 1989; 92:73-79.
117. Carter SR, Slomiany A, Gwozdzinski K et al. Enzymatic sulfation of mucus glycoprotein in gastric mucosa. J Biol Chem 1988; 263:11977-11984.
118. Kasinathan C, Liau YH, Murty VLN et al. Identification of gastric mucus glycoprotein sulfotransferase. Biochemistry International 1991; 24:43-49.
119. Slomiany A, Okazaki K, Piotrowski J et al. In vitro sulfation of sublingual salivary gland mucin; structures of ^{35}S-labelled oligosaccharides. Biochem Int 1989; 18:589-597.
120. Goso Y, Hotta K. Characterization of a rat corpus sulfotransferase for the 6-0-sulfation of β-D-N-acetylglucosamine residues on oligosaccharides. Glycoconj J 1993; 10:226.
121. Goso Y, Hotta K. Types of oligosaccharide sulfation, depending on mucus glycoprotein source, corpus or antral, in rat stomach. Biochem J 1989; 264:805-812.
122. Lo-Guidice J, Périni J-M, LaFitte J-J, et al. Characterization of a sulfotransferase from human airways responsible for the 3-0-sulfation of terminal galactose in N-acetyllactosamine-containing mucin carbohydrate chains. J Biol Chem 1995; 270:27544-27550.
123. Komatsu K, Driscoll W, Koh Y et al. A P-loop related motif (GxxGxxK) highly conserved in sulfotransferases is required for binding the activated sulfate donor. Biochem Biophys Res Comm 1994; 204:1178-1185.
124. Superti-Furga A, Hästbacka J, Wilcox WR et al. Achondrogenesis type IB is caused by mutations in the diastrophic dysplasia sulphate transporter gene. Nat Gen 1996; 12:100-102.

CHAPTER 5

Structures of β-N-Acetylglucosamine–Asn-Linked Oligosaccharides (N-Glycans)

N-glycans are common on soluble and membrane bound glycoproteins and are often cell-, tissue-, and species-specific and may vary among different glycoproteins synthesized in the same cell, and between glycosylation sites of the same protein.

For structural analysis, N-glycans may be released by hydrazine or by reductive amination, and more selectively by endoglycosidases such as endo-β-N-acetylglucosaminidase H (releasing high mannose structures) or endo-β-N-acetylglucosaminidase F (releasing most N-glycans). Radioactive or fluorescent labels may be introduced at the reducing end for chromatographic identification by comparison with oligosaccharide standards. In contrast to O-glycans, all N-glycans have one common core structure shown below, although they also exhibit great microheterogeneity. Thus glycoproteins, often present as broad bands after gel electrophoresis, are resolved as sharper bands after removal of the carbohydrate. The branched trimannosyl core structure:

```
Manα1-6
   \
    Manβ1-4GlcNAcβ1-4GlcNAcβ-
   /
Man α1-3
```

may be extended by Man residues (of high Man structures), or by Man residues at the Mana1-6-arm and at least one GlcNAc residue at the Mana1-3-arm (hybrid structures) (Fig. 3). Hybrid structures often have a bisecting GlcNAc residue b1-4-linked to the internal Manb1-4 residue. Complex N-glycans have at least two GlcNAc residues attached to the

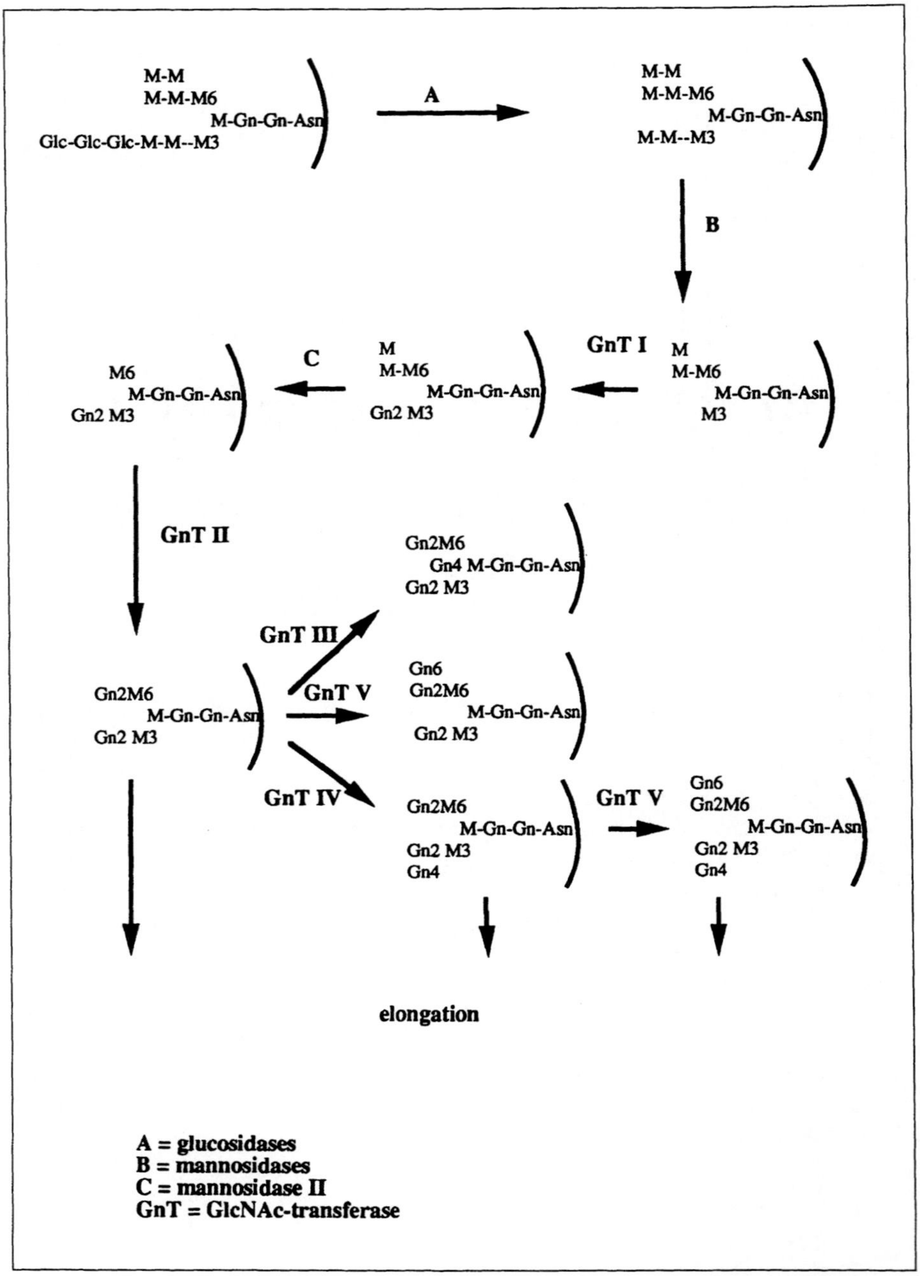

Fig. 3. N-glycan biosynthetic pathways. After preassembly of a high mannose chain with 3 Glc residues on dolichol-pyrophosphate, the glycan chain is transferred to Asn residues of Asn - x- Thr/Ser sequences of the peptide by oligosaccharyltransferase. The N-glycan chains are then processed. The three Glc residues are removed by glucosidases I and II (step A) and Man residues are cleaved by mannosidases in the ER and Golgi (step B) to form the Man_5 structure. Man_5 is the natural substrate for the first branching enzyme, GlcNAc-transferase I (GnT I), localized to the medial Golgi compartment. The product of the GnT I catalyzed reaction is a substrate for the medial Golgi enzyme mannosidase II (step C), and also for GnT III, α6-Fuc-transferase and possibly other enzymes. After removal of the two Man residues, complex N-glycans can be assembled by GnT II, III, IV and V. The GlcNAc-terminating structures may be elongated to form "antennae" that carry glycans important for cell adhesion and other biological phenomena. Gn, GlcNAc; M, mannose.

core; these are classified as bi- tri-, tetra- or penta-antennary structures and may also carry a bisecting GlcNAc residue. The most highly branched complex N-glycan, a bisected penta-antennary structure (Fig. 3) is found in hen ovomucoid and in eggs from other species. These complex N-glycans are thought to present their elongated, branched, often antigenic carbohydrate chains as antennae to their biological environment, implying that the presentation of these sugars relates to specific cellular functions.

With few exceptions, N-glycan chains may be elongated in a similar fashion as O-glycans. Single or repeating Galβ1-3/4-GlcNAcβ1-3 residues (poly N-acetyllactosaminoglycans, blood group i antigen, Table 1), or additional GlcNAcβ1-6 branches at the Gal residues (blood group I antigen, Table 1) are found. Sugars at the nonreducing termini may be sialic acid, Gal, GalNAc, Fuc or GlcNAc in a number of different linkages. Sialic acid is commonly found in α2-6-linkage but may also occur in other types of linkages. Phosphate groups are found esterified to Man in lysosomal enzymes. Blood group A, B, H and the Lewis determinants (Table 1) may occur at the nonreducing end of N-glycans.

Sulfate may be added to the core or the antennae of hybrid and complex N-glycan chains. Structural and metabolic studies have documented these sulfate linkages on a variety of glycoproteins. In avian adipocytes, lipoprotein lipase is degraded before secretion. Metabolic labelling studies established that inhibition of sulfation by chlorate reduces the secretion of lipoprotein lipase in a dose-dependent manner. This effect may be due to the undersulfation of proteoglycans which are essential for the degradation process. Sulfate is attached to GlcNAc N-linked to Asn-45 of adipocyte lipoprotein lipase.[1]

Sulfate on N-glycans of glycoprotein hormones such as bovine lutropin may be in a characteristic 4-O-sulfated GalNAcβ1-4 structure.[2] N-linked structures from calf thyroglobulin contain Galβ1-4GlcNAc sequences with sulfate attached to the 3-position of terminal Gal residues.[3] Porcine thyroglobulin carries complex N-linked oligosaccharides with sialylated and sulfated termini, with sulfate attached the 3-position of Gal and the 6-position of GlcNAc.[4] Sulfated N-glycans present in recombinant human tissue plasminogen activator were characterized by the presence of a Neu5Acα2-3 (6-sulfated) Gal unit.[5] The functions of these sulfated structures are largely unknown. Sulfate may protect sugars from degradation by glycosidases. In addition, sulfate may be involved in determining the half life of glycoprotein hormones.[2]

References

1. Hoogewerf AJ, Cisar LA, Evans DC et al. Effect of chlorate on the sulfation of lipoprotein lipase and heparan sulfate proteoglycans. Sulfation of heparan sulfate proteoglycans affects lipoprotein lipase degradation. J Biol Chem 1991; 266:16564-16571.
2. Green E, Morishima C, Boime I et al. Structural requirements for sulfation of asparagine linked oligosaccharides of lutropin. Proc Natl Acad Sci USA 1985; 82:7850-7854.
3. Kato Y, Spiro R. Characterization of a thyroid sulfotransferase responsible for the 3-0-sulfation of terminal β-D-galactosyl residues in N-linked carbohydrate units. J Biol Chem 1989; 264:3364-3371.
4. de Waard P, Koorevaar A, Kamerling J et al. Structure determination by H NMR spectroscopy of sulfated sialylated N-linked carbohydrate chains released from porcine thyroglobulin by peptide-N^{4}-(N-acetyl -β-glucosylaminyl) asparagine Amidase-F*. J Biol Chem 1991; 266:4237-4243.
5. Pfeiffer G, Strube K, Geyer R. Biosynthesis of sulfated glycoprotein N-glycans present in recombinant human tissue plasminogen activator. Biochem Biophys Res Comm 1992; 189:1681-1685.

CHAPTER 6

Biosynthesis of N-Glycans

The initial steps in the biosynthesis of N-glycans have been are preserved throughout evolution and are similar in lower and higher species. In contrast to O-glycans, N-glycans are pre-assembled as a dolichol-pyrophosphate-(Dol-P-P-) intermediate and then transferred to protein by the action of oligosaccharyltransferase in the ER. Subsequent processing includes the trimming of Glc and Man residues and the addition of various sugars by Golgi glycosyltransferases (Fig. 3).

6.A. BIOSYNTHETIC REACTIONS IN THE ENDOPLASMIC RETICULUM

Biosynthetic reactions that take place on the cytoplasmic side of the ER membrane utilize nucleotide-sugar donors, and reactions that take place on the lumenal side utilize the dolichol derivatives. The GlcNAc residues and five of the mannosyl residues in the $Glc_3Man_9GlcNAc_2$ precursor oligosaccharide are derived by glycosyl transfer from nucleotide sugars on the cytoplasmic side of the ER membrane. Following translocation of the heptasaccharide $Man_5GlcNAc_2$-P-P-Dol across the ER membrane, the last four mannose and three glucosyl residues are transferred from Dol-linked intermediates on the lumenal side of the ER membrane.[1] The oligosaccharide is then transferred to appropriate Asn residues in nascent polypeptide acceptors on the lumenal side of the microsomal membrane. The presence of completed oligosaccharide chains on the lumenal side of the ER membrane requires the translocation of hydrophilic carbohydrates across the hydrophobic barrier of the membrane. The lipophilic dolichol moiety functions to anchor the activated sugar residues, as well as the growing oligosaccharide chains, in membranes of the ER. The long chain polyisoprenes are probably also involved in the translocation of the activated carbohydrates across membranes.

The first steps in the assembly of dolichol derivatives up to the $Man_5GlcNAc_2$-P-P-Dol structure take place at the cytoplasmic surface of the ER membrane by transfer of GlcNAc and Man from cytosolic UDP-GlcNAc and GDP-Man. The initiation of N-glycan synthesis, the synthesis of Dol-P-P-GlcNAc by GlcNAc-1-phospho-transferase (Table 2), is inhibited by the antibiotic tunicamycin, a structural analog of UDP-GlcNAc, which has the ability to enter cells and block the entire N-glycan biosynthetic pathways (Table 3). GlcNAc-1-phospho transferase has been cloned in yeast and mammalian cells and is an integral multiple-ER membrane spanning protein.[2,3]

The β4-Man-transferase adding the first Man residue to $GlcNAc_2$-P-P-Dol is possibly a type 1 membrane protein with the active site at the carboxyl-terminal oriented towards the cytoplasm.[4,5] After translocation of $Man_5GlcNAc_2$-P-P-Dol across the ER membrane by an unknown mechanism, Man and Glc residues are added at the lumenal membrane surface by transfer from Dol-P-Man and Dol-P-Glc. Dol-P-Man is synthesized by Dol-P-Man synthase (Table 2), an ER membrane protein with the catalytic site oriented towards the cytoplasm.[6] The activated mannose residue must then be translocated to the luminal side of the ER membrane. The mechanism of this translocation is still unknown but it is likely that the dolichyl moiety promotes the translocation due to the destabilizing effects of the long chain polyisoprenol on phospholipid bilayer properties.[7]

Dol-P-Man synthase has been cloned from yeast and expressed as an active enzyme in *E. coli*.[6] More recently the gene was inserted in a high expression vector that allowed the purification of the enzyme to homogeneity[8] although the mammalian enzyme had previously been highly purified;[9] both enzymes have unique requirements for phospholipids for optimal catalytic activity. Yeast Dol-P-Glc synthase (Table 2) has significant homology with Dol-P-Man synthase. Dol-P-Glc synthase has been cloned, and was also expressed as active enzyme in *E. coli*.[10] Eight different enzymes in the dolichol cycle have been identified.[11,12] The activities appear to be regulated by the composition of the membrane environment. For example, Dol-P-Man synthase from rat liver has a characteristic lipid requirement for optimal activity. Products of phospholipid metabolism have a marked effect on activity when these are added to the enzyme reconstituted in phosphatidylcholine or phosphatidylethanolamine vesicles.[13]

The final product of these reactions, $Glc_3Man_9GlcNAc_2$, is transferred to protein by an oligosaccharyl transferase complex.[14] Several proteins are associated with the enzyme which appears to be intimately associated with membrane bound ribosomes. One of the proteins essential for oligosaccharyl transferase activity has been identified in yeast as a WBP1 protein.[15] The enzyme is specific for the Asn-X-Ser/Thr-X sequon in the glycoprotein substrate (where X cannot be Pro).[16] Interestingly, the tripeptide Asn-x-Thr is effectively glycosylated and transported, yet not all Asn-X-Ser/Thr sequences of glycoproteins carry N-glycans.[17] The genes encoding oligosaccharyltransferase subunits in yeast are subject to control during the growth cycle.[18]

Upon transfer, the N-glycan chains are processed by the removal of α2- and α3-linked Glc residues by glucosidase I and II, respectively, and of one α2-linked Man residue by a specific ER α-mannosidase.[19] These cleavage steps are known to be blocked by inhibitors such as the antibiotic deoxynojirimycin and the plant alkaloid castanospermine which inhibit glucosidase, and deoxymannojirimycin which inhibits mannosidase (Table 3).

6.B. GOLGI PROCESSING REACTIONS

The trimmed N-glycoprotein may retain high Man structures or may be processed further to hybrid and complex structures after its transport into the Golgi. Cis-Golgi localized α-mannosidase I is an important enzyme for this processing since it may cleave Man residues to expose the Manα1-3 arm for the addition of GlcNAc. Golgi α-mannosidase I is cell type-, glycoprotein- and site-specific; the action of the enzyme may be regulated by the accessibility or the conformation of the carbohydrate chain, which may be a result of interaction with or folding of the protein moiety of substrates. The enzyme is a type II membrane bound glycoprotein; it is dependent on Ca^{2+} and has homology to Ca^{2+}-binding proteins.[20,21] Further processing is catalyzed by sequentially arranged glycosyltransferases in the Golgi apparatus.

6.B.1. Synthesis of Hybrid and Complex N-Glycans

The pathways of GlcNAc additions to the N-glycan core are guided by the substrate specificities and relative activities of the enzymes involved. Many of the possible pathways in the assembly of complex N-glycans remain to be elucidated. After processing of the Man residues to the Man_5 structure (Fig. 3), the first GlcNAc is added to the Manα1-3 arm by β2-GlcNAc-transferase I, which has been localized to the medial Golgi.[22] This biosynthetic step is an essential prerequisite for the conversion of high mannose to hybrid or complex N-glycan structures and initiates the formation of antennae. The α1-3- and α1-

Table 3. Inhibitors of glycosylation

Name	Enzymatic step inhibited	Type of inhibition	Chapter/ Reference
Tunicamycin	synthesis of Dol-P-GlcNAc	strong	8/45,46
Castanospermine	processing glucosidase I and II	strong	8/45,46
Deoxynojirimycin	processing glucosidase I and II	strong	8/45,46
Deoxymannojirimycin	processing α-mannosidase I	strong	8/45,46
Swainsonine	α-mannosidase II	strong	8/45,46
Manα6(6-5-iodoacetamido-pentyl Manα3) Man-R	GlcNAc-T I	irreversible	8/37
Manα6(6-4,4azopentyl Manα3) Man-R		irreversible	
Manα6(6-deoxy-Manα3) Man-R		competitive	
GlcNAcβ2(2-deoxy-)Manα6 (Manα3)Man-R	GlcNAc-T II	competitive	8/55
GlcNAcβ2(3-O-pentyl-) Manα6(Manα3)Man-R		competitive	
GlcNAcβ2Manα6Glc-R derivatives	GlcNAc-T V		8/53,54
GlcNAcβ2(4,6dimethyl) Manα6Glc-R	GlcNAc-T V	weak	8/58
GlcNAcβ2(4-methyl)Manα6 Glc-R	GlcNAc-T V	competitive	8/57
GalNAcα-aryl	O-glycan extension	competitive	8/49,50
4-azido-UDP-GalNAc/ GlcNAc	UDP-GalNAc/GlcNAc-binding transferases	irreversible	8/47,48
CMP-sialic acid fluoro derivative	sialyltransferases		8/56
Thio-substrate derivative	β4-Gal-T		8/59
Galβ3GalNAc-p-nitrophenyl	core 2 β6GlcNAc-T	strong irreversible	8/59
deoxy-substrate derivatives Gal-T	GlcNAc-T	competitive	8/53,54
3-deoxy-Galβ3GalNAc-R	α3-sialyl-T	weak competitive	4/29
Galβ3(6-deoxy)GalNAc-R VIII/53	core 2 β6-GlcNAc-T	weak competitive	4/29
2-deoxy-Galβ3GlcNAcβ-R	α2-Fuc-T	weak competitive	8/53
Galβ3(4-deoxy)GlcNAcβ-R	α4-Fuc-T	weak competitive	8/53
3-deoxy-Galβ3GlcNAcβ-R	α3-sialyl-T (N)	weak competitive	8/53
Bisubstrate analog (GDP linked to Galβ-phenyl)	α2-Fuc-T	competitive	8/52

6-linked Man residues of the Manα1-6 arm are then removed by mannosidase II, a medial Golgi enzyme inhibited by swainsonine and requiring the prior action of GlcNAc-transferase I. Further branching by GlcNAc-transferases II to VI may occur, and Gal, GlcNAc and terminal sugars as well as sulfate may be added to form multi-antennary complex N-glycan structures presenting a variety of epitopes. The addition of Fucα1-6 to the Asn-linked GlcNAc of the core, and the action of the branching enzymes GlcNAc-transferase II, III and IV and α-mannosidase II (Fig. 3) all depend on the GlcNAcβ1-2Manα1-3 moiety synthesized by GlcNAc-transferase I for substrate binding.

GlcNAc-transferase I (Table 2, Fig. 3) has been purified and cloned in several species,[22] and has been localized to human chromosome 5q35. The enzyme appears to be present in all mammalian cells, although a phytohaemagglutinin-resistant mutant cell line LEC1[23] has been shown to be deficient in the enzyme. Knockout mouse embryos live for about 10 days, indicating that GlcNAc-transferase I is essential for a developing organism.[24,25] This implies a specific function of hybrid or complex N-glycans in cell-cell interactions which are not essential for cells in culture.

After the action of GlcNAc-transferase I and mannosidase II, the Manα1-6 residue may be substituted by the next branching enzyme, β2-GlcNAc-transferase II (Table 2, Fig. 3), to form bi-antennary N-glycans. The enzyme has been purified and cloned and localized to human chromosome 14q21.[22] The activity is ubiquitous with the rare exception of patients with a carbohydrate deficient glycoprotein syndrome (chapter 16).[26] In these patients, the lack of GlcNAc-transferase II is associated with severe disturbances of neural development. GlcNAc-transferase I and II show little sequence homology although they both bind UDP-GlcNAc, Mn^{2+} and substrates of similar structure, and synthesize GlcNAcβ1-2Man linkages, suggesting that the binding sites are similar but may be composed of amino acids from distant parts of the protein.

At various stages of hybrid or complex-type N-glycan branching, a "bisecting" GlcNAc can be linked to the internal Manβ-residue by β4-GlcNAc-transferase III (Table 2, Fig. 3).[27] This step has special significance in that it inhibits many of the otherwise possible elongation and branching reactions, either due to lack of accessibility or due to the peculiar conformation of bisected structures, which may not be favorable for the recognition of processing enzymes.[28] The enzyme has been purified and cloned and has been localized to human chromosome 22q.13.1.[29,30] GlcNAc-transferase III is expressed in a cell type-specific fashion and appears to be regulated in leukemia[31] and in rat hepatoma.[32,33]

β4-GlcNAc-transferase IV adds a GlcNAcβ1-4 residue to the Manα1-3 arm and is thus involved in the synthesis of tri-antennary structures (Table 2, Fig. 3).[34] Like other branching GlcNAc-transferases, the enzyme appears to be regulated during differentiation.[35,36] Although the synthesis of tri-antennary chains appears to be efficient in vivo, for example in the synthesis of serum glycoproteins in the liver, the activity measured in vitro is always very low. This may indicate that the intracellular organization of glycosyltransferases, which is important for efficient glycan synthesis, is disturbed in in vitro assays.

β6-GlcNAc-transferase V acts after GlcNAc-transferase II and adds the GlcNAcβ1-6 branch to the Manα1-6 arm of bi- and tri-antennary substrates to form tri- and tetra-antennary N-glycans, respectively (Table 2, Fig. 3).[22] The enzyme does not need divalent metal ions for activity and can function in the presence of EDTA. The enzyme has been implicated in cancer, transformation and metastasis (see below). Transfection of the GlcNAc-transferase III gene into mouse melanoma cells showed that the metastatic potential can be reduced by the overexpression of bisected structures which is expected to suppress the formation of GlcNAc-transferase V products.[37] Direct transfection of the GlcNAc-transferase V gene into epithelial cells demonstrated reduced adhesion properties.[38] The activity appears to be regulated by TGF β[39] and by phosphorylation through protein kinases.[40] GlcNAc-transferase V has been isolated from hamster kidney.[41] The cDNA has been cloned and the gene has been localized to human chromosome 2q21.[42] The gene has an unusually high number of 17 exons and codes for a protein with similar domain structure but much longer peptide chain than other glycosyltransferases.

GlcNAc-transferase VI adds the final branch to N-glycans to form penta-antennary structures in birds and other non-mammalian species.[43] The enzyme requires the prior actions of GlcNAc-transferases II and V (Table 2, Fig. 3). An enzyme with similar specificity was detected in hen oviduct microsomes and termed GlcNAc-transferase VI'.[44] This enzyme does not require the prior action of GlcNAc-transferase V. GlcNAc-transferase VI synthesizes the intersected structure with β1-2-, β1-4, and β1-6-linked GlcNAc residues on the Manα1-6 arm while GlcNAc-transferase VI' synthesizes the branched β1-2- and β1-4-Man structure on the Manα1-6 arm, similar to the product of GlcNAc-transferase IV on the Manα1-3 arm. GlcNAc-transferases VI and VI' have not yet been demonstrated in human tissues.

6.B.2. Addition of the Mannose-6-Phosphate (Man-6-P) Marker

Lysosomal enzymes carry a Man-6-P group which allows them to bind to Man-6-P receptors with subsequent targeting to lysosomal compartments.[45] The synthesis of the Man-6-P group takes place in several steps. A GlcNAc-1-phosphate residue is first transferred to carbon 6 of a pen–ultimate Man residue of high Man structures by a lysosomal enzyme, GlcNAc-1-phos–photransferase, (Table 2) in the cis-Golgi.[46] The enzyme recognizes a specific domain present in lysosomal enzymes.[47] Subsequently, GlcNAc is cleaved by a specific hexosaminidase, exposing Man-6-P. Man-6-P receptors are recirculated and internalized based on the recognition of a Tyr-Ser-Lys-Val sequon by proteins in coated pits of the endocytic pathways. Other receptors may have similar recognition signals, for example, the Tyr-Thr-Arg-Phe sequence of the transferrin receptor.[48] A deficiency in the ability to add the Man-6-P marker leads to a general deficiency in lysosomal enzymes and accumulation of undegraded material in the lysosomes in I cell disease.[49]

6.B.3. Elongation Reactions

Elongation reactions in the synthesis of N-glycans involve the addition of poly-n-acetyl-lactosamine chains to the terminal GlcNAc residues with the exception of the bisected GlcNAc residue which remains unsubstituted. The enzymes synthesizing Galβ1-3/4 GlcNAcβ1-3 repeating units (i antigen) or GlcNAcβ1-6 branches (I antigen) are likely to be the same as those elongating O-glycans. N-glycans of glycoprotein hormones, in addition, may be substituted by GalNAcβ1-4 residues (instead of Galβ1-4) which then becomes sulfated. The β4-GalNAc-transferase has been characterized in a number of tissues (see section 6.B.5)

6.B.4. Terminal Reactions

The additions of most terminal sugars to N-glycans chains are catalyzed by the enzymes that have been described for O-glycans. However, several terminal structures are apparently unique for N-glycans. These include Fucα1-6-linked to GlcNAc attached to Asn and sialic acid α2-6-linked to Gal residues. Sialic acid α2-3 linked to Gal occurs on both O- and N-glycans but it has not been established if these linkages are synthesized by the same enzyme (Fig. 2, Table 2).

The α6-Fuc-transferase that adds Fuc to the core GlcNAc requires an unsubstituted GlcNAcβ1-2Manα1-3 arm, synthesized by GlcNAc-transferase I, for recognition but it does not require peptide. The enzyme has been purified from human skin fibroblasts.[50]

Insects such as honey bees possess an α3-Fuc-transferase which synthesizes the

+/-Fucα1-6 (Fucα1-3) GlcNAcβ-Asn structure.[51] This enzyme is absent from humans who may become sensitized and develop anti Fucα1-3 antibodies.

The α6- and α3-sialyltransferases acting on Gal residues of N-glycans have been purified and cloned.[52-56] α2-3 linked sialic acid is preferably added to Galβ1-3 GlcNAc- termini and not to type 2 chains. The cDNA encoding α3-sialyltransferase has been cloned from human placenta[56] and rat liver.[53] It is possible that this enzyme is involved in the regulation of sialyl-Le^x and sialyl-Le^a expression. Another α3-sialyltransferase from human placenta acts on type 2 chains rather than type 1, and also can utilize Galβ1-3 GalNAc substrates.[57,58]

The expression of α6-sialyltransferase is subject to tissue-specific and growth-dependent regulation (see below). Human colon cancer tissue has high levels of the activity.[59] The invasiveness of oncogene transformed rat fibroblasts is correlated with α6-sialyltransferase levels.[60] The gene is regulated by at least three promoters.[61-63]

α8-sialyltransferases synthesizing the polysialic acid chains of NCAM are expressed in a tissue- and stage-specific fashion. One can assume that these α8-sialyltransferases act on termini of both N- and O-glycans. The cDNAs encoding several enzyme species including fetal human brain, have been cloned.[64]

The enzyme that synthesizes the sialylα2-6 GlcNAc- linkage found in human milk and other glycoproteins was characterized from human placenta[65] and other sources. This GlcNAc α6-sialyltransferase acts on sialylα2-3 Galβ1-4 GlcNAc- structures; the α3-sialyltransferase (N) therefore has to act before the α6-sialyltransferase, similar to the sequence in the O-glycan pathways.

6.B.5. Biosynthesis of Sulfated N-Glycans

The PAPS substrate requirements for N-glycan sulfation are the same as those described for sulfated O-glycans. It is not yet known if sulfotransferases acting on O-glycans also act on N-glycans, and the number of different enzymes that exist. The addition of sulfate to form 4-sulfated GalNAcβ1-4 N-glycans is catalyzed by an oligosaccharide-specific sulfotransferase.[66] Incorporation of sulfate into GalNAc-R substrate was stimulated by the presence of basic proteins, including deglycosylated lutropin, suggesting that the protein portions of glycoprotein substrates may modulate sulfo-transferase activity.[67] Other enzymes, including chondroitin-6-sulfotransferase and chondroitin-4-sulfo-transferase, have been shown to be similarly responsive to polycationic stimulation.[68,69]

A GalNAc-transferase and GalNAc 4-sulfo-transferase are present in bovine submaxillary gland. The sulfotransferase has similar or identical kinetic parameters as the pituitary enzyme and in addition can sulfate terminal GalNAc residues on carbonic anhydrase VI. A different, non-sulfated glycoform of carbonic anhydrase VI is found in parotid gland which lacks sulfotransferase but possesses β4-GalNAc-transferase.[70] This suggests that the sulfotransferase is not hormone-specific but acts on various oligosaccharides with terminal GalNAcβ1-4 residues.

The addition of sulfate to the N-glycan core occurs after some processing, at least by glucosidases.[71] Castanospermine treatment which inhibits glucosidase I results in sulfate-free glycans of Glc_3 $Man_{7-9}GlcNAc_2$ types. Thus, removal of glucose units appears to be necessary before sulfate can be added.

Modifications of sulfated glycans can be induced by changes in the membrane transport of sulfate,[72,73] alterations in PAPS biosynthesis by specific inhibitors such as chlorate[74-76] and certain drugs,[77,78] or by sulfatases.[79,80]

References

1. Hirschberg CB, Snider MB. Topography of glycosylation in the rough endoplasmic reticulum and Golgi apparatus. Ann Rev Biochem 1987; 56:63-87.
2. Zhu X, Lehrman MA. Cloning, sequence, and expression of a cDNA encoding hamster UDP-GlcNAc:dolichol phosphate N-

acetylglucosamine-1-phosphate transferase. J Biol Chem 1990; 265:14250-14255.
3. Scocca JR, Krag SS. Sequence of a cDNA that specifies the uridine diphosphate N-acetyl-D-glucosamine:dolichol phosphate N-acetylglucosamine-1-phosphate transferase from chinese hamster ovary cells. J Biol Chem 1990; 265:20621-20626.
4. Albright CF, Robbins PW. The sequence and transcript heterogeneity of the yeast gene ALG1 and essential mannosyltransferase involved in N-glycosylation. J Biol Chem 1990; 265:7041-7049.
5. Couto JR, Huffaker TC, Robbins PW. Cloning and expression in Escherichia coli of a yeast mannosyltransferase from the asparagine-linked glycosylation pathway. J Biol Chem 1983; 259:378-382.
6. Orlean P, Albright C, Robbins PW. Cloning and sequencing of the yeast gene for dolichol phosphate mannose synthase, an essential protein. J Biol Chem 1988; 263:17499-17507.
7. Schutzbach JS, Jensen JW, Lai C-S et al. Membrane structure and mannosyltransferase activities:the effects of dolichols on membranes. Chemica Scripta 1987; 27:109-118.
8. Schutzbach JS, Zimmerman JW, Forsee WT. The purification and characterization of recombinant yeast dolichyl-phosphate-mannose synthase. J Biol Chem 1993; 268:24190-24196.
9. Jensen JW, Schutzbach JS. Activation of dolichyl-phospho-mannose synthase by phospholipids. Eur J Biochem 1985; 153:41-48.
10. te Heesen S, Lehle L, Weissmann A et al. Isolation of the ALG5 locus encoding the UDP-Glucose:dolichyl-phosphate glucosyltransferase from Saccharomyces cerevisiae. Eur J Biochem 1994; 224:71-79.
11. Kukuruzinska MA, Bergh MLE, Jackson BJ. Protein glycosylation in yeast. Ann Rev Biochem 1987; 56:915-944.
12. Herscovics A, Orlean P. Glycoprotein biosynthesis in yeast. FASEB J 1993; 7:540-550.
13. Jensen JW, Schutzbach JS. Phospholipase-induced modulation of dolichyl-phosphomannose synthase activity. Biochemistry 1989; 28:851-855.
14. Verbert A. From $Glc_3Man_9GlcNAc_2$-protein to $Man_5GlcNAc_2$-protein:transfer 'en bloc' and processing. In:Glycoproteins. Montreuil J, Vliegenthart JFG, Schachter H eds. New York:Elsevier Pub 1995; 145-152.
15. te Heesen S, Janetzky B, Lehle L et al. The yeast WBP1 is essential for oligosaccharyltransferase activity in vivo and in vitro. EMBO J 1992; 11:2071-2075.
16. Bause E. Structural requirements of N-glycosylation of proteins. Studies with proline peptides as conformational probes. Biochem J 1983; 209:331-336.
17. Wieland FT, Gleason ML, Serafini TA et al. The rate of bulk flow from the endoplasmic reticulum to the cell surface. Cell 1987; 50:289-300.
18. Lennon K, Pretel R, Kesselheim J et al. Proliferation-dependent differential regulation of the dolichol pathway genes in Saccharomyces cerevisiae. Glycobiol 1995; 5:633-642.
19. Bischoff J, Moremen K, Lodish HF. Isolation, characterization, and expression of cDNA encoding a rat liver endoplasmic reticulum alpha-mannosidase. J Biol Chem 1990; 265:17110-17117.
20. Lal A, Schutzbach JS, Forsee WT et al. Isolation and expression of murine and rabbit cDNAs encoding an α1,2-mannosidase involved in the processing of asparagine-linked oligosaccharides. J Biol Chem 1994; 269:9872-9881.
21. Schutzbach JS, Forsee WT. Calcium ion activation of rabbit liver α1,2-mannosidase. J Biol Chem 1990; 265:2546-2549.
22. Schachter H. Biosynthesis 2c: Glycosyltransferases involved in the synthesis of N-glycan antennae. In:Montreuil J, Vliegenthart JFG, Schachter H eds. Glycoproteins. Elsevier pub. 1995:153-199.
23. Stanley P, Narasimhan S, Siminovitch L et al. Chinese Hamster Ovary cells selected for resistance to the cytotoxicity of phytohemagglutinin are deficient in a UDP N-acetylglucosaminyl glycoprotein N-acetylglucosaminyltransferase activity. Proc Natl Acad Sci USA 1975; 72:3323-3327.
24. Metzler M, Gertz A, Sarkar M et al. Complex asparagine-linked oligosaccharides are required for morphogenic events during post-implan-

tation development. EMBO J 1994; 13:2056-2065.

25. Ioffe E, Stanley P. Mice lacking N-Acetylglucosaminyltransferase I activity die at mid-gestation, Revealing an essential role for complex or hybrid N-linked carbohydrates. Proc Natl Acad Sci USA. 1994; 91:728-732.
26. Tan J, Dunn J, Jaeken J. Carbohydrate-deficient glycoprotein syndrome type II. An autosomal recessive disease due to point mutations in the coding region of the N-acetylglucosaminyltransferase II gene. Glycoconj J 1995; 12:478.
27. Narasimhan S. Control of glycoprotein synthesis VII. UDP-GlcNAc:glycopeptide β4-N-acetylglucosaminyltransferase III, an enzyme from hen oviduct which adds GlcNAc in β1-4 linkage to the β-linked mannose of the trimannosyl core of N-glycosyl oligosaccharides. J Biol Chem 1982; 257:10235-10242.
28. Brisson JR, Carver JP. The relation of three-dimensional structure to biosynthesis in the N-linked oligosaccharides. Can J Biochem Cell Biol 1993; 61:1067-1078.
29. Nishikawa A, Ihara Y, Hatakeyama M et al. Purification, cDNA cloning, and expression of UDP-N-acetylglucosamine:β-D-mannoside β-1,4N-acetylglucosaminyltransferase III from rat kidney. J Biol Chem 1992; 267:18199-18204.
30. Ihara Y, Nishikawa A, Tohma T et al. cDNA cloning, expression, and chromosomal localization of human N-acetylglucosaminyltransferase III (GnT-III). J Biochem 1993; 113:692-698.
31. Yoshimura M, Nishikawa A, Ihara Y et al. High expression of UDP-N-acetylglucosamine:β-D-mannoside β-1,4-N-acetylglucosaminyltransferase III (GnT III) in chronic myelogenous leukemia in blastic crisis. Int J Cancer 1995; 60:443-449.
32. Narasimhan S, Schachter H, Rajalakshmi S. Expression of N-acetylglucosaminyltransferase III in hepatic nodules during rat liver carcinogenesis promoted by orotic acid. J Biol Chem 1988; 263:1273-1281.
33. Miyoshi E, Nishikawa A, Ihara Y et al. N-acetylglucosaminyltransferase III and V messenger RNA levels in LEC rats during hepatocarcinogenesis. Cancer Res 1993; 53:3899-3902.
34. Gleeson PA, Schachter H. Control of glycoprotein synthesis. VIII. UDP-GlcNAc: GnGn (GlcNAc to Manα1-3) β4-N-acetylglucosaminyltransferase IV, an enzyme in hen oviduct which adds GlcNAc in β1-4 linkage to the α1-3-linked Man residue of the trimannosyl core of N-glycosyl oligosaccharides to form a tri-antennary structure. J Biol Chem 1983; 258:6162-6173.
35. Koenderman A, Wijermans P, van den Eijnden D. Changes in the expression of N-acetylglucosaminyltransferase III, IV, V associated with the differentiation of HL60 cells. FEBS Lett 1987; 222:42-46.
36. Brockhausen I, Romero P, Herscovics A. Glycosyltransferase changes upon differentiation of CaCo-2 human colonic adenocarcinoma cells. Cancer Res 1991; 51:3136-3142.
37. Yoshimura M, Nishikawa A, Ihara Y. Suppression of lung metastasis of B16 mouse melanoma by N-acetylglucosaminyltransferase III gene transfection. Proc Natl Acad Sci. USA 1995; 92:8754-8758.
38. Demetriou M, Nabi IR, Coppolino M et al. Reduced contact-inhibition and substratum adhesion in epithelial cells expressing GlcNAc-Transferase V. J Cell Biol 1995; 130:383-392.
39. Miyoshi E, Nishikawa A, Ihara Y et al. Transforming growth factor β up-regulates expression of the N-acetylglucosaminyltransferaseV gene in mouse melanoma cells. J Biol Chem 1995; 270:6216-6220.
40. Ju T-Z, Chen H-L, Gu J-X. Regulation of N-acetylglucosaminyltransferase V by protein kinases. Glycoconj J 1995; 12:767-772.
41. Shoreibah MG, Hindsgaul O, Pierce M. Purification and characterization of rat kidney UDP-N-acetylglucosamine:α-6-D-mannoside β-1,6-N-acetylglucosaminyltransferase. J Biol Chem 1992; 267:2920-2927.
42. Saito H, Nishikawa A, Gu JG. CDNA cloning and chromosomal mapping of human N-acetylglucosaminyltransferase V. Biochem Biophys Res Commun 1994; 198:318-327.
43. Brockhausen I, Hull E, Hindsgaul O et al. Control of glycoprotein synthesis. Detection and characterization of a novel branching

enzyme from hen oviduct, UDP-N-acetylglucosamine:GlcNAcβ1-6(GlcNAcβ1-2)Manα-R (GlcNAc to Man) β-4-N-acetylglucosaminyltransferase VI. J Biol Chem 1989; 264:11211-11221.

44. Brockhausen I, Möller G, Yang JM et al. Control of glycoprotein synthesis. Characterization of (1→4)-N-acetyl-beta-D-glucosaminyltransferases acting on the alpha-D-(1→3)-and alpha-D-(1→6)-linked arms of N-linked oligosaccharides. Carbohydr Res 1992; 236:281-299.
45. Kornfeld S. Lysosomal enzyme targeting. Biochem Soc Trans 1990; 18:367-374.
46. Ketcham CM, Kornfeld S. Characterization of UDP-N-acetylglucosamine:glycoprotein N-acetylglucosamine-1-phosphotransferase from Acanthamoeba castellanii. J Biol Chem 1992; 267:11654-11659.
47. Baranski T, Koelsch G, Hartsuck JA et al. Mapping and molecular modelling of a recognition domain for lysosomal enzyme targeting. J Biol Chem 1991; 266:23365-23372.
48. Jadot M, Canfield WM, Gregory W et al. Characterization of the signal for rapid internalization of the bovine mannose-6-phosphate/insulin-like growth factor-II receptor. J Biol Chem 1992; 267:11069-11077.
49. Reitman ML, Varki A, Kornfeld S. Fibroblasts from patients with I-cell disease and pseudo-Hurler polydystrophy are deficient in uridine 5'-diphosphate-N-acetylglucosamine:glycoprotein N-acetylglucosaminylphosphotransferase. J Clin Invest 1981; 67:1574-1579.
50. Voynow JA, Kaiser RS, Scanlin TF et al. Purification and characterization of GDP-L-fucose-N-acetyl β-D-glucosaminide α1-6 fucosyltransferase from cultured human skin fibroblasts. Requirement of a specific biantennary oligosaccharide as substrate. J Biol Chem 1991; 266:21572-21577.
51. Staudacher E, Altmann F, Glössl J. GDP-fucose:beta-N-acetylglucosamine (Fuc to (Fucalpha1—>6GlcNAc)-Asn-peptide) alpha1—>3-fucosyltransferase activity in honeybee (Apis mellifica) venom glands—The difucosylation of asparagine-bound N-acetylglucosamine. Eur J Biochem 1991; 199:745-751.
52. Weinstein J, DeSouza-e-Silva U, Paulson JC. Purification of a Galβ1-4GlcNAc α2-6-sialyltransferase and a Galβ1-3(4)GlcNAc α2-3-sialyltransferase to homogeneity from rat liver. J Biol Chem 1982; 257:13835-13844.
53. Wen DX, Livingston BD, Medzihradszky KF et al. Primary structure of Galβ1, 3(4)GlcNAc α,3-sialyltransferase determined by mass spectrometry sequence analysis and molecular cloning. Evidence for a protein motif in the sialyltransferase gene family. J Biol Chem 1992; 267:21011-21019.
54. Lee EU, Roth J, Paulson JC. Alteration of terminal glycosylation sequences on N-linked oligosaccharides of chinese hamster ovary cells by expression of β-galactoside α2,6-sialyltransferase. J Biol Chem 1989; 264:13848-13855.
55. Weinstein J, Lee EU, McEntee K et al. Primary structure of β-galactoside α2,6-sialyltransferase. Conversion of membrane-bound enzyme to soluble forms by cleavage of the NH_2-terminal signal anchor. J Biol Chem 1987; 262:17735-17743.
56. Kitagawa H, Paulson JC. Cloning and expression of human Galbeta1,3(4)GlcNAc alpha2,3-sialyltransferase. Biochem Biophys Res Commun 1993; 194:375-382.
57. Kitagawa H, Paulson JC. Cloning of a novel α2-3-sialyltransferase that sialylates glycoprotein and glycolipid carbohydrate groups. J Biol Chem 1994; 269:1394-1401.
58. Nemansky M, Van den Eijnden DH :Enzymatic characterization of CMP-NeuAc:Galbeta1-4GlcNAc-R alpha(2-3)-sialyltransferase from human placenta. Glycoconj J 1993; 10:99-108.
59. Dall'Olio F, Malagolini N, Di Stefano G et al. Increased CMP-NeuAc:Galβ1-4GlcNAc-R α2-6 sialyltransferase activity in human colorectal cancer tissues. Int J Cancer 1989; 44:434-439.
60. Le Marer N, Stéhelin D. High alpha-2,6-sialylation of N-acetyllactosamine sequences in ras-transformed rat fibroblasts correlates with high invasive potential. Glycobiology 1995; 5:219-226.
61. Aasheim HC, Aas-Eng DA, Deggerdal A. Cell-specific expression of human beta-galactoside alpha2,6-sialyltransferase tran-

scripts differing in the 5' untranslated region. Eur J Biochem 1993; 213:467-475.
62. Wang X-C, O'Hanlon TP, Young RF et al. Rat β-galactoside α2,6-sialyltransferase genomic organization:alternate promoters direct the synthesis of liver and kidney transcripts. Glycobiology 1990; 1:25-31.
63. Harduin-Lepers A, Recchi MA, Delannoy P. 1994, The year of sialyltransferases. Glycobiology 1995; 5:741-758.
64. Nakayama J, Fukuda MN, Fredette B et al. Expression cloning of a human polysialyltransferase that forms the polysialylated neural cell adhesion molecule present in embryonic brain. Proc Natl Acad Sci USA 1995; 92:7031-7035.
65. de Heij HT, Koppen PL, van den Eijnden DH. Biosynthesis of sialylated β-D-galactopyranosyl-(1-3)-2-deoxy-β-D-glucopyranosyl oligosaccharide chain. Identification of a β-D-galactoside ST(3)N-1 α-(2-3)- and a 2-acetamido-2-deoxy-β-D-glucoside α-(2-6)-sialyltransferase in regenerating rat liver and other tissues. Carbohydr Res 1986; 149:85-99.
66. Hooper LV, Hindsgaul O, Baenziger JU. Purification and characterization of the GalNAc-4-sulfotransferase responsible for sulfation of GalNAcβ1,4GlcNAc-bearing oligosaccharides. J Biol Chem 1995; 270:16327-16332.
67. Skelton T, Hooper L, Srivastava V et al. Characterization of a sulfotransferase responsible for the 4-0-sulfation of terminal β-N-Acetyl-D-galactosamine on asparagine-linked oligosaccharides of glycoprotein hormones. J Biol Chem 1991; 266:17142-17150.
68. Habuchi O, Miyata K. Stimulation of glycosaminoglycan sulfotransferase from chick embryo cartilage by basic proteins and polyamines. Biochim Biophys Acta 1980; 616:208-217.
69. Habuchi O, Miyashita N. Separation and characterization of chondroitin-6-sulfotransferase and chondroitin-4-sulfotransferase from chick embryo cartilage. Biochim Biophys Acta 1982; 717:414-421.
70. Hooper LV, Beraneck MC, Manzella SM et al. Differential expression of GalNAc-4-sulfotransferase and GalNActransferase results in distinct glycoforms of carbonic anhydrase VI in parotid and submaxillary glands. J Biol Chem 1995; 270:5985-5993.
71. Merkle R, Elbein A, Heifetz A. The effect of swainsonine and castanospermine on the sulfation of the oligosaccharide chains of N-linked glycoproteins. J Biol Chem 1985; 260:1083-1089.
72. Humphries D, Silbert C, Silbert J. Glycosaminoglycan production by bovine aortic endothelial cells cultured in sulfate depleted medium. J Biol Chem 1986; 261:9122-9127.
73. van der Kraan P, de Vries B, Vitters E et al. The effect of low sulfate concentrations on the glycosaminoglycan synthesis in anatomically intact articular cartilage of the mouse. J Orthoped Res 1989; 7:645-653.
74. Baeuerle P, Huttner W. Chlorate-a potent inhibitor of sulfation in intact cells. Biochem Biophys Res Comm 1986; 141:870-877.
75. Dekker J, Van Beurden-Lamers W, Strous G. Biosynthesis of gastric mucus glycoprotein of the rat. J Biol Chem 1989; 264:10431-10437.
76. Mintz K, Fisher L, Grzesik W et al. Chlorate-induced inhibition of tyrosine sulfation on bone sialoprotein synthesized by a rat osteoblast-like cell line (UMR 106-01 BSP). J Biol Chem 1994; 269:4845-4852.
77. Forstner J, Roomi N, Khorasani R et al. Effect of reserpine on the histochemical and biochemical properties of rat intestinal mucin. Exp Mol Path 1991; 54:129-143.
78. Mawhinney T. Effects of reserpine pretreatment on secretion of sulfated and sialylated glycoproteins by the rat trachea in vivo and in vitro. In:Martinez J, Barbero G, eds, Animal Models for Cystic Fibrosis:The Reserpine Treated Rat. San Francisco:San Francisco Press, Inc, 1985:1235-1237.
79. Roberton A, McKenzie C, Sharfe N et al. A glycosulfatase that removes sulfate from mucus glycoprotein. Biochem J 1993; 293:683-689.
80. Corfield A, Wagner S, Clamp J. Detection of a carbohydrate sulfatase in human fecal extracts. Biochem Soc Trans 1987; 15:1089.

CHAPTER 7

Less Common O-Linked Carbohydrates of Glycoproteins

7.A. O-GlcNAc

O-linked GlcNAc β-linked to Ser/Thr, where GlcNAc remains unsubstituted, is found in a number of cytoplasmic, nuclear and nuclear envelope proteins in pecies from yeast to man, including transcription factors, crystallins, kinases and viral glycoproteins. These O-GlcNAc residues have been detected by using O-GlcNAc containing glycoproteins as a substrate for bovine milk β4-Gal-transferase with subsequent release of Galβ1-4GlcNAc-OH by beta-elimination. The functions of O-GlcNAc have been suggested to be assembly and maintenance of protein complexes, regulation of protein synthesis and control of protein phosphorylation.[1]

The O-GlcNAc residue is transferred by a β-GlcNAc-transferase from UDP-GlcNAc[2] and removed by a β-hexosaminidase which preferably acts on O-GlcNAc.[3] Similar to other O-glycosylation reactions, no specific peptide sequon for O-GlcNAc addition has been identified. O-GlcNAc appears to be turned over faster than the protein moiety and this turnover appears to be growth stimulation-dependent. For example, activation of T-lymphocytes shifts the amount of O-GlcNAc from the cytosolic to nuclear glycoproteins.[1] Most O-GlcNAc proteins may also be phosphorylated. There is a reciprocal relationship of O-GlcNAc modification with protein phosphorylation; this suggests a regulatory function for O-GlcNAc, possibly because O-GlcNAc modification may sterically block phosphorylation sites.[4]

7.B. O-MAN

O-linked Man is common in yeast and is synthesized by a Man-transferase requiring Dol-P-Man as a sugar donor.[5,6] Pro residues near the possible glycosylation site are favorable, as has been found for GalNAc-

Ser/Thr O-glycosylation. Recently, the cytoplasmic enzyme phosphoglucomutase (PGM) from yeast, rat cortical synaptosomes and pheochromocytoma PC12 cells was shown to be glycosylated[7,8] with a Man residue O-linked to Ser. It has been shown that, in rat liver, the Man residue may be substituted at the 6-position with Glcα-1-phosphate due to the action of a Glcα-1-phosphotransferase with subsequent removal of the Glc residue by a cytosolic Glcα-1-phosphodiesterase.[9] PGM is the only cytoplasmic protein that has been shown to carry this unusual linkage which is believed to function in the interactions with ER membranes. The Man-transferase may therefore be very specific for the peptide sequence of PGM.

O-linked Man has also been reported in rat brain proteoglycans.[10] The biosynthesis of the mammalian type of O-Man glycans has not yet been elucidated and it is not known which sugar donor may be utilized. Both GDP-Man or Dol-P-Man sugar donors are accessible to glycosyltransferases in the cytoplasm. It is likely that the proteoglycan O-Man linkage is not synthesized in the cytoplasm but rather in the ER or the Golgi apparatus.

7.C. O-FUC AND O-GLC

Two unusual O-glycans, based on Fucα-O-Ser/Thr (O-Fuc glycans) and Glcβ-O-Ser (O-Glc glycans) have recently been identified in the epidermal growth factor domain of several proteases involved in thrombus formation or dissolution.[11] Tissue plasminogen activator (tPA) converts plasminogen to plasmin and is therefore an important factor for the dissolution of fibrin clots; it has been used therapeutically for fibrinolysis in myocardial infarctions and other cardiovascular diseases. Natural human tPA has N- and O-glycans and one O-Fuc glycan and one O-Glc glycan.[11,12] Human hepatoma cells HepG2 bind recombinant tPA but not the de-O-fucosylated species.[13] The function of the O-Fuc may therefore involve the regulation of the half life of glycoproteins. CxxGGS/TC sequences are homologous sequences at O-Fuc glycosylation sites. The O-Fuc based glycan chain of human blood coagulation factor IX may be extended to form sialic acidα2-6Galβ1-4GlcNAcβ1-3Fuc- chains.[14] The Ser/Thr α-Fuc-transferase has not yet been described and probably uses GDP-Fuc as a sugar donor. The β3-GlcNAc-transferase is also likely to be novel since its specificity does not match with those of known GlcNAc-transferases. The sialylα2-6Galβ1-4 chains are probably added by the known α6-sialyltransferase (acting on N-glycans) and β4-Gal-transferase, based on their specificities.

In addition or instead of O-Fuc glycans, many serine proteases such as coagulation factors VII and IX carry Xylα1-3 Xylα1-3 Glcβ- chains or their apparent biosynthetic intermediates linked via the β-Glc residue to Ser. While it has been suggested that O-Glc glycans are modulating coagulation,[15] nothing is known about their biosynthesis. The consensus sequence for O-glucosylation appears to be CxSxPC. Both UDP-Glc or Dol-P-Glc are possible sugar donors for the glucosylation reaction which is yet unknown. Since N- and O-glycans may also be present on these molecules, the enzymes synthesizing O-Fuc and O-Glc chains are probably localized in the ER or in the Golgi in various cell types including the liver and endothelial cells. The urinary form of tPA (human urokinase) carries N-glycans with unusual GalNAcβ1-4 (Fucα1-3) GlcNAcβ1-2 terminal structures, resembling those found in glycoprotein hormones.[16]

References

1. Hayes BK, Hart GW. Novel forms of protein glycosylation. Curr Opin Struc Biol 1994; 4:692-696.
2. Haltiwanger RS, Blomberg MA, Hart GW. Glycosylation of nuclear and cytoplasmic proteins. J Biol Chem 1992; 267:9005-9013.
3. Dong LYD, Hart GW. Purification and characterization of an O-GlcNAc selective N-acetylglucosaminidase from rat spleen cytosol. J Biol Chem 1994; 269:19321-19330.
4. Hart GW, Greis KD, Dong LY et al. O-linked N-acetylglucosamine: the yin-yang

of Ser/Thr phosphorylation? Nuclear and cytoplasmic glycosylation. Adv Exp Med Biol 1995; 376:115-123.

5. Strahl-Bolsinger S, Tanner W. Protein O-glycosylation in Saccharomyces cerevisiae. Purification and characterization of the dolichyl-phosphate-D-mannose-protein O-D-mannosyltransferase. Eur J Biochem 1991; 196:185-190.
6. Gentzsch M, Strahl-Bolsinger S, Tanner W. A new Dol-P-Man:protein O-D-mannosyltransferase activity from Saccharomyces cerevisiae. Glycobiol 1995; 5:77-82.
7. Marchase RB, Bounelis P, Brumley LM et al. Phosphoglucomutase in Saccharomyces cerevisiae is a cytoplasmic glycoprotein and the acceptor for a Glc-phosphotransferase. J Biol Chem 1993; 268:8341-8349.
8. Veyna NA, Jay JC, Srisomsap C et al. The addition of glucose-1-phosphate to the cytoplasmic glycoprotein phosphoglucomutase is modulated by intracellular calcium in PC12 cells and rat cortical synaptosomes. J Neurochem 1994; 62:456-464.
9. Srisomsap C, Richardson KL, Jay JC et al. An α-glucose-1-phosphate phosphodiesterase is present in rat liver cytosol. J Biol Chem 1989; 264:20540-20546.
10. Krusius T, Reinhold V, Margolis RK et al. Structural studies on sialylated and sulphated O-glycosidic mannose-linked oligosaccharides in the chondroitin sulphate proteoglycan of brain. Biochem J 1987; 245:229-234.
11. Harris RJ, Spellman MW. O-linked fucose and other posttranslational modifications unique to EGF modules. Glycobiology 1993; 3:219-224.
12. Bergwerff AA, van Oostrum J, Asselbergs FAM et al. Primary structure of N-linked carbohydrate chains of a human chimeric plasminogen activator K_2tuPA expressed in Chinese Hamster Ovary cells. Eur J Biochem 1993; 212:639-656
13. Hajjar K, Reynolds C. α-Fucose mediated binding and degradation of tissue type plasminogen activator by HepG2 cells. J Clin Invest 1994; 93:703-710.
14. Nishimura H, Takao T, Hase S et al. Human factor X has a tetrasaccharide O-glycosidically linked to Serine 61 through the fucose residue. J Biol Chem 1992; 267:17520-17525.
15. Bjoern S, Foster DC, Thim L et al. Human plasma and recombinant factor VII. Characterization of O-glycosylations at serine residues 52 and 60 and effects of site-directed mutagenesis of serine 52 to alanine. J Biol Chem 1991; 266:11051-11057.
16. Bergwerff AA, Thomas-Oates JE, van Oostrum J et al. Human urokinase contains GalNAcβ(1-4) [Fucα(1-3)] GlcNAcβ(1-2) as a novel terminal element in N-linked carbohydrate chains. FEBS Lett 1992; 314:389-394.

CHAPTER 8

Control Mechanisms in the Biosynthesis of N- and O-Glycans

Although N- and O-glycans exhibit great heterogeneity in their structures, they vary in a cell-type-, species- and developmental stage- and growth-specific fashion. In addition, different glycoproteins synthesized in the same cell may be differently glycosylated, and a glycoprotein molecule may carry different glycans characteristic of the individual glycosylation site. This indicates that the biosynthesis is tightly controlled. Changes in glycoprotein structures during disease reflect a reprogramming of these complex control mechanisms.

Many factors are known to affect the relative amounts and structures of glycoprotein-bound oligosaccharides. Genes for glycosyltransferases and substrates are often expressed in a cell-type- and growth-specific fashion. The transport of substrates and products through the endomembrane system and into the Golgi apparatus, and the localization of glycosyltransferases are important for proper glycosylation. The relative activities and the substrate specificities of glycosyltransferases determine the structures and relative amounts of individual glycan chains. The kinetic properties and the specificity of glycosyltransferases are known to be regulated by many different factors such as metal ion and other cofactor concentrations, the membrane composition and possibly post-translational modifications of enzymes. All of these factors may change in disease.

8.A. GENETIC CONTROL OF GLYCOSYLTRANSFERASES

Most often, glycosyltransferases with similar specificities exist as families of proteins with homologous regions. The genes for glycosyltransferases may have one exon (e.g. α3-Fuc-transferases, core 2 β6-GlcNAc-transferase, GlcNAc-transferases I and II) or multiple exons (e.g. α6-sialyltransferase, β4-Gal-transferase). Several mRNA transcripts may be present and more than one promoter may control the gene in a tissue-specific fashion. The genes encoding α6-sialyltransferase (N) and β4-Gal-transferase have been studied extensively.

The α6-sialyltransferase gene in the rat contains eleven exons,[1] and at least three (one constitutive and several tissue-specific) promoters control transcription.[2,3] Five or more mRNA transcripts may be produced.[2,4-6] Liver specific transcription factors may bind to the liver-specific promoter to produce mRNA transcripts characteristic for this organ. Different promoters are used in the kidney and other tissues resulting in varying stability of mRNA. Some of the mRNA species may code for catalytically inactive proteins. Constitutive expression produces a low level of enzyme activity which may be augmented by upregulation using different promoters. Dexamethasone or ras oncogene transfection may elicit this effect in rat fibroblasts,[7,8] while thyroid stimulating hormone may cause a downregulation of mRNA expression in rat thyroid cells.[9] In situ hybridization suggests that during the development of thymocytes the expression of one of the mRNA species appears to increase since cortical regions of immature cells show much lower levels of expression than medullary regions of mature thymocytes.[10]

The gene encoding bovine β4-Gal-transferase has two distinct promoters translating into two sets of mRNA of different lengths, coding for two proteins that differ by 13 amino acids in the cytoplasmic domain but having the same topology and enzyme activity.[11,12] A single copy gene for the enzyme exists on chromosome 9.[13] It appears that the upstream promoter controls the expression of the long transcript in a constitutive fashion, leading to a basic low level of enzyme activity. The other promoter may be subject to regulation in cells with high levels of activities. Negative control elements have also been identified that appear to block the production of the short transcript. During lactation the negative control is deactivated and larger amounts of the short form transcript and high activities are produced. At least in the mouse, male germ cells utilize one promoter which may be germ cell-specific and leads to the long transcript.[14,15]

8.B. INTRACELLULAR LOCALIZATION OF PROCESSING ENZYMES

Enzymes involved in the assembly of glycoprotein glycans appear to be arranged in Golgi compartments in a sequential order from cis to trans Golgi compartments. For example, the first enzyme of the O-glycosylation pathway, polypeptide α-GalNAc-transferase, has been localized to the cis-Golgi by immunoelectronmicroscopy.[16] GlcNAc-transferase I and mannosidase II which are acting at intermediate stages of N-glycan biosynthesis have been localized to the medial Golgi[17] and the late acting enzyme α6-sialyltransferase (N) to the trans Golgi and trans-Golgi-network.[18]

There is also evidence for the existence of glycosyltransferase complexes where several transferases act with high efficiency to construct complex carbohydrate chains. The latter model is supported by the fact that glycosyltransferase activities responsible for the synthesis of abundant glycans may be very low in vitro where the intracellular organization is disturbed, but must be high in vivo. In addition, the mixing of different glycosyltransferases in vitro may cause an enhancement of activities. This has been shown by the increase in N-acetyl-lactosamine synthesizing activity of bovine milk β4-Gal-transferase when mixed with soluble bovine colostrum GlcNAc-transferase I. In addition, complex formation between these two enzymes was demonstrated.[19]

ER processing enzymes are integral membrane proteins with a variety of topological orientations and no common domain structure. In contrast, most Golgi processing enzymes are of the type II membrane glycoprotein structure with a short N-terminal end directed towards the cytoplasm, an uncleaved membrane-bound anchor region and a stem region thought to expose the C-terminal globular peptide containing the catalytic site. Proteolytic cleavage at the stem region, for example by the acute phase response or during enzyme purification, destroys Golgi targeting signals and causes secretion of glycosyltransferases.[20,21] The factors responsible for localizing a glycosyltransferase in the Golgi apparatus include the sequence, hydrophobicity, three-dimensional arrangement of the membrane anchor and adjacent peptides, as well as the spacing between these adjacent regions. The association with other proteins and the formation of aggregates may also play an important role in retaining proteins in the appropriate compartment.[22] In addition, the Golgi membrane structure may be instrumental in sorting proteins with membrane anchor regions suitable for retention. Ionophores like monensin and the drug Brefeldin A disturb the intracellular distribution of Golgi enzymes and thus seriously affect N- and O-glycan processing. During differentiation or in disease the intracellular localization of glycosyltransferases may also change. This was shown in HT29 human colon cancer cells where Golgi proteins are redistributed compared to normal colonic tissue.[23] Substrates for glycosyltransferases may thus be inappropriately placed in diseased cells, and this could result in the synthesis of abnormal glycans, although enzyme activities may be unchanged. Incomplete synthesis of chains may also occur when there is an increased rate of secretion, for example, in certain hypersecretory diseases.

Glycosyltransferases have also been found on cell surfaces. A certain proportion of the β4-Gal-transferase, for example, is localized with the active site facing towards the extracellular space. Here it does not act as an enzyme since the nucleotide sugar donor is lacking. It has been suggested that these cell surface glycosyltransferases can act as lectins and are involved in interactions with other cells or with the extracellular matrix.[24-26] Overexpression of the enzyme in mouse 3T3 fibroblasts leads to slower growth than in cells with reduced expression of the enzyme. Thus, cell surface Gal-transferase may interact with and modulate the EGF receptor thereby affecting EGF-dependent cell proliferation.[27]

8.C. FACTORS AFFECTING ENZYME ACTIVITIES

Glycosyltransferase activity levels determine major and minor pathways of glycoprotein biosynthesis. There are often crossroads in the biosynthetic pathways where the relative rates of glycosyltransferase activities competing for a common substrate will determine the ratio of oligosaccharide products. Glycosyltransferase activities are influenced by substrate and cofactor concentrations as well as by the membrane environment and possibly post-translational modifications of the enzyme protein. Most glycosyltransferases require divalent metal ions with the exception of Fuc-transferases, sialyltransferases, and β6-GlcNAc-transferases, although certain metal ions may have a strong inhibitory or stimulatory effect. In vitro measured glycosyltransferase activities are often optimal at metal ion concentrations outside of the physiological ranges.

β4-Gal-transferase activity may be regulated by a phosphorylation mechanism.[28] The enzyme binds to α-lactalbumin in the mammary gland which changes kinetic properties and specificity and lead to the synthesis of lactose instead of poly-n-acetyl-lactosamine chains.[29] Another glycosyltransferase-binding protein is the yeast protein WBP1, which is required for oligosaccharyltransferase action in yeast and may be a component of the enzyme.[30] A soluble activation factor specific for α6-sialyltransferase has been reported in rat colon[31] which might be involved in the developmental regulation of the enzyme activity. Other activator proteins analogous to those associated with glycosidases have not been identified for glycosyltransferases.[32]

8.D. GLYCOSYLTRANSFERASE SUBSTRATE SPECIFICITIES

Glycoproteins may carry a number of sugars in various linkages, branches and substitution patterns. Although the theoretical number of possible complex glycan structures is inconceivably high, only several thousand of these chains are actually being synthesized in nature due to the restricted range of glycosyltransferase specificities.

Enzymes have a strict requirement for certain partial structures in the substrate, but allow binding to a variety of substrates differing in other parts of the molecule. Sugar residues may convert an active substrate to an inactive substrate by introducing a STOP signal that blocks further processing steps. For example, the addition of an α2-6-linked sialic acid to GalNAc of O-glycans is a STOP signal for the synthesis of core 1 to 4 structures. On the other hand, the insertion of a critical residue may be required as an essential GO signal for other enzymes. Sialic acid α2-3-linked to the Gal residue of core 1 is a GO signal for the addition of sialic acid in α2-6 linkage to GalNAc. Fuc added in α1-2-linkage to the Gal residue of O-glycan core 1 represents a GO signal for the formation of blood group A and B but is a STOP signal for the conversion to O-glycan core 2.

The bisecting GlcNAc of N-glycans is a STOP signal for further branching by GlcNAc-transferases IV and V[33,34] and inhibits trimming by α-mannosidase II which explains why most of the hybrid structures analyzed are bisected.[35] GlcNAc-transferase III activity is reduced when the substrate is galactosylated, therefore only a small number of bisected, galactosylated structures have been demonstrated.

GlcNAc-transferase I initiates trimming by mannosidase II, the actions of GlcNAc-transferase II, III and IV and α6-Fuc-transferase. This suggests that all of these enzymes have a binding site for the GlcNAcβ1-2Manα1-3- moiety of N-glycans. The ability of GlcNAc-transferase I to bind this moiety has been demonstrated by detailed specificity studies.[36,37] Although the enzyme acts on the 2-hydroxyl of the Manα1-3 residue, it has an absolute requirement for the 2- and 4-hydroxyls of the Manβ1-4 residue in the axial and equatorial configurations, respectively. Omission of the 4- and 6-hydroxyls of Manα1-3 decreases activity significantly while there is little structural requirement for the Manα1-6 arm.

The peptide portion controls peptide glycosylation as well as glycan processing. Examples for reactions that are known to involve protein recognition are the addition of the Man-6-P marker to lysosomal enzymes, the addition of GalNAc in β1-4-linkage to GlcNAc of glycoprotein hormones, the transfer of oligosaccharides from a dolichol derivative to the Asn-X-Ser/Thr sequon of proteins, and the transfer of GalNAc to Ser/Thr-containing protein.

The transfer of GalNAc to Thr protein as the first step in O-glycan synthesis depends on the sequence as well as the glycosylation of peptide substrates[38-41] although no specific sequon is required. Other O-glycan processing enzymes are also greatly influenced by the peptide structures of substrates although they do not directly use these as substrates.[42,43] This supports the idea of site-directed processing.

It appears that glycosyltransferases not only recognize the primary structures of their substrates but also specific conformations. These conformations may be a result of the interactions of the peptide moiety with the carbohydrate. Experimental support came from studies by Lindh and Hindsgaul,[44] who synthesized substrate analogues for GlcNAc-transferase V. Some of the analogues contain bulky groups that restrict rotation and thus stabilize certain conformations, and concomitantly show increased enzyme activity.

Glycosyltransferase inhibitors may be useful to examine oligosaccharide functions, the role of individual processing reactions, to map enzyme binding sites, as affinity ligands for glycosyltransferase purifications, and to develop therapeutic drugs to prevent the synthesis of specific carbohydrate structures in disease.

Several inhibitors for N-glycan processing steps are known that turn off the entire N-glycan synthesis (tunicamycin) or inhibit processing glycosidases (e.g. nojirimycin, mannojirimycin, swainsonine) thus resulting in unprocessed N-glycans (Table 3).[45,46] UDP-GalNAc analogs[47,48] may generally inhibit GalNAc-transferases and thus inhibit initiation of O-glycosylation. Since the most common O-glycan core structures, core 1 to 4, can also be formed on non-glycoprotein substrates such as GalNAc-phenyl in vitro, it has been possible to inhibit endogenous mucin glycosylation by these synthetic compounds which diffuse into cells due to their hydrophobic nature and act as competitive substrates in vivo. GalNAcα-aryl derivatives[49,50] were shown to inhibit mucin synthesis by competition with O-glycan extension, and the synthesis of Le^x structures.

Recently, specific inhibitors for a number of glycosyltransferases became available (Table 3).[37,51-56] Deoxy derivatives of substrates lacking the hydroxyl that the enzyme acts upon may be weak competitive inhibitors for about half of the glycosyltransferases tested.[53,55] Substitution of hydroxyls by methyl groups may create specific inhibitors.[54,57,58] Replacement of the 6-hydroxyl by a sulfhydryl group leads to a competitive inhibitor for β4-Gal-transferase.[59] Irreversible inhibitors may contain reactive groups such as iodoacetamides and uv reactive nitrophenyl groups.[55,60] A bisubstrate inhibitor, containing a nucleotide-sugar and acceptor analog linked covalently and thought to represent a transition state analog, has been shown to inhibit α2-Fuc-transferase.[52] It is hoped that some of these specific inhibitors may be able to cross membranes to reach the site of glycosyltransferase action, but this has yet to be demonstrated experimentally.

8.E. FACTORS AFFECTING THE AVAILABILITY OF SUBSTRATES

The expression of individual glycoproteins is variable under different growth conditions and in disease. Since glycosylation is directed by the peptide moiety, it is conceivable that glycosylation changes follow alterations of glycoprotein expression. This may be significantly influencing primary glycosylation (see chapters 4.B and 6.A) as well as elongation and termination reactions (chapters 4 and 6) since these are dependent on the presentation of the appropriate presursor structures. Regulation of early acting enzymes controls substrate availability for later acting enzymes. Thus the concentration of substrates as well as their structures and glycosylation influence the effectiveness of glycosylation reactions.

Multiple factors affect the availability of glycosylation donor and acceptor substrates for glycoprotein synthesis. For example, the expression of MUC mucin genes is altered in disease (see chapter 12.A). The message of MUC genes is polydisperse as demonstrated on Northern blots, possibly due to alternative splicing. Polydispersity of mRNA has been noted to increase in disease, for example, in cystic fibrosis.[61]

Glycosylation reactions may also be affected by the transport and metabolism of nucleotide sugars. Regulation of nucleotide sugar transport from the cytoplasm into the Golgi occurs by specific antiport-transport systems which control sugar-donor substrate and nucleotide inhibitor availability and thus, glycosyltransferase activities. Nucleotide metabolism may also be modulated by free sugars,[62] by gangliosides,[63] and by pyrophosphatases.[64] Variable pool sizes of nucleotide sugars are found in human colon cancer HT29 cells, which have been induced to differentiate. The intracellular accumulation of UDP-GlcNAc and UDP-GalNAc has been correlated with an inability of HT-29 human colonic cancer cells to differentiate, suggesting that the nucleotide-sugar levels are involved in the regulation of differentiation.[65]

References

1. Svensson EC, Soreghan B, Paulson JC. Organization of the β-galactoside α2,6-sialyltransferase gene. Evidence for the transcriptional regulation of terminal glycosylation. J Biol Chem 1990; 265:20863-208688.
2. Wang X-C, O'Hanlon TP, Young RF et al. Rat β-galactoside α2,6-sialyltransferase genomic organization:alternate promoters direct the synthesis of liver and kidney transcripts.

Glycobiology 1990; 1:25-31.
3. Aasheim HC, Aas-Eng DA, Deggerdal A. Cell-specific expression of human beta-galactoside alpha2,6-sialyltransferase transcripts differing in the 5' untranslated region. Eur J Biochem 1993; 213:467-475.
4. Paulson JC, Weinstein J, Schauer A. Tissue-specific expression of sialyltransferases. J Biol Chem 1989; 264:10931-10934.
5. Wen DX, Svensson EC, Paulson JC. Tissue-specific alternative splicing of the β-galactoside α2,6 sialyltransferase gene. J Biol Chem 1992; 267:2512-2518.
6. O'Hanlon TP, Lau KM, Wang XC et al. Tissue-specific expression of β-galactoside α–2,6sialyltransferase. J Biol Chem 1989; 264:17389-17394.
7. LeMarer N, Laudet V, Svensson EC et al. The c-Ha-ras oncogene induces increased expression of β-galactoside α-2,6 sialyltransferase in rat fibroblast (FR3T3) cells. Glycobiology 1992; 2:49-56.
8. LeMarer N, Stehelin D. High alpha-2,6-sialylation of N-acetyllactosamine sequences in ras-transformed rat fibroblasts correlates with high invasive potential. Glycobiology 1995; 5:219-226.
9. Grollman EF, Doi SQ, Weiss P et al. Hyposialylated thyroglobulin in a patient with congenital goiter and hypothyroidism. J Clin Endocrin Metab 1992; 74:43-48.
10. Gillespie W, Paulson J, Kelm S et al. Regulation of α2,3-sialyltransferase expression correlates with conversion of peanut agglutinin (PNA)+ to PNA- phenotype in developing thymocytes. J Biol Chem 1993; 268:3801-3804.
11. Russo RN, Shaper NL, Shaper JH. Bovine beta1-4-galactosyltransferase:Two sets of mRNA transcripts encode two forms of the protein with different amino-terminal domains. In vitro translation experiments demonstrate that both the short and the long forms of the enzyme are type II membrane-bound glycoproteins. J Biol Chem 1990; 265:3324-3331.
12. Harduin-Lepers A, Shaper JH, Shaper NL. Characterization of two cis-regulatory regions in the murine β1,4-galactosyltransferase gene:evidence for a negative regulatory element that controls initiation at the proximal site. J Biol Chem 1993; 268:14348-14359.
13. Shaper NL, Shaper JH, Meuth JL et al. Bovine galactosyltransferase:a clone identified by direct immunological screening of a cDNA expression library. Proc Natl Acad Sci USA 1986; 83:1573-1577.
14. Shaper NL, Wright WW, Shaper JH. Murine beta1,4-galactosyltransferase:Both the amounts and structure of the mRNA are regulated during spermatogenesis. Proc Natl Acad Sci USA 1990; 87:791-795.
15. Shaper NL, Harduin-Lepers A, Shaper JH. Male germ cell expression of murine beta 4-galactosyltransferase–A 796-base pair genomic region, containing two cAMP-responsive element (CRE)-like elements, mediates male germ cell-specific expression in transgenic mice. J Biol Chem 1994; 269:25165-25171.
16. Roth J, Wang Y, Eckhardt AE et al. Subcellular localization of the UDP-N-acetyl-D-galactosamine:polypeptide N-acetylgalactosaminyltransferase-mediated O-glycosylation reaction in the submaxillary gland. Proc Natl Acad Sci USA 1994; 91:8935-8939.
17. Paulson JC, Colley KJ. Glycosyltransferases. Structure, localization, and control of cell type specific glycosylation. J Biol Chem 1989; 264:17615-17618.
18. Taatjes DJ, Roth J, Weinstein J et al. Post-Golgi apparatus localization and regional expression of rat intestinal sialyltransferase detected by immunoelectron microscopy with polypeptide epitope-purified antibody. J Biol Chem 1988; 263:6302-6309.
19. Moscarello MA, Mitranic MM, Vella G. Stimulation of bovine milk galactosyltransferase activity by bovine colostrum N-acetylglucosaminyltransferase I. Biochim Biophys Acta 1985; 831:192-200.
20. Colley KJ, Lee EU, Adler B et al. Conversion of a Golgi apparatus sialyltransferase to a secretory protein by replacement of the NH2-terminal signal anchor with a signal peptide. J Biol Chem 1989; 264:17619-17622.
21. Lammers G, Jamieson JC. Cathepsin D-like activity in the release of Galbeta1-4GlcNAcα2-6sialyltransferase from mouse

and guinea pig liver Golgi membranes during the acute phase response. J Comp Biochem Physiol 1990; 95:327-334.

22. Gleeson PA, Teasdale RD, Burke J. Targeting of proteins to the Golgi apparatus. Glycoconj J 1994; 11:381-394.
23. Egea G, Francí C, Gambús G et al. Cis-Golgi resident proteins and O-glycans are abnormally compartmentalized in the RER of colon cancer cells. J Cell Sci 1993; 105:819-830.
24. Eckstein DJ, Shur BD. Cell surface beta-1,4-galactosyltransferase is associated with the detergent-insoluble cytoskeleton on migrating mesenchymal cells. Exp Cell Res 1992; 201:83-90.
25. Barcellos-Hoff M. Mammary epithelial reorganization on extracellular matrix is mediated by cell surface galactosyltransferase. Exp Cell Res 1992; 201:225-234.
26. Shur BD. Cell surface β1,4-galactosyltransferase: twenty years later. Glycobiology 1991; 1:563-575.
27. Hinton DA, Evans SC, Shur BD. Altering the expression of cell surface β1,4-galactosyltransferase modulates cell growth. Exp Cell Res 1995; 219:640-649.
28. Bunnell BA, Adams DE, Kidd VJ. Transient expression of a p58 protein kinase cDNA enhances mammalian glycosyltrans–ferase activity. Biochem Biophys Res Comm 1990; 171:196-203.
29. O'Keeffe ET, Mordick T, Bell JE. Bovine galactosyltransferase:interaction with α-lactalbumin and the role of α-lactalbumin in lactose synthase. Biochemistry 1980; 19:4962-4966.
30. te Heesen S, Janetzky B, Lehle L et al. The yeast WBP1 is essential for oligosaccharyltransferase activity in vivo and in vitro. EMBO J 1992; 11:2071-2075.
31. Nagpurkar A, Hunt D, Mookerjea S. A low molecular weight protein factor in rat colon specifically activates α2-6 sialyltransferase. Int Carbohydr Symp, Paris 1992, Abstract B062.
32. Sandhoff K, Klein A. Intracellular trafficking of glycosphingolipids:role of sphingolipid activator proteins in the topology of endocytosis and lysosomal digestion. FEBS Lett 1994; 346:103-107.
33. Gleeson PA, Schachter H. Control of glycoprotein synthesis. VIII. UDP-GlcNAc:GnGn (GlcNAc to Manα1-3) β4-N-acetylglucosaminyltransferase IV, an enzyme in hen oviduct which adds GlcNAc in β1-4 linkage to the α1-3-linked Man residue of the trimannosyl core of N-glycosyl oligosaccharides to form a tri-antennary structure. J Biol Chem 1983; 258:6162-6173.
34. Narasimhan S. Control of glycoprotein synthesis VII. UDP-GlcNAc:glycopeptide β4-N-acetylglucosaminyltransferase III, an enzyme from hen oviduct which adds GlcNAc in β1-4 linkage to the β-linked mannose of the trimannosyl core of N-glycosyl oligosaccharides. J Biol Chem 1982; 257:10235-10242.
35. Allen SD, Tsai D, Schachter H. Control of glycoprotein synthesis. X. The in vitro synthesis by hen oviduct membrane preparations of hybrid Asn-linked oligosaccharides containing five mannose residues. J Biol Chem 1984; 259:6984-6990.
36. Möller G, Reck F, Paulsen H et al. Control of glycoprotein synthesis:substrate specificity of rat liver UDP-GlcNAc:Manα3R β2-N-acetylglucosaminyl-transferase I using synthetic substrate analogues. Glycoconj J 1992; 9:180-190.
37. Reck F, Springer M, Meinjohanns E et al. Synthetic substrate analogues for UDP-GlcNAc:Manα1-3R β1-2-N-acetylglucosaminyltransferase I. Substrate specificity and inhibitors for the enzyme. Glycoconj J 1995; 12:747-754.
38. Wang Y, Abernethy JL, Eckhardt AE et al. Purification and characterization of a UDP-GalNAc:polypeptide N-acetylgalactosaminyltransferase specific for glycosylation of threonine residues. J Biol Chem 1992; 267:12709-12716.
39. O'Connell BC, Hagen FK, Tabak LA. The influence of flanking sequence on the O-glycosylation of threonine in vitro. J Biol Chem 1992; 267:25010-25018.
40. Brockhausen I, Toki D, Brockhausen J et al. Specificity of O-glycosylation by bovine colostrum UDP-GalNAc:polypeptide α-N-acetylgalactosaminyltransferase using synthetic glycopeptide substrates. Glycoconj J 1996; in press.
41. Hughes RC, Bradbury AF, Smyth DG.

Substrate recognition by UDP-N-acetyl-α-D-galactosamine:polypeptide N-acetyl-α-D-galactosaminyltransferase. Effects of chain length and disulphide bonding of synthetic peptide substrates. Carbohydr Res 1988; 178:259-269.

42. Brockhausen I, Möller G, Merz G et al. Control of glycoprotein synthesis:The peptide portion of synthetic O-glycopeptide substrates influences the activity of O-glycan core 1 Uridine 5'-diphospho-galactose:N-acetylgalactosamineα-R β3-galactosyl-transferase. Biochemistry 1990; 29:10206-10212.
43. Granovsky M, Bielfeldt T, Peters S et al. O-glycan core 1 UDP-Gal:GalNAc β3-galactosyltransferase is controlled by the amino acid sequence and glycosylation of glycopeptide substrates. Eur J Biochem 1994; 221:1039-1046.
44. Lindh I, Hindsgaul O. Synthesis and enzymatic evaluation of two conformationally restricted trisaccharide analogues as substrates for N-acetylglucosaminyltransferase V. J Am Chem Soc 1991; 113:216-223.
45. Schwarz RT, Datema R. Inhibitors of trimming:new tools in glycoprotein research. TIBS 1984:9:32-34.
46. Kaushal GP, Elbein AD. Glycosidase inhibitors in study of glycoconjugates. Methods Enzymol 1994; 230:316-329.
47. Drake RR, Elbein AD. Photoaffinity labelling of glycosyltransferases. Glycobiology 1992; 2:279-284.
48. Hatanaka K, Slama JT, Elbein AD. Synthesis of a new inhibitor of the UDP-GalNAc:polypeptide galactosaminyl transferase. Biochem Biophys Res Comm 1991; 175:668-672.
49. Kuan S, Byrd J, Basbaum C et al. Inhibition of mucin glycosylation by Aryl-N-acetyl-α-galactosaminides in human colon cancer cells. J Biol Chem 1989; 264:19271-19277.
50. Kojima N, Handa K, Newman W et al. Inhibition of selectin-dependent tumor cell adhesion to endothelial cells and platelets by blocking O-glycosylation of these cells. Biochem Biophys Res Comm 1992; 182:1288-1295.
51. Kuhns W, Rutz V, Paulsen H et al. Processing O-glycan core 1, Galβ1-3GalNAcα-R. Specificities of core 2, UDP-GlcNAc:Galβ1-3GalNAc-R(GlcNAc to GalNAc) β6-N acetylglucosaminyltransferase and CMP-sialic acid:Galβ1-3GalNAc-R α3-sialyltransferase. Glycoconj J 1993; 10:381-394.
52. Palcic MM, Heerze LD, Srivastava OP et al. A bisubstrate analog inhibitor for α(1→2)-fucosyltransferase. J Biol Chem 1989; 264:17174-17181.
53. Hindsgaul O, Kaur KJ, Srivastava G et al. Evaluation of deoxygenated oligosaccharide acceptor analogs as specific inhibitors of glycosyltransferases. J Biol Chem 1991; 226:17858-17862.
54. Khan SH, Matta KL. Recent advances in the development of potential inhibitors of glycosyltransferases. In: Allen HJ, Kisailus EC eds. Glycoconjugates:Composition, Structure, and Function. New York:Marcel Dekker, Inc. 1992; 361-378.
55. Reck F, Meinjohanns E, Springer M et al. Synthetic substrate analogues for UDP-GlcNAc:Manα1-6R β(1-2)-N-acetylglucosaminyltransferase II. Substrate specificity and inhibitors for the enzyme. Glyconj J 1994; 11:210-216.
56. Vijima-Suda I, Miyamato Y, Toyoshima S et al. Inhibition of experimental pulmonary metastasis of mouse colon adenocarcinoma 26 sublines by a sialic acid: nucleoside conjugate having sialyltransferase inhibiting activity. Cancer Res 1986; 46:858-862.
57. Khan SH, Crawley SC, Kanie O et al. A trisaccharide acceptor analog for N-acetylglucosaminyltransferase V which binds to the enzyme but sterically precludes the transfer reaction. J Biol Chem 1993; 268:2468-2473.
58. Brockhausen I, Reck F, Kuhns W et al. Substrate specificity and inhibition of UDP-GlcNAc:GlcNAcβ1-2Manα1-6R β1,6-N-acetylglucosaminyl-transferase V using synthetic substrate analogues. Glycoconj J 1995; 12:371-379.
59. Kajihara Y, Hashimoto H, Kodama H. Methyl-3-O-(2-acetamido-2-deoxy-6-thio-β-D-glucopyranosyl)-β-D-galactopyranoside:a slow reacting acceptor-analogue which inhibits glycosylation by UDP-D-galactose-N-acetyl-D-glucosamine-(1-4)-β-D-galactosyl-

transferase. Carbohydr Res 1992; 229:C5-C9.
60. Toki D, Granovsky MA, Reck F et al. Inhibition of UDP-GlcNAc:Galβ1-3GalNAc-R (GlcNAc to GalNAc) β6-N-acetylglucosaminyltransferase from acute myeloid leukaemia cells by photoreactive nitrophenyl substrate derivatives. Biochem Biophys Res Comm 1994; 198:417-423.
61. Verma M, Davidson EA. Mucin genes: structure, expression and regulation. Glycoconj J 1994; 3:172-179.
62. Krug E, Zweibaum A, Schulz-Holstege C et al. D-glucosamine-induced changes in nucleotide metabolism and growth of colon-carcinoma cells in culture. Biochem J 1984; 217:701-708.
63. Merritt WD, Morré DJ. Glycolipids modulate glycosyl transfer to endogenous protein acceptors. Biochim Biophys Acta 1980; 620:261-269.
64. Jentoft N, Cheng P, Carlson D. Glycosyltransferases and glycoprotein biosynthesis. In:Martonosi A, ed. The Enzymes of Biological Membranes. Volume 2. Biosynthesis of Cell Components. New York NY:Plenum Press, 1976:343-383.
65. Wice BM, Trugnan G, Pinto M et al. The intracellular accumulation of UDP-N-acetylhexosamines is concomitant with the inability of human colon cancer cells to differentiate. J Biol Chem 1985; 260:139-146.

CHAPTER 9

Glycoproteins and Cell Adhesion Functions

Oligosaccharide structures of glycoproteins differ between species, organs and cell types and are associated with a wide range of biological functions. Glycosylation patterns often fluctuate during growth, development and differentiation. The role of carbohydrate has been studied following the removal or modifications of carbohydrate by enzymic or chemical methods such as: with lectins and anticarbohydrate antibodies, by inhibition of certain steps in the biosynthetic pathways, by the use of mutant cell lines lacking enzymes involved in carbohydrate processing, by the introduction of genes or knocking out of genes or mRNA coding for glycosyltransferases, or by the expression of mammalian glycoproteins in systems differing in glycosylation.

The role of the oligosaccharide heterogeneity encountered in most glycoproteins remains an unresolved mystery. In addition to the complexity of primary structures, oligosaccharides exist in different conformations and these may contain essential signals required for interaction with other molecules. Postulated functions of N- and O-glycans include the following:

1. contribution to the physical and conformational properties of a protein;
2. protection from proteolysis;
3. role in intracellular transport, targeting and secretion;
4. contribution to antigenic and chemical properties of a glycoprotein;
5. cell adhesion;
6. binding of microbes;
7. biological activity of enzymes, hormones and receptors;
8. role in fertilization;
9. functions in the immune system;
10. involvement in growth, differentiation, development;
11. involvement in programmed cell death (apoptosis);
12. signal transduction; and
13. quality control of newly synthesized proteins.

9.A. CELL ADHESION MOLECULES

Cell adhesion events are essential in a multicellular organism and are an integral part of growth, the immune system or the invasion of cancer cells. Cell adhesion molecules (CAM) may be categorized as: (a) cadherins; (b) integrins; (c) immunoglobulin-related molecules; (d) selectins; and (e) lectins including galectins. Selectins, galectins and integrins will be discussed more fully in chapters to follow. Cell surface glycosyltransferases such as Gal-transferase may function as lectin-like molecules in cell adhesion. Alternatively, they may possess receptor functions whereby substrate binding and/or glycosylation induces responses in target cells.[1]

CAM may function cooperatively with other molecules via carbohydrates to facilitate critical adhesion events. CAM contain variable numbers of potential N-glycosylation sites and exist as glycosylation variants.[2] NCAM, the most abundant and well characterized of the known CAM of the immunoglobulin superfamily involved in neural recognition events, was found to possess a sialic acid content of 10-30% by weight. NCAM also contains sulfated carbohydrate residues that have been implicated in cell-cell adhesiveness. Most sialic acids consist of linear polymers of about 5-55 sialic acid residues in α2-8 linkages attached via N-glycans at three glycosylation sites in the central domain of NCAM. The effects of polysialic acid (PSA) upon cell functions have been studied using a specific endo-N-neuraminidase and antibodies prepared against different forms of the oligosaccharide. These and other studies have suggested that PSA may function as a negative modulator of cell adhesion by NCAM.[3]

PSA is also involved in NCAM and N-cadherin adhesion associated with muscle development. Maturing myotubes contain decreasing levels of these adhesins, but they change from non-sialylated to lipid-linked membrane protein containing PSA.[4] In mature tissue, PSA forms of NCAM are restricted to regions separated from myotube surfaces.

A PSA-transferase has been used to study the expression of PSA in neural tissue. The amount of PSA-transferase transcript has been well correlated with PSA in various tissues, and is abundant in fetal human brain but not in adult brain. HeLa cells in monolayer culture, stably expressing α8-polysialyltransferase and NCAM, promoted overlaying chicken embryo neurite outgrowth and spreading. The results suggest that plasticity in neural cells is regulated by PSA transferase which biosynthesizes polysialylated embryonic NCAM.[5]

The structural diversity of sialic acids can influence cell adhesion mediated by a number of molecules. Examples of this include sialoadhesin, CD22 and the selectins. Sialoadhesins are found in macrophages, hematopoietic cells and B-lymphocytes and bind to cell surface glycoconjugates terminating in Neu5Acα2-3Galβ1-3GalNAc and/or Neu5Acα2-3 Galβ1-3/4 GlcNAc. The sialylated ligands are found on developing myeloid cells in the bone marrow, and on lymphocyte subsets in lymphoid tissue.[6] The selective adhesions thus contribute to cellular maturation and also provide a means by which macrophage scavenger functions are carried out.

Sialoadhesins have homology to CD22, a B cell specific receptor related to adhesion and signal transduction.[7] It has been shown that the 9-0-acetyl group of Neu5,9Ac_2 and the N-glycolyl group of Neu5Gc interfere with sialoadhesion binding. On the other hand, CD22 binds more firmly to Neu5Gc compared to Neu5Ac. Sialoadhesin and CD22 have thus slightly different binding specificities towards synthetic sialic acids.[8,9] Recognition by the glycerol side chain of sialic acid is important for binding with CD22 but not E-, L-, or P selectins. Modifications of sialoglycoconjugates can greatly influence cell-cell interactions mediated by adhesins such as sialoadhesin of macrophages and lymphocyte CD22.[6,8]

Sialoglycoproteins of endothelial cells including human vascular adhesion protein 1 mediate the binding of lymphocytes to endothelial cells. Glycosidase digestion experiments have shown that the sialic acid moities are critical for this function.[10]

N-glycans may block cell adhesion interactions. Osteonectin is a multiply N-glycosylated component of the bone matrix. Site directed mutagenesis of osteonectin at a specific Asn residue reduces its interaction with collagen V.[11] Similarly, human tissue plasminogen activator (tPA) requires complete N-glycosylation for normal tPA-fibrin binding.[12] Loss of N-glycan chains results in higher fibrinolytic compared to fibrinogenolytic activity.

Transfection of GlcNAc-transferase V in immortalized lung epithelial cells reduces their adhesiveness on surfaces coated with fibronectin and collagen type IV while increasing the apparent sizes of cell surface glycoproteins.[13] The effect is prevented by swainsonine which inhibits the formation of complex N-glycans. Transfected cells injected into nude mice produce significantly more tumors than control cells.

Carbohydrate-carbohydrate interactions have been proposed as an important force mediating some forms of cell adhesion; these reactions are calcium-dependent. They include Le^x-Le^x interaction in the case of teratocarcinoma cells,[14] GM_3-GM_3 ganglioside association in melanoma and lymphoma cells[15] and adhesion proteoglycan self-binding in marine sponges.[16]

9.B. MAMMALIAN LECTINS

Mammalian lectins may be cytosolic, secreted or cell surface bound; they may recognize various sugar structures and may differ in regard to calcium dependence. The carbohydrate binding activity generally resides in a limited portion of the molecule named carbohydrate recognition domain (CRD)[17]. Studies of galectins derived from *C. elegans* have shown that all of the critical amino acids involved in carbohydrate binding have been conserved, despite the phylogenetic distance between nematodes and vertebrates.[18] CRDs are found in P-type (Man-6-phosphate receptors), C-type (Ca^{2+}-binding) and S-type (Galβ-binding) lectins.

P-type lectins function as mannose-6-phosphate receptors which mediate the targeting of lysosomal enzymes to lysosomes.[19] These lectins recognize Man-6-phosphate groups on high mannose N-glycans of lysosomal enzymes.

C-type animal lectins are diverse and include the selectins, discussed separately below. The sugar binding specificity can be altered by minor amino acid sequence changes.[20] A well characterized lectin is the hepatic Gal/GalNAc-specific asialoglycoprotein receptor which functions in the clearance of glycoproteins from the circulation.[21] The hepatic asialoglycoprotein receptor in parenchymal hepatocytes and HepG2 cells binds glycoproteins with terminal Gal, GalNAc and related galactosides.[22,23] The asialo receptors of macrophages differ from those in hepatic cells in the spatial arrangements of their sugar combining sites and in subunit organization.[24] As a mechanism for glycoprotein turnover, interactions with asialoglycoprotein receptors serve to immobilize, internalize and finally degrade glycoproteins. The lifespan of circulating erythrocytes depends upon masking of cell surfaces with sialic acid. Removal of sialic acid results in erythrocyte clearance within a few hours since desialylated erythrocytes and other cells are taken up via asialo receptors on Kupffer cells in the liver and various macrophages.[25] The binding specificity of the hepatic asialoglycoprotein receptor has been demonstrated for N-glycan derivatives. Tri-antennary N-glycans with Gal or GalNAc termini are bound efficiently by the receptor but bi-antennary N-glycans with terminal Gal are not. In contrast, bi-antennary chains containing GalNAc termini target receptors with much greater efficiency than Gal terminated tri-antennary oligosaccharide.[26]

Fucose receptors have also been described in Kupffer cells derived from rat liver,[27] and in the hepatoma HepG2 cell line.[28] These may bind and determine the half life of O-Fuc containing glycoproteins such as tPA. A C-type lectin, NKR-P1 protein on natural killer (NK) cells, binds a diversity of oligosaccharide ligands with high affinity, including those with sialylated, sulfated and mannosylated termini.[29] This is important since NK cells display a multitude of functions including destruction of tumor and virally infected cells, transplantation reactions, and control of cell numbers in hematopoiesis.

A further example of C-type lectin is the mannose-binding receptor on some macrophages and also on endothelial cells.[30] The mannose-binding receptor of macrophages is important for the clearance of pathogens via calcium-dependent mannose recognition. It differs from other lectins of its type in that it contains CRDs in a single polypeptide chain; these CRDs can form a ternary complex with sugar and calcium.[31] This facilitates internalization of glycoproteins containing N-linked high mannose structures such as β-glucuronidase and yeast invertase. One function proposed for these receptors may be to scavenge misrouted secreted lysosomal enzymes. Collectins, C-type CRDs associated with collagens, frequently possess high affinities for mannose, and can mediate defense by way of opsonization and complement fixation.[17]

9.B.1. Galectins

S-type lectins are now classified as galectins.[32] Members of the galectin family are categorized by shared characteristic amino acid sequences and affinity for β–galactoside ligands. Galectins-1 and -2 are similar peptides containing approximately 130 amino acids whereas galectins-3 and -4 possess additional domains. The shared domain contains the carbohydrate binding sequence. Some recent additional but less well characterized galectins are termed galectins-5, -6 and-7. Galectin-1, galectin-2 and galectin-4 form dimers and thus have two carbohydrate binding sites which may crosslink molecules or cell surfaces. Galectin-3 can form aggregates and thus gain carbohydrate binding sites.[32] The binding sites of galectin-3 have been mapped in detail by Rini et al (Fig. 4). Galectins are primarily cytosolic but are also secreted, although this seems to occur in the absence of a typical secretion signal peptide. Nuclear galectins have been described; studies on galectin-depleted nuclear extracts from HeLa cells reconstituted with recombinant galectins suggest that these may be required for pre mRNA splicing.[33] Galectins possess distinct but somewhat overlapping distribution in tissues: for example, galectin-1 is abundant in muscle, neuronal tissue, thymus, kidney and placenta while galectin-3 is highest in activated macrophages, mast cells, some epithelial cells and some neurons. Their presence on the endothelium of lymph nodes depends on the state of activation of this tissue.[34]

Glycoproteins that contain poly-n-acetyl-lactosamines are especially good ligands for galectins. CD45 is one example and is believed to be a counter-receptor for galectin-1.[35] Blood group structures have also been shown to be reactive with galectins.[36] It is thought that galectins could act by cross-linking carbohydrate

Fig. 4. (opposite page) Carbohydrate binding site of galectin-3. The schematic drawing represents galectin-3 molecules bound to cell surface ligands. The insert indicates the binding domain of human galectin-3 in complex with N-acetyl-lactosamine. Galectin-3 is a carbohydrate binding protein whose three-dimensional structure contains a galactose-binding site pocket. Adhesive proteins such as laminin, carcinoembryonic antigen and lysosomal-associated membrane proteins possess N-acetyl-lactosamine ligands which can bind to galectin-3. Upregulation of galectin-3 can occur in cancers such as colon carcinoma where its expression is associated with tumor invasiveness and neoplastic progression. Galectin is secreted as a monomeric protein but may aggregate. The initial event in this process may be the binding of tumor cells to endothelium via galectin-3 and its counter-receptors. Affinity for β-galactosides and sequence similarity in the carbohydrate-binding site are requirements for inclusion in the galectin superfamily.[32] As seen in the X-ray crystal structure, His 158, Asn 160 and Arg 162 make direct hydrogen bond interactions with the C4 hydroxyl group of galactose. Reprinted with the permission of Dr. James Rini, Department of Biochemistry, University of Toronto.

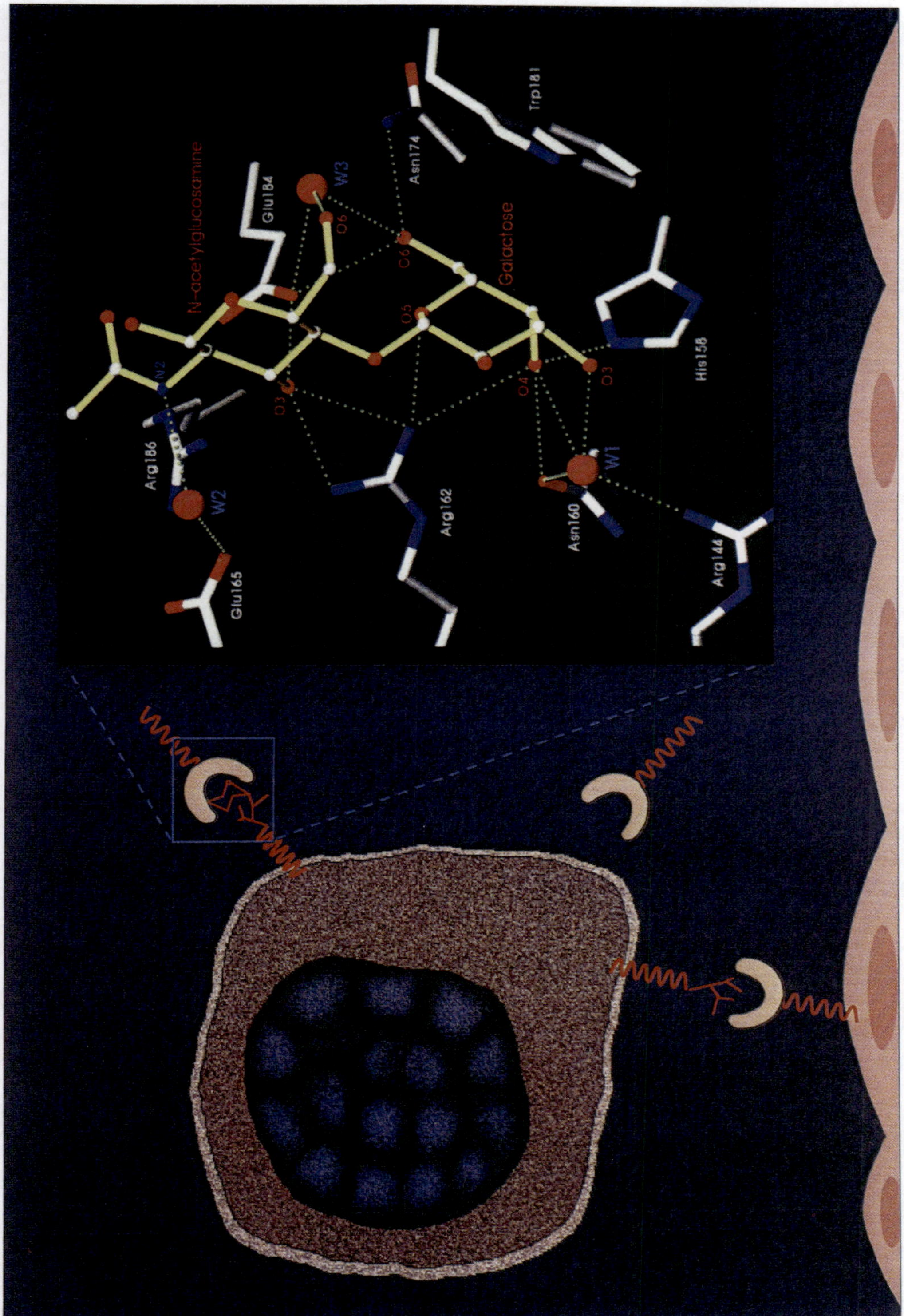
N-acetylglucosamine
Galactose
Glu184
W3
Asn174
Trp181
O6
C6
O5
N2
O3
O4
O3
Arg186
W2
Glu165
Arg162
Asn160
W1
Arg144
His158

chains on cell surfaces or the extracellular matrix. Interactions of galectin-1 and CD45 may initiate critical cell functions as illustrated by the following experiment. Activated human T-cells containing recombinant galectin-1 are converted to apoptotic cell forms, a phenomenon which requires the expression of CD45. Glycosylation is critical since the effect is reversed by blocking N-glycan elongation with swainsonine, and can be enhanced by inhibiting O-glycan elongation with GalNAcα-benzyl.[35]

The functions of galectins may be expressed through their interactions with cell surface components such as laminin,[37] integrin,[38] carcinoembryonic antigen[39] and membrane components involved in cell adhesion and cell migration.

Since galectins are expressed on microvascular endothelial cells, their presence in such locations could play a role in adhesive interactions in tumor cell metastases.[40] Galectin-3 expression is significantly increased in high grade dysplasia and early invasive colon cancers in comparison with the adenomatous tissues from which they evolved.[41] Similarly, galectin-3 is upregulated in HTLV-1 infected human T-cells and cells derived from anaplastic large cell lymphoma.[42] Studies of transgenic mice in which galectin-1 gene had been deleted showed no apparent abnormalities when compared with the normal; these findings suggest that alternate mechanisms could assume functions ordinarily ascribed to galectins.[43]

9.C. SELECTINS

Three homologous glycoprotein types termed L- E- and P-selectins comprise the selectin family. The term selectin indicates their selective expression and function and also denotes the presence of lectin-like domains. All three types were identified by the use of specific antibodies prepared against cell surface antigens.[44-46] Selectins have the CRD at the amino-terminus and have overlapping sugar binding specificities. Adjacent to this is a domain similar to a repeat first described in epidermal growth factor (EGF). The areas which follow are a series of repeats which resemble complement regulatory proteins.

Selectins function in cell adhesion and act in synchrony with other CAMs such as leukocyte integrins, vascular cell adhesion molecule-1 and intracellular adhesion molecule (ICAM-1), and establish cohesion between platelets, leukocytes and endothelial cells. The appearance of selectins on cell surfaces can be triggered by inflammatory agents; for example, lipopolysaccharide derived from Gram negative bacteria can greatly increase levels of E-selectins derived from endothelium.[47] Selectins are particularly important in leukocyte rolling on vessel walls, a reversible phenomenon which can precede firm attachment and movement of cells into tissue during inflammation. The extent to which P-selectin extends its CRD above membranes plays a role in neutrophil attachment under shear forces.[48] L-selectin is considered a leukocyte homing receptor. It is believed to transmit intracellular signals, including tyrosine phosphorylation and activation of protein kinases, and these events may contribute to the activation of neutrophils during adhesion.[49] L-selectin-deficient mice produced by gene targeting exhibit an impaired primary T cell response and cytokine production.[50]

There are subtle differences in the sugar binding specificities of selectins. The carbohydrate ligands for selectins fall into several classes of oligosaccharides: sialyl Le^x and sialyl Le^a related oligosaccharides, phosphorylated saccharides and sulfated oligosaccharides. The best ligands are sialylated or sulfated Lewis antigens.

Sulfated and sialylated O-linked carbohydrate chains were found on an endothelial ligand for L-selectin.[51] The gene encoding a mucin-like glycoprotein ligand,[52] GlyCAM 1 (glycosylation-dependent cell adhesion molecule 1), is similar to that of glycophorin and CD34, but differs from that of leukosialin which has only one exon.[53] Endothelial cell GlyCAM 1 and CD34 both function as ligands through the presentation of highly clustered O-linked chains to L-selectin on leukocyte cell surfaces. Independent of overall sialylation and fucosylation on GlyCAM 1, sulfation is necessary for ligand binding as shown by inhibition experiments using the sulfation inhibitor chlorate.[54] Sulfated structures of GlyCAM1 important for selectin binding were identified as N-acetyl-lactosamine chains with 6-sulfated Gal and 6-sulfated GlcNAc residues.[55,56]

Some insight into the molecular nature of ligand binding has been derived from a combination of X-ray crystallography and mutagenesis of lectin/epidermal growth factor-like domains of human E-selectin. The application of these techniques has resulted in modelling a three-dimensional structure of the E-selectin ligand binding region involving seven highly conserved basic amino acid residues on the surface of the domain which could interact with a negatively charged carbohydrate ligand. Calcium dependence of such binding may be the result of the Ca^{2+} binding properties of a specific amino acid (Asn 82) which in turn can act as a hydrogen donor to a sugar ligand hydroxyl group.[17,57]

Knowledge of the molecular chemistry of selectins and their ligands has been applied to study the specific behavior of cell types. T lymphocyte attachment and rolling on P selectin is abrogated by *P. hemolytica* O-glycoprotease (cleaving peptide at certain sialylated O-glycosylation sites) and by neuraminidase. The effects of glycoprotease upon cell binding to E-selectin are negligible but binding is abolished by neuraminidase. E-selectin binding has been correlated strictly with sialyl-Le^x structures whereas P-selectin binding is related to glycoforms shared with T lymphocytes and myeloid cells.[58,59] Activated B cells but not resting cells bind to purified E- and P-selectins and this activity is paralleled by the induction of different carbohydrate epitopes. Thus, phorbol ester induced B cells react with a panel of monoclonal antibodies produced against Le^x, sialyl-Le^x and sulfated glucuronyl ceramides, in contrast to their non-activated counterparts.[60] Binding can be reduced by treatment with sialidase.

In vitro transfection experiments of Chinese Hamster Ovary (CHO) cells with human α3-Fuc-transferase cDNA demonstrate that α3-Fuc-tranferase regulates the expression of E-selectin ligands and ELAM-1 dependent endothelial cell adhesion.[61,62] Specifically, α3-Fuc-transferase VII may be involved in the synthesis of E- and P-selectin ligands in the mouse.[63]

The role of N-glycosylation in the expression of E-selectin on endothelial cells was studied using interleukin-1 stimulated human umbilical vein endothelial cells (HUVEC) cultured with or without tunicamycin or castanospermine. E-selectin expression is significantly reduced by tunicamycin but only slightly reduced by castanospermine; however, the deglycoslated forms of E-selectin are fully capable of mediating adhesion of HT29 cells in vitro.[64a]

Several colon cancer cell lines expressing mucin or non-mucin ligands react with E-, P- and L-selectins. L- and P-selectins react primarily with mucin-type ligands on colon cancer cells, whereas E-selectin recognizes mucin and non-mucin ligands.[64b] Both sialyl Le^a and sialyl Le^x play significant roles in the adhesion of human cancer cells to vascular endothelium.[65-68] With epithelial cell lines, malignant colon and pancreas cell adhesion to E-selectin is dependent mostly on sialyl Le^a, but binding of other cell lines including those from lung and liver is mediated to a great extent by sialyl Le^x structures. The efficacy of binding of colon carcinoma cells has been correlated with their metastatic potential since highly metastatic cells express more lysosomal associated membrane proteins (lamp) and sialyl Le^x on their surfaces than those with low metastatic potential.[66] Adhesion is prevented by soluble lamp-1 or leukosialin carrying sialyl Le^x termini but inhibition is not seen when these compounds lack sialyl Le^x. Mimics and analogues of sialyl Le^x structures have been developed as potential anti-inflammatory agents and will be discussed in chapter 27.[69]

9.D. EXTRACELLULAR MATRIX AND INTEGRINS

Extracellular matrix (ECM) is a complex mixture produced by the cells embedded in it, acting as adhesive and protective macromolecules, transmitters of information between cells, and modifiers of gene expression, mitogenic activity and DNA synthesis.[70-73] ECM components include collagens, proteoglycans such as heparan, chondroitin and hyaluronic acid, and glycoproteins (e.g. fibronectin, laminin).[72] Some of these molecules possess trans-membrane domains and can bind cytokines or growth factors.[74] A considerable number of matrix components share a tripeptide RGD motif capable of binding in a calcium-dependent fashion to membrane-bound integrin receptors.[74-77]

Many proteoglycans and their receptors contain acidic and sulfated repeating disaccharide chains, as well as N-linked and O-linked oligosaccharides.[78-80] These sugar chains are believed to function as inhibitors of proteolytic degradation within the matrix.

Cell-matrix interactions are dependent on oligosaccharides from both ligand and receptor. This has been shown by an analysis of laminin-integrin interactions. Deletion of integrins in mice results in peri-implantation lethality and morphogenetic defects, underscoring their vital role in development.[81-84] Analysis of α and β integrin subunits of mouse melanoma cells shows that both components possess complex sialylated N-glycan chains which are involved in cell adhesion to laminin. Modulation of integrin-mediated cell adhesion and spreading can be brought about by altering the glycosylation of its subunits. Gp120/140, an $\alpha_6\beta_1$ integrin from murine melanoma cells can bind to laminin if desialylated; however, no binding is observed following treatment of desialylated structures with α–galactosidase or anti-α-galactosidase antibody. Spreading of whole cells on laminin can be modulated by pre-treatment with swainsonine.[83]

The binding of $\alpha_5\beta 1$ integrins to fibronectin has been examined in a ricin-resistant variant (Ric14) of baby hamster kidney (BHK) cells lacking GlcNAc-transferase I and therefore unable to synthesize complex or hybrid glycans. Although there is reduced binding by the variant cells and also by cells treated with 1-deoxymannojirimycin, relatively normal levels of integrins are present at the cell surface, which suggests that binding in this model is not dependent on complete N-glycan processing.[84] Inhibition of N-glycan processing can yield disparate results, perhaps depending on the cell line, the drug sensitivity or the functions of individual glycoproteins.[85,86]

A functional role for N-glycosylation of $\alpha_5\beta_1$ integrin has been shown in studies of Zheng et al.[87] Normally the two subunits are tightly associated with each other. N-glycosidase F treatment results in the dissociation of these chains, shown by immunoprecipitation experiments. N-glycosidase treated integrin subunits are only precipitated by antibody to the specific subunit. Adhesion of K562 cells to fibronectin plates which is mediated by $\alpha_5\beta_1$ integrin is inhibited when cells are treated with N-glycosidase F. In addition, binding of purified $\alpha_5\beta_1$ integrin incorporated into liposomes to fibronectin-coated plates depends on N-glycosylation. This suggests that N-glycosylation is necessary for both integrin subunit association and optimal binding to fibronectin.

Glycosylation changes occur on integrin chains in the course of T-cell ontogeny. Immature thymocytes have a greater content of N-glycosylated sugar chains and fewer sialic acid residues on β_1 chains than do mature thymocytes and peripheral lymphocytes; similarly, α-chains are undersialylated on immature forms. The undersialylated forms show better binding to laminin and fibronectin.[88]

Altered glycosylation of integrins in transformed cells can affect their invasive behavior as shown in colon cancer cells treated with glycosylation inhibitors.[89] Comparisons between β_1 integrins in a highly metastatic mouse B16F1 melanoma cell line and its weakly metastatic mutant showed that both forms of B16 cells possess high mannose-type and bi-, tri- and tetra-antennary N-glycans. However, less sialylated forms are present in the mutant along with Le^x structures, mainly on tri- and tetra-antennary structures, while the parental strain does not exhibit Le^x. These differences are thought to account for the reduction in metastatic potential in the mutant and decreased adhesion to fibronectin and laminin.[90] A possibility exists that metastasis is initiated by sphingolipid (GM3/lactosylceramide) interaction, as judged from dynamic flow studies in which B16 melanoma cells were employed.[91]

In a system utilizing HL60 cells, integrin-dependent adhesion to fibronectin is thought to be regulated by O-linked sugars. GalNAc-α-benzyl has been used to block elongation of O-glycans leading to an accumulation of GalNAc-O-Ser/Thr. No adhesion of HL60 cells to fibronectin occurs in the presence of elongated O-linked glycans, but significant adhesion is observed following the appearance of truncated glycans in the presence of GalNAc-α-benzyl.[92] This suggests that certain O-glycan chains may inhibit cell attachment to the extracellular matrix.

9.E. FERTILIZATION

The events of fertilization consist of four major activities: contact/recognition between sperm and egg, regulation of sperm entry into the egg, fusion of sperm/egg genes, and activation of egg metabolism to begin development.[93] Glycoproteins and glycosyltransferases,[94-97] as well as glycosidases[98] are involved in this process. Preparatory events include increased expression of MUC-1 mucin mRNA following ovulation.[99] Porcine zona pellucida glycoproteins of mature ova are rich in neutral and acidic branched N- and O-glycans; the N-glycans consist of bi-, tri- and tetra-antennary structures.[100-103]

Porcine zona pellucida glycoprotein ZP3 inhibits sperm-egg binding in in vitro competition assays. N-glycans of porcine ZP3 are believed to play a major role in sperm binding. Elimination of these structures with N-glycanase treatment decreases the inhibitory action of ZP3, while elimination of O-linked chains by alkali treatment has little effect on binding.[104] In contrast, in the invertebrate *Unio elongatus*, O-glycans but not N-glycans of the vitellin coat glycoprotein are involved in sperm binding to eggs.[94]

A discrete region of the mouse zona pellucida ZP3 possesses sperm receptor activity.[105] The removal of O-glycans by alkali treatment destroys sperm receptor activity. Galα1-3 residues at the non-reducing terminus of O-linked glycans of ZP3 are thought to be involved in sperm binding.[106-108] In the mouse, a Galα1-3 residue has been identified as the ligand for sperm interaction[107] and recent studies have shown that oligosaccharides containing terminal Galα1-3Gal moieties inhibit sperm-egg binding in a dose-dependent manner in the mouse.[108] However, other epitopes may be involved since it has been found that normal mouse litters can be derived from parent mice lacking the α3-Gal-transferase gene.[109] Fucose linked to O-glycans also seems to be involved in invertebrate species.[94]

Not only the Galα1-3Gal structure but also terminal GlcNAc has been identified as an essential part of the ligand for sperm binding on mouse egg ZP3. Gal-transferase is present in seminal plasma and epididymal fluid.[110] It appears that the expression of cell surface bound β4-Gal-tranferase is regulated in a stage-specific fashion during spermatogenesis and the enzyme plays an important role in fertilization by binding to glycoconjugates of the zona pellucida glycoprotein of the egg.[111] Inhibition of sperm egg binding in the mouse is achieved by adding anti β4-Gal-transferase antibody to sperm. The addition of UDP-Gal, which leads to the conversion of terminal GlcNAc residues to Galβ1-4 GlcNAc termini of ZP3, also inhibits sperm binding to the egg.[112] Primary gamete binding sites can be removed by digestion of ZP3 with N-acetylhexosaminidase which cleaves GlcNAc residues; presumably this process occurs following egg activation to prevent polyspermy.

Overexpression of β4-Gal-transferase in transgenic mice reduces its binding to radiolabelled ZP3. This may be explained by the excessive production of glycans which may mask ligands.[113] Other carbohydrates and transferases have also been reported to play a role in fertilization. Removal of sialic acid from the zona pellucida of the rabbit ovum results in an inability of the sperm to penetrate.[114] Sperm-zona interactions may also be influenced by activities of Fuc-transferase[115] and sialyltransferase.[116]

References

1. Roseman S. The synthesis of complex carbohydrates by multiglycosyltransferase systems and their potential function in intercellular adhesion. Chem Phys Lipids 1970; 5:270-297.
2. Pigott R, Power C. The Adhesion Molecule Factsbook. Academic Press, San Diego CA 1993:115-120.
3. Rutishauser U. Neural cell adhesion molecule and polysialic acid. In:McDonald J, Mecham R, eds. Receptors for Extracellular Matrix. San Diego CA:Academic Press, 1991:131-156.
4. Fredette B, Rutishauser U, Landmesser L. Regulation and activity-dependence of N-cadherin, NCAM isoforms and polysialic acid on chick myotubes during development. J Cell Biol 1993; 123:1867-1888.
5. Nakayama J, Fukuda MN, Fredette B et al. Expression cloning of a human polysialyltransferase that forms the polysialylated neural cell adhesion molecule present in embryonic brain. Proc Natl Acad Sci USA 1995; 92:7031-7035.
6. Crocker P, Mucklow S, Bouckson V et al. Sialoadhesion, a macrophage sialic acid binding receptor for hematopoietic cells with 17 immunoglobulin-like domains. EMBO J 1994; 13:4490-4503.
7. Clark E. CD22, a B cell-specific receptor, mediates adhesion and signal transduction. J Immun 1993; 150:4715-4718.
8. Kelm S, Schauer R, Manuguerra J et al. Modifications of cell surface sialic acids modulate cell adhesion mediated by sialoadhesion and CD22. Glycoconj J 1994; 11:576-585.
9. Powell L, Jain R, Matta K et al. Characterization of sialyloligosaccharide binding by recombinant soluble and native cell-associated CD22. J Biol Chem 1995; 270:7523-7532.
10. Salmi M, Jalkanen S. Human vascular adhesion protein 1 (VAP-1) is a unique sialoglycoprotein that mediates carbohydrate-dependent binding of lymphocytes to endothelial cells. J Exp Med 1996; 183:569-579.
11. Xie R, Long G. Role of N-linked glycosylation in human osteonectin. J Biol Chem 1995; 270:23212-23217.
12. Hansen L, Blue Y, Barone K et al. Functional effects of asparagine-linked oligosaccharide on natural and variant human tissue-type plasminogen activator. J Biol Chem 1988; 263:15713-15719.
13. Demetriou M, Nabi IR, Coppolino M et al. Reduced contact-inhibition and substratum adhesion in epithelial cells expressing GlcNAc-Transferase V. J Cell Biol 1995; 130:383-392.
14. Eggens I, Fenderson B, Tokoyuni T et al. Specific interaction between Le x and Le x determinants. J Biol Chem 1989; 264:9476-9484.
15. Kojima N, Hakomori S. Specific interaction between gangliotriaosylceramide (Gga) and sialosyllactosylceramide (GM3) as a basis for specific cellular recognition between lymphoma and melanoma cells. J Biol Chem 1989; 264:20159-20162.
16. Misevic G, Burger M. Carbohydrate-carbohydrate interactions of a novel acidic glycan can mediate sponge cell adhesion. J Biol Chem 1993; 268:4922-4929.
17. Drickamer K. Three-dimensional view of a selectin cell adhesion molecule. Glycobiology 1994; 4:245-248.
18. Hirabayashi J, Ubukata T, Kasai K. Purification and molecular characterization of a novel 16 kDa galectin from the nematode caenorhabditis elegans. J Biol Chem 1996; 271:2497-2505.
19. Baranski T, Koelsch G, Hartsuck JA et al. Mapping and molecular modelling of a recognition domain for lysosomal enzyme targeting. J Biol Chem 1991; 266:23365-23372.

20. Iobst ST, Drickamer K. Binding of sugar ligands to Ca^{2+}-dependent animal lectins. II. Generation of high affinity galactose binding by site-directed mutagenesis. J Biol Chem 1994; 269:15512-15519.
21. Ashwell G, Harford J. Carbohydrate-specific receptors of the liver. Ann Rev Biochem 1982; 51:531-554.
22. Baenziger J, Maynard Y. Human hepatic lectin. Physiochemical properties and specificity. J Biol Chem 1980; 255:4607-4613.
23. Schwartz A, Rup D. Biosynthesis of the human asialoglycoprotein receptor. J Biol Chem 1983; 258:11249-11255.
24. Ozaki K, Lee R, Lee Y et al. The differences in structural specificity for recognition and binding between asialoglycoprotein receptors of liver and macrophages. Glycoconj J 1995; 12:268-274.
25. Schauer R. Chemistry, metabolism and biological functions of sialic acids. Adv Carbohydr Chem Biochem 1982; 40:131-234.
26. Chiu M, Tamura T, Wadhwa M et al. In vivo targeting function of N-linked oligosaccharides with terminating galactose and N-acetylgalactosamine residues. J Biol Chem 1994; 269:16195-16202.
27. Hoyle G, Hill R. Molecular cloning and sequencing of a cDNA for a carbohydrate binding receptor unique to rat Kupffer cells. J Biol Chem 1988; 263:7487-7492.
28. Hajjar K, Reynolds C. α-Fucose mediated binding and degradation of tissue type plasminogen activator by HepG2 cells. J Clin Invest 1994; 93:703-710.
29. Bezouska K, Yuen C-T, O'Brien J et al. Oligosaccharide ligands for NKR-P1 protein activate NK cells and cytotoxicity. Nature 1994; 372:150-157.
30. Stahl P. The macrophage mannose receptor:current status. Am J Respir Cell Molec Biol 1990; 2:317.
31. Mullin N, Hall K, Taylor M. Characterization of ligand binding to a carbohydrate-recognition domain of the macrophage mannose receptor. J Biol Chem 1994; 269:28405-28413.
32. Barondes SH, Cooper DNW, Gitt MA et al. Galectins. Structure and function of a large family of animal lectins. J Biol Chem 1994; 269:20807-20810.
33. Dagher S, Wang S, Wang L et al. Nuclear galectins are functionally redundant in their activity in pre mRNA splicing. Glycoconj J 1995; 12:545.
34. Baum L, Seilhamer J, Pang M et al. Synthesis of an endogenous lectin, galectin-1, by human endothelial cells is up-regulated by endothelial cell activation. Glycoconj J 1995; 12:63-68.
35. Perillo NL, Pace KE, Seilhamer JJ et al. Apoptosis of T cells mediated by galectin-1. Nature 1995; 378:736-739.
36. Abbott WM, Hounsell EF, Feizi T. Further studies of oligosaccharide recognition by the soluble 13 kDa lectin of bovine heart muscle. Ability to accomodate the blood-group-H and -B-related sequences. Biochem J 1988; 252:283-287.
37. Zhou Q, Cummings R. L-14 recognition of laminin and its promotion of in vitro cell adhesion. Arch Biochem Biophys 1993; 300:6-17.
38. Cooper D, Cannon V, Farrell E et al. Galectin-1 binds to integrins on vascular smooth muscle cells and inhibits their migration. Glycoconj J 1995; 12:546.
39. Ohannesian DW, Lotan D, Thomas P et al. Carcinoembryonic antigen and other glycoconjugates act as ligands for galectin-3 in human colon carcinoma cells. Cancer Res 1995; 55:2191-2199.
40. Lotan R, Ito H, Yasui W et al. Expression of a 31-kDa lactoside-binding lectin in normal human gastric mucosa and in primary and metastatic gastric carcinomas. Int J Cancer 1994; 56:474-480.
41. Schoeppner H, Raz A, Ho S et al. Expression of an endogenous galactose-binding lectin correlated with neoplastic progression in the colon. Cancer 1995; 75:2818-2826.
42. Konstantinov KN, Robbins BA, Liu F-T. Galectin-3, a β-galactoside-binding animal lectin, is a marker of anaplastic large-cell lymphoma. Am J Pathol 1996; 148:25-30.
43. Poirier F, Robertson E. Galectin is not necessary for animal life? "knock-out" mouse tells the truth. Development 1993; 119:1229-1236.
44. Lasky L. Selectins:interpreters of cell-specific carbohydrates information during inflammation. Science 1992; 258:964-970.

45. Bevilacqua M, Nelson R. Selectins. J Clin Invest 1993; 91:379-387.
46. Larsen GR, Sako D, Ahern TJ et al. P-selectin and E-selectin:distinct but overlapping leukocyte ligand specificities. J Biol Chem 1992; 267:11104-11110.
47. Kawamura N, Imanishi N, Koike H et al. Lipoteichoic acid-induced neutrophil adhesion via E-selectin to human umbilical vein endothelial cells (HUVECs). Biochem Biophys Res Comm 1995; 217:1208-1215.
48. Patel K, Nollert M, McEver R. P-Selectin must extend a sufficient length from the plasma membrane to mediate rolling of neutrophils. J Cell Biol 1995; 131:1893-1902.
49. Waddell T, Fialkow L, Chan C et al. Signalling functions of L-selectins. J Biol Chem 1995; 270:15403-15411.
50. Xu J, Grewal IS, Geba GP et al. Impaired primary T cell responses in L-selectin-deficient mice. J Exp Med 1996; 183:589-598.
51. Imai Y, Rosen S. Direct demonstration of heterogeneous sulfated O-linked carbohydrate chains on an endothelial ligand for L-selectin. Glycoconj J 1993; 10:34-39.
52. Lasky LA, Singer MS, Dowbenko D et al. An endothelial ligand for L-selectin is a novel mucin-like molecule. Cell 1992; 69:927-938.
53. Dowbenko D, Andalibi A, Young PE et al. Structure and chromosomal localization of the murine gene encoding GLYCAM 1. A mucin-like endothelial ligand for L selectin. J Biol Chem 1993; 268:4525-4529.
54. Imai Y, Lasky L, Rosen S. Sulfation requirement for GlyCAM-1, an endothelial ligand for L-selectin. Nature 1993; 361:555-557.
55. Hemmerich S, Bertozzi CR, Leffler H et al. Identification of the sulfated monosaccharides of GlyCAM-1, an endothelial derived ligand for L-selectin. Biochemistry 1994; 33:4820-4829.
56. Hemmerich S, Rosen SD. 6'-sulfated sialyl Lewis x is a major capping group of GlyCAM-1. Biochemistry 1994; 33:4830-4835.
57. Graves BJ, Crowther RL, Chandran C et al. Insight into E-selectin/ligand interaction from the crystal structure and mutagenesis of the lec/EGF domains. Nature 1994; 367:532-538.
58. Alon R, Rossiter H, Wang X et al. Distinct cell surface ligands mediate T lymphocyte attachment and rolling on P- and E-selectin under physiological flow. J Cell Biol 1994; 127:1485-1495.
59. Bennett TA, Schammel CMG, Lynam EB et al. Evidence for a third component in neutrophil aggregation:potential roles of O-linked glycoproteins as L-selectin counter structures. J Leukoc Biol 1995; 58:510-518.
60. Postigo A. B lymphocyte binding to E and P-selectins is mediated through the de novo expression of carbohydrates on in vitro and in vivo activated human B cells. J Clin Invest 1994; 94:1585-1596.
61. Lowe JB, Stoolman LM, Nair RP et al. ELAM-1-dependent cell adhesion to vascular endothelium determined by a transfected human fucosyltransferase cDNA. Cell 1990; 63:475-484.
62. Gersten KM, Natsuka S, Trinchera M et al. Molecular cloning, expression, chromosomal assignment, and tissue-specific expression of a murine α-(1,3)-fucosyltransferase locus corresponding to the human ELAM-1 ligand fucosyl transferase. J Biol Chem 1995; 270:25047-25056.
63. Smith PL, Gersten KM, Petryniak B et al. Expression of the α(1,3)fucosyltransferase Fuc-TVII in lymphoid aggregate high endothelial venules correlates with expression of L-selectin ligands. J Biol Chem 1996; 271:8250-8259.
64a. Pahlsson P, Strindhall J, Srinivas U et al. Role of N-linked glycosylation in expression of E-selectin on human endothelial cells. Eur J Immunol 1995; 25:2452-2459.
64b. Mannori G, Crottet P, Cecconi O et al. Differential colon cancer cell adhesion to E-, P-, and L-selectin:role of mucin-type glycoproteins. Cancer Res 1995; 55:4425-4430.
65. Takada A, Ohmori K, Yoneda T et al. Contribution of carbohydrate antigens sialyl Lewis a and sialyl Lewis x to adhesion of human cancer cells to vascular endothelium. Cancer Res 1993; 53:354-361.
66. Saitoh O, Wang W-C, Lotan R et al. Differential glycosylation and cell surface expression of lysosomal membrane glycoproteins in sublines of a human colon cancer exhibiting distinct metastatic potentials. J Biol Chem 1992; 267:5700-5711.

67. Sawada R, Tsuboi S, Fukuda M. Differential E-selectin-dependent adhesion efficiency in sublines of a human colon cancer exhibiting distinct metastatic potentials. J Biol Chem 1994; 269:1425-1431.
68. Sawada T, Ho JJL, Chung Y-S et al. E-selectin binding by pancreatic tumor cells is inhibited by cancer sera. Int J Cancer 1994; 57:901-907.
69. Rao BNN, Anderson MB, Musser JH et al. Sialyl Lewis X mimics derived from a pharmacophore search are selectin inhibitors with anti-inflammatory activity. J Biol Chem 1994; 269:19663-19666.
70. Wang D, Birkenmeier T, Yang J et al. Release from quiescence stimulates the expression of integrin $\alpha_5\beta_1$ which regulates DNA synthesis in human fibrosarcoma HT1080 cells. J Cell Physiol 1995; 164:499-508.
71. Hynes R. Integrins:versatility, modulation, and signaling in cell adhesion. Cell 1992; 69:11-25.
72. Ayad S, Boot-Handford R, Humphries M et al. The Extracellular Matrix FactsBook. San Diego CA:Academic Press 1994:10-15.
73. Fukushima K, Watanabe H, Takeo K et al. N-linked sugar chain structure of recombinant human lymphotoxin produced by CHO cells:the functional role of carbohydrate as to its lectin-like character and clearance velocity. Arch Biochem Biophys 1993; 304:144-153.
74. Ruoslahti E, Pierschbacher M. New perspectives in cell adhesion:RGD and integrins. Science 1987; 238:491-497.
75. Ruoslahti E, Yamaguchi Y. Proteoglycans as modulators of growth factor activities. Cell 1991; 64:867-869.
76. Haas T, Plow E. Ligand-ligand interactions:a year in review. Curr Opin Cell Biol 1994; 6:656-662.
77. Wight T, Heinegard D, Hascall V. Proteoglycans:structure and function. In:Hay E ed. Cell Biology of Extracellular Matrix. New York:Plenum Press; 1991:51-52.
78. Nilsson B, De Luca S, Lohmander S. Structures of N-linked and O-linked oliogosaccharides on proteoglycan monomer isolated from the swarm rat chondrosarcoma. J Biol Chem 1982; 257:10920-10927.
79. Hounsell E, Feeney J, Scudder P et al. NMR studies at 500 MHz of a neutral disaccharide and sulfated di-, tetra, hexa, and larger oligosaccharides obtained by endo-β-galactosidase treatment of keratan sulfate. Eur J Biochem 1986; 157:375-384.
80. Akiyama F, Stevens RL, Hayashi S et al. The structures of N- and O-glycosidic carbohydrate chains of a chondroitin sulfate proteoglycan isolated from the media of the human aorta. Arch Biochem Biophys 1987; 252:574-590.
81. Stephens L, Sutherland A, Klimanskaya I et al. Deletion of β1 integrins in mice results in inner cell mass failure and peri-implantation lethality. Gene Develop 1995; 9:1883-1895.
82. Fässler R, Meyer M. Consequences of lack of β1 integrin gene expression in mice. Gene Develop 1995; 9:1896-1908.
83. Chammas R, Veiga S, Travassos L et al. Functionally distinct roles for glycosylation of α and β integrin chains in cell-matrix interactions. Proc Natl Acad Sci USA 1993; 90:1795-1799.
84. Koyama T, Hughes RC. Functional integrins from normal and glycosylation-deficient baby hamster kidney cells. J Biol Chem 1992; 267:25939-25944.
85. Akiyama S, Yamada S, Yamada K. Analysis of the role of the human fibronectin receptor. J Biol Chem 1989; 264:18011-18018.
86. Elbein A, Legler G, Tlusty A et al. The effect of deoxymannojirimycin on the processing of the influenza viral glycoproteins. Arch Biochem Biophys 1984; 235:579-588.
87. Zheng M, Fang H, Hakomori S. Functional role of N-glycosylation in α5β1 integrin receptor. J Biol Chem 1994; 269:12325-12331.
88. Wadsworth S, Halvorson M, Chang A et al. Multiple changes in VLA protein glycosylation, expression, and function occur during mouse T cell ontogeny. J Immunol 1993; 150:847-857.
89. Lampe B, Stallmach A, Riecken EO. Altered glycosylation of integrin adhesion molecules in colorectal cancer cells and decreased adhesion to the extracellular matrix. Gut 1993; 34:829-836.
90. Kawano T, Takasaki S, Tao T et al. Altered glycosylation of β_1 integrins associated with reduced adhesiveness to fibronectin and

laminin. Int J Cancer 1993; 53:91-96.

91. Kojima N, Shiota M, Sadahira Y et al. Cell adhesion in a dynamic flow system as compared to static system. J Biol Chem 1992; 267:17264-17270.
92. Kojima N, Saito M, Tsuji S. Role of cell surface O-linked oligosaccharides in adhesion of HL60 cells to fibronectin:regulation of integrin-dependent cell adhesion by O-linked oligosaccharide elongation. Exp Cell Res 1994; 214:537-542.
93. Gilbert S. Developmental Biology. 3rd ed. Sunderland MA:Sinauer Associates Inc, 1991:32-33.
94. Focarelli R, Rosati F. The 220-kDa vitelline coat glycoprotein mediates sperm binding in the polarized egg Unio elongatulus through O-linked oligosaccharides. Development Biol 1995; 171:606-614.
95. Shur B. Expression and function of cell surface galactosyltransferase. Biochim Biophys Acta 1989; 988:389-409.
96. Dell A, Morris HR, Easton RL et al. Structural analysis of the oligosaccharides derived from glycodelin, a human glycoprotein with potent immunosuppressive and contraceptive activities. J Biol Chem 1995; 270:24116-24126.
97. Miller D, Macek B, Shur B. Complementarity between sperm surface β1-4 galactosyltransferase and egg coat ZP3 mediates sperm-egg binding. Nature 1992; 357:589-593.
98. Tulsiani D, Skudlarek M, Araki K et al. Purification and characterization of two forms of β-D-galactosidase from rat epididymal fluid:evidence for their role in the modification of sperm plasma membrane glycoprotein(s). Biochem J 1995; 305:41-50.
99. Hey N, Graham R, Seif, R et al. The polymorphic epithelial mucin MUC1 in human endometrium is regulated with maximal expression in the implantation phase. J Clin Endocrin Metab 1994; 78:337-342.
100. Mori E, Takasaki S, Hedrick J et al. Neutral oligosaccharide structures linked to asparagines of porcine zona pellucida glycoproteins. Biochemistry 1991; 30:2078-2087.
101. Noguchi S, Nakano M. Strucure of the acidic N-linked carbohydrate chains of the 55-kDa glycoprotein family (PZP3) from porcine zona pellucida. Eur J Biochem 1992; 209:883-894.
102. Hokke C, Damm J, Penninkhof B et al. Structure of the O-linked carbohydrate chains of porcine zona pellucida glycoproteins. Eur J Biochem 1994; 221:491-512.
103. Yurewicz E, Pack B, Sacco A. Porcine oocyte zone pellucida Mr 55,000 glycoproteins:identification of O-glycosylated domains. Molec Reprod Develop 1992; 33:182-188.
104. Yonezawa N, Aoki H, Hatanaka Y et al. Involvement of N-linked carbohydrate chains of pig zona pellucida in sperm-egg binding. Eur J Biochem 1995; 233:35-41.
105. Rosiere T, Wassarman P. Identification of a region of mouse zona pellucida glycoprotein mZP3 that possesses sperm receptor activity. Develop Biol 1992; 154:309-317.
106. Bleil JD, Wassarman PM. Galactose at the nonreducing terminus of O-linked oligosaccharides of mouse egg zona pellucida glycoprotein ZP3 is essential for the glycoprotein's sperm receptor activity. Proc Natl Acad Sci USA 1988; 85:6778-6782.
107. Wassarman P. Zone pellucida glycoproteins. Ann Rev Biochem 1988; 57:415-442.
108. Litscher E, Juntunen K, Seppo A et al. Oligosaccharide constructs with defined structures that inhibit binding of mouse sperm to unfertilized eggs in vitro. Biochemistry 1995; 34:4662-4669.
109. Thall A, Maly P, Lowell J. Oocyte Galα1-3Gal epitopes implicated in sperm cell adhesion to the zona pellucida glycoprotein ZP3 are not required for fertilization in the mouse. J Biol Chem 1995; 270:21437-21440.
110. Ross P, Vigneault N, Provencher S et al. Partial characterization of galactosyltransferase in human seminal plasma and its distribution in the human epididymis. J Reprod Fertil 1993; 98:129-137.
111. Scully N, Shur B. Stage-specific increase in cell surface galactosyltransferase activity during spermatogenesis in mice bearing t alleles. Develop Biol 1988; 125:195-199.
112. Lopez L, Bayna E, Litoff D et al. Receptor function of mouse sperm surface galactosyltransferase during fertilization. J Cell Biol 1985; 101:1501-1510.
113. Youakim A, Hathaway H, Miller D et al. Overexpressing sperm surface β1-4-galacto-

syltransferase in transgenic mice affects multiple aspects of sperm-egg interactions. J Cell Biol 1994; 126:1573-1583.

114. Lambert H, Le A. Possible involvement of a sialylated component of the sperm plasma membrane in sperm-zona interaction in the mouse. Gamete Res 1984; 10:153-163.

115. Cardullo R, Armant D, Millette C. Increased Fuc-transferase activity between isolated mouse spermatocytes and spermatids. J Cell Biol 1986; 103:81a.

116. Durr R, Shur S, Roth S. Sperm associated sialyltransferase activity. Nature 1977; 265:547-548.

CHAPTER 10

Role of Glycoproteins of the Immune and Blood Coagulation Systems

10.A. IMMUNE SYSTEM

Immunoglobulins and immunoglobulin-like cell surface domains are the most pervasive structures of the immune system. The latter group comprises over 70 members of the immunoglobulin superfamily which control cell behavior by acting as matrix binders, intercellullar adhesion molecules and/or signal transducing molecules. Many glycoproteins are found on the surfaces of cells of the immune system and their functions may be greatly influenced by glycosylation.[1] For example, terminal sialic acid on lymphocytes helps to maintain normal homing patterns to tissues and organs; this function is altered following treatment of cells with neuraminidase.[2] Neuraminidase also greatly reduces the colony forming ability of bone marrow stem cells.[3] Selectin-mediated adhesion of leukocytes and tumor cells expressing sialyl Le^x to endothelium is believed to antecede cell migration to ectopic sites of inflammation.[4]

Different cell types of the immune system possess characteristic carbohydrate structures, as well as glycosyltransferases which can modify cell function. The enzymatic activity of α6-sialyltransferase acting on N-glycans has been found to generate four different sialylated B lymphocyte differentiation antigens.[5] Sialylation of B lymphocyte glycoprotein CD22 by α6-sialyltransferase (N) produces significant T cell rosetting which may be abrogated with neuraminidase treatment. This suggests that α6-sialyltransferase might serve as a regulator of adhesion in developing B lymphocytes.[6]

A role for N-glycans in B cell maturation has been suggested by studies in which swainsonine and castanospermine treatment were shown to suppress adhesion between treated B cells and untreated T cells, but not between untreated B cells and treated T cells.[7] N-glycosylation is required for human CD2 immunoadhesion to CD58, a determinant of antigen presenting cells and other cell types, since mutant CD2 transmembrane receptor lacking an N-glycosylation site as well as deglycosylated recombinant CD2 are no longer able to bind CD58.[8] The three-dimensional structure of a glycosylated domain of human CD2 was studied by NMR. This suggests a model whereby residues important for CD58 binding could be identified as well as those affecting conformational stability.[9]

B cells and cell lines are rich in GlcNAc-transferase III activity, in contrast to T cells and cell lines which contain negligible enzyme activities.[10] The biological significance of introducing the bisecting GlcNAc residue may be the control of further N-glycan processing.

N-glycans change upon T-cell activation. Marked increases of L-PHA lectin binding structures characterized by GlcNAcβ1-6 branching of complex N-glycans occurs in both T4 and T8 cells following lectin or phorbol ester activation.[11] N-glycan branching and sialylation of Thy-1 antigen is altered during differentiation of mouse T lymphocytes. Thy-1 glycoprotein exhibits a decreased tri-antennary to bi-antennary chain ratio on mature thymocytes compared with immature cells, but N-glycans from immature cell Thy-1 possess more sialic acid.[12]

The surfaces of granulocytes (PMN) and other hematopoietic cells are characterized by leukosialin, (CD43), a glycoprotein which contains ample amounts of O-glycans but only a single N-linked chain.[13,14] Differences are observed between myeloid HL60 and erythroid K562 cell O-glycans; O-glycan core 2 structures are commonly expressed on myeloid glycans, whereas erythroid structures seem to possess more O-glycan core 1 structures. Leukosialin is rich in poly-n-acetyl-lactosamines which are important for the adhesive properties of cell surfaces; they can also serve as acceptors for glycosyltransferases that add antigenic or ligand structures of importance for immune functions.

Differential regulation of O-linked CD43 glycoforms on CD4 and CD8 T lymphocytes has been observed in graft-versus-host disease (GVHD). A 115 kDa glycoform carries primarily O-glycan core 1 tetrasaccharide, sialylα2-3 Galβ1-3 (sialylα2-6) GalNAc-, and is a marker for all T cells; a 130 kDa glycoform has mainly disialylated O-glycan core 2 hexasaccharides and is associated with T cell activation. In GVHD, T lymphocyte subsets were examined for CD43 expression; both glycoforms are upregulated in CD4 T cells, whereas in CD8 T cells, the O-glycan core 1 tetrasaccharides of CD43 are downregulated while the core 2 hexasaccharides are markedly upregulated.[15]

Activation of human T lymphocytes by anti-CD3 and interleukin results in a conversion of the O-glycan core 1 tetrasaccharide to the core 2 O-glycan hexasaccharide (described below) compared to resting T-lymphocytes.[16] This striking difference in structure has been attributed to an increase in O-glycan core 2 β6-GlcNAc-transferase as well as a decrease in sialyltransferase acting on asialo-bovine submaxillary mucin. In the Wiskott-Aldrich syndrome, characterized by a defect in antigen processing or recognition, this regulation is lost and is accompanied by a fixed increase in sialylated core 2 hexasaccharide and core 2 GlcNAc-transferase,[17,18] and defective cell surface sialoglycoproteins.[19]

Cytotoxic T-lymphocytes recognize target cell surface ligands via carbohydrate structures which are under control of the major histocompatibility complex (MHC). Glycopeptides have been shown to inhibit this effect but only if they were derived from MHC that was the same as the MHC of target cells.[20] High mannose oligosaccharides have been shown to directly inhibit antigen driven T cell responses, suggesting that such structures may possess an immunomodulating role.[21]

Carbohydrate chains on immunoglobulin molecules are generally found on heavy chain C regions, with possible exceptions when an N-glycosylation sequon Asn-X-Ser/Thr was found in the V region.[22] N-glycosylation of newly synthesized IgA and IgE is necessary for normal secretion to occur, since inhibition of secretion occurs following tunicamycin treatment of immunoglobulin secreting plasma cells.[23] Regulation of IgE secretion occurs by way of IgE binding protein; depending on the occupancy of glycosylation sequons, this protein either enhances (IgE potentiating factor) or inhibits (IgE suppressing factor) IgE synthesis.[24,25]

Glycoform differences in Fc and Fab fragments of IgG have significant effects upon antibody properties such as self-association, aggregation and cryoprecipitation.[26] High mannose type oligosaccharides on the non-binding Fab arm of IgG may function as ligands for the serum mannan-binding protein, with specificities for mannose and GlcNAc, and possibly GalNAc[27] and thus activate the classical pathway of complement.[28] Innate forms of immunity may be effected via mannose-binding lectins which are known to be present in normal sera. Cells infected with influenza virus which contains high mannose oligosaccharides show calcium-dependent binding to the mannose binding lectin, in contrast to non-infected cells. Lysis of infected cells then occurs in the presence of complement.[29] Serum conglutinin can serve this function via Man-binding domains and also by mediating contacts between complement coated bacteria or immune complexes and effector cells.[30]

NK lymphocytes, also termed null cells because they fail to express T or B cell receptors, show the selective expression of a developmentally regulated poly-N-acetyllactosamine epitope associated with a mucin-type glycoprotein; it has been stated that this may protect such cells from their own cytolytic activities.[31] Consistent with this is the finding that exogenous mucins[32] and O-glycan rich glycophorin[33] inhibit NK cell killing. Selected target carbohydrate structures may function as high affinity ligands for NK cells in the presence of calcium, particularly for the NKR-P1 protein.[34] NKR-P1 protein has a higher affinity of binding to β-D-GalNAc than to β-D-GlcNAc and a lower affinity towards α-L-fucose. This binding may play a role in target recognition by NK cells. However, the possible involvement of N-linked glycans in NK killing is controversial and may relate to the choice of cell lines used for studies.[35-37]

Maternal immune tolerance to the fetus is a well known phenomenon, but its mechanism is not understood. Pregnant women display this effect which is not due to immunological incompetence since they exhibit otherwise normal immune responses. A recent study has indicated that extracts derived from syncytiotrophoblast membrane glycocalyx of term placenta inhibit thymidine incorporation into peripheral blood mononuclear leukocytes. The inhibition can be overcome by the addition of free oligosaccharides obtained by hydrazinolysis of placental extracts. Immunosuppressive activity is associated with N-linked oligosaccharides, including high mannose, hybrid and incompletely processed complex bi-antennary structures.[38]

10.B. BLOOD GROUPS

Blood group antigens are present on glycolipids and glycoproteins of erythrocyte membranes (Table 1). According to the genetically determined secretor status, ABO and other blood groups may also be found in mucins and other glycoproteins in secretions and on the membranes of non-erythrocytic cells. Blood groups elicit the formation of antibody in incompatible hosts and are related to disease susceptibility, although the mechanism in the latter case is not well understood. Apart from their role as antigens or modifiers of antigens, glycoproteins carrying blood groups are associated with mitosis and migration, development, differention and tumor cell growth.[39-41] They play significant roles in the structural organization and cell shape of the erythrocyte cytoskeleton,[42] as membrane transport proteins, integrin-like receptors, and enzymes,[43] receptors for adhesion molecules[44,45] and ligands for parasitic and microbial adhesion.[46]

M and N blood groups reside on glycophorin A and B. Variants within this system are based on amino acid heterogeneity and include the rare Miltenberger antigens. The M and N antigens are specified by disialylated O-glycan core 1 structures of glycophorin (Table 1). The effects of O-glycosylation upon cell surface expression of glycophorin A were studied following transfection of cDNA encoding a Miltenberger mutant glycophorin A into the CHO ldlD cell line which has a 4-epimerase defect and allows manipulation of O- and N-glycosylation. A marked decrease in cell surface expression of glycophorin is noted when cells are cultured in the absence of Gal and GalNAc, conditions leading to the absence of O-glycans, or in the presence of Gal alone which restores N-glycans only.[47] This shows that O-glycans are important in the cell surface appearance of this glycoprotein.

The Rh blood group antigens reside on a complex of glycosylated and non-glycosylated erythrocyte proteins. Interaction between glycophorin B and Rh glycoprotein and peptides appears to be required for appropriate cell surface Rh expression. The N-glycan chains of Rh glycoprotein of glycophorin B-deficient cells are smaller than normal.[48] Similarly, En(a-) erythrocytes which lack the major sialoglycoprotein express band 3 molecules with different electrophoretic mobility and sugar composition.[49] In these systems, the lack of one glycoprotein does not seem to cause ectopic expression of the missing glycan structures on other glycoproteins as a compensation mechanism.

P system antigens are based on Galα1-4Gal or GalNAcβ1-3Galα1-4 structures. They appear to reside primarily on glycolipids but are also found on mucins. Glycoproteins with Galα1-4Gal appear to be absent from red cell membranes.[50] Considerable structural diversity exists on membranes to allow for binding to many types of bacteria and toxins.[51] Erythrocyte P antigen functions as a receptor for B19 parvovirus. Hematopoietic target cells can be protected from infection and toxicity by preincubation with monoclonal antibody to globoside carrying this blood group P determinant.[52]

10.C. BLOOD THROMBUS FORMATION AND DISSOLUTION

Normal physiological responses to vascular injury primarily involve single platelet adhesion to disrupted vascular endothelium and subendothelium, followed by platelet activation, aggregation and secretion of thrombogenic compounds.[53] Platelet activation is accompanied by degranulation and the appearance of surface glycoproteins.

Glycoproteins GPIIb/IIIa of activated platelets are modified post-translationally from high mannose to complex N-glycan structures; a primary function of this complex is to bind fibrinogen, another glycoprotein with a variety of sialylated N-glycans, in the formation of blood clots.[54,55] Synthesis and secretion of fibrinogen do not appear to be affected by N-glycosylation as judged from tunicamycin treatment of cultured cells,[56] although an inverse relationship between sialic acid content of fibrinogen and fibrin cross-linking has been described.[57]

Platelet activation events are mediated by platelet membrane integrin glycoprotein receptors IIb/IIIa as well as GPIb/IX which can bind such molecules as fibronectin, fibrinogen, vitronectin and von Willebrand factor (vWF).[58] GPIb/IX is present on resting platelets and amplifies the platelet activation response to thrombin.[59] vWF can bridge vascular subendothelium and a receptor on GPIb and is thus important as a critical initial step in hemostasis.[60] The thrombin binding site on GPIbα involves a negatively charged sequence near a highly O-glycosylated domain.[61,62]

The cell-surface mucin glycocalicin is derived from GP1b and has many O-glycans with sialylated core 2 structures as well as tetrasialylated monofucosylated tetra-antennary N-glycans.[63,64] Removal of GlcNAc but not Gal or sialic acid by glycosidases results in reduced von Willebrand factor binding.[65] Patients with Bernard-Soulier syndrome, a severe hereditary bleeding disorder, lack the platelet GPIb glycoprotein complex.[53] It is thought that glycocalicin may function as a protectant by preventing contact of macromolecules with the surface of resting platelets.[66]

Human vWF has N- and many O-glycans and is sulfated. The carbohydrate moieties appear to control the binding activity. The affinities towards GP1b of several glycosylated forms of vWF produced by expression of its cDNA in CHO cells are reduced when compared with a non-glycosylated vWF produced in *E. coli*.[67] Human endothelial cells cultured in the presence of tunicamycin fail to secrete vWF. High mannose carbohydrate chains are thought to be important for human pro-vWF to achieve the conformation needed for dimerization and secretion. The effect may be species-specific since tunicamycin treated bovine endothelial cells process and secrete vWF normally.[68] All of the later processing steps of vWR can be inhibited by monensin which is known to disrupt Golgi function.[69]

A number of other soluble coagulation factors have been shown to possess carbohydrate moieties, presumably as regulators for their clearance from the circulation. Coagulation factor VIII from pig tissue and plasma contains sialylated bi-, tri- and tetra-antennary complex N-glycans with a significant proportion of bi-antennary chains containing the bisecting GlcNAc residue. Factor VIII in the circulation is complexed with vWF which masks Galα1-3Gal residues present on the pig glycoprotein. This removes a potential hazard entailed in the use of porcine Factor VIII in human treatment, since the masked disaccharide does not react with the naturally occurring anti-Galα1-3 Gal antibodies present in human blood.[70]

Fibrinolysis, the dissolution of fibrin clots, is accomplished by the plasminogen-plasmin system. Plasminogen is a glycoprotein with a number of glycoforms differing in sialic acid content of O-linked glycans.[71] Tissue plasminogen activator activates plasminogen by proteolytic cleavage. The catalytic efficiency of tPA in activating these glycoforms appears to be lower in the plasminogen glycoforms with a higher proportion of sialic acid.

Recombinant tPA produced in a number of cell types is modified by O-linked α-fucose.[72] Enzymatic removal of Fuc from tPA reduces its binding to HepG2 cells, although binding is unaffected by removal of galactose or N-linked sugars. Similarly, specific inhibition of binding can be accomplished by Fuc or Fuc-derivatives.[73] Results suggest that binding and degradation of tPA may be related to O-Fuc via fucose-binding hepatic receptors. Macrophages apparently use two independently regulated mechanisms for the uptake and degradation of tPA, via a mannose binding receptor and a low density lipoprotein receptor.[74]

Recombinant antithrombin produced by two cell lines is heterogenously glycosylated, and the increased carbohydrate content of a high proportion of these molecules causes a reduction in antithrombin affinity for heparin.[75] A weaker binding glycoform contains a higher proportion of tri- and tetra-antennary N-glycans which may modulate heparin-antithrombin binding.

References

1. Dustin ML, Staunton DE, Springer TA. Supergene families meet in the immune system. Immunol Today 1988; 9:213-215.
2. Gesner B, Ginsburg V. Effect of glycosidases on the fate of transfused lymphocytes. Proc Natl Acad Sci USA 1964; 52:750-755.
3. Tonelli Q, Meints R. Sialic acid:a specific role in hematopoietic spleen colony formation. J Supramolec Struct 1978; 8:67-78.
4. Hakomori S. Le x and related structures as adhesion molecules. Histochem J 1992; 24:771-776.
5. Keppler O, Moldenhauer G, Oppenländer M et al. Human Golgi β-galactoside α-2-6-sialyltransferase generates a group of sialylated B lymphocyte differentiation antigens. Eur J Immun 1992; 22:2777-2781.
6. Braesch-Andersen S, Stamenkovic I. Sialylation of the B lymphocyte molecule CD22 by α2,6sialyltransferase is implicated in the regulation of CD22-mediated adhesion. J Biol Chem 1994; 269:11783-11786.

7. Karasuno T, Kanayama Y, Nishiura T et al. Glycosidase inhibitors (castanospermine and swainsonine) and neuraminidase inhibit pokeweed mitogen-induced B cell maturation. Eur J Immun 1992; 22:2003-2008.
8. Recny M, Luther M, Knoppers M et al. N-glycosylation is required for human CD2 immunoadhesion functions. J Biol Chem 1992; 267:22428-22434.
9. Withka J, Wyss D, Wagner G et al. Structure of the glycosylated adhesion domain of human T lymphocyte glycoprotein CD2. Curr Biol 1993; 1:69-81.
10. Narasimhan S, Lee J, Cheung R et al. β-1,4-mannosyl glycoprotein β-1,4-N-acetylglucosaminyl transferase III in human B and T lymphocyte lines and in tonsillar B and T lymphocytes. Biochem Cell Biol 1988; 66:889-900.
11. Lemaire S, Derappe C, Michalski J et al. Expression of β1-6 branched N-linked oligosaccharides is associated with activation in human T4 and T8 cell populations. J Biol Chem 1994; 269:8069-8074.
12. Carlsson S. Changes in glycan branching and sialylation of the Thy-1 antigen during normal differentiation of mouse T lymphocytes. Biochem J 1985; 226:519-525.
13. Carlsson S, Sasaki H, Fukuda M. Structural variations of O-linked oligosaccharides present in leukosialin isolated from erythroid, myeloid, and T-lymphoid cell lines. J Biol Chem 1986; 261:12787-12795.
14. Fukuda M. Leukosialin, a major O-glycan-containing sialoglycoprotein defining leukocyte differentiation and malignancy. Glycobiology 1991; 1:347-356.
15. Ellies L, Jones A, Williams M et al. Differential regulation of CD43 glycoforms on CD4 and CD8 lymphocytes in graft versus host disease. Glycobiology 1994; 4:885-893.
16. Piller F, Piller V, Fox R et al. Human T lymphocyte activation is associated with changes in O-glycan biosynthesis. J Biol Chem 1988; 263:15146-15150.
17. Higgins E, Siminovitch K, Zhuang D et al. Aberrant O-linked oligosaccharide biosynthesis in lymphocytes and platelets from patients with the Wiskott-Aldrich syndrome. J Biol Chem 1991; 266:6280-6290.
18. Piller F, Le Deist F, Weinberg KI et al. Altered O-glycan synthesis in lymphocytes from patients with Wiscott-Aldrich syndrome. J Exp Med 1991; 173:1501-1510.
19. Mentzer S, Remold-O'Donnell E, Crimmins M et al. Sialophorin, a surface sialoglycoprotein defective in the Wiskott-Aldrich syndrome, is involved in T-lymphocyte proliferation. J Exp Med 1987; 165:1383-1392.
20. Pimlott N, Miller R. Glycopeptides inhibit allospecific cytotoxic T lymphocyte recognition in an MHC-specific manner. J Immun 1986; 136:6-11.
21. Muchmore A, Sathyamoorthy N, Decker J et al. Evidence that specific oligosaccharides block early events necessary for the expression of antigen-specific proliferation by human lymphocytes. J Immun 1980; 125:1306-1311.
22. Jeske DJ, Capra JD. Immunoglobulins: Structure and Function. Paul W ed. Fundamental Immunology. New York: Raven Press 1984; 131.
23. Hickman S, Kulczycki A, Lynch R et al. Studies of the mechanism of tunicamycin inhibition of IgA and IgE secretion by plasma cells. J Biol Chem 1977; 252:4402-4408.
24. Huff T, Uede T, Iwata M et al. Modulation of the biologic activities of IgE-binding factors III. Switching of a T-cell hybrid clone from the formation of IgE-suppressive factor to the formation of IgE-potentiating factor. J Immun 1983; 131:1090-1095.
25. Huff T, Jardieu P, Ishizaka K. Regulatory effects of human IgE binding factors on the IgE response of rat lymphocytes. J Immun 1986; 136:955-962.
26. Rademacher T, Parekh R, Dwek R. Glycobiology. Ann Rev Biochem 1988; 57:792-794.
27. Aoki N, Furukawa K, Iwatsuki K et al. A bovine IgG heavy chain contains N-acetylgalactosaminylated N-linked sugar chains. Biochem Biophys Res Comm 1995; 210:275-280.
28. Ikeda K, Sannoh T, Kawasaki N et al. Serum lectin with known structure activates complement through the classical pathway. J Biol Chem 1987; 262:7451-7454.
29. Reading PC, Hartley CA, Ezekowitz AB et al. A serum mannose-binding lectin mediates complement-dependent lysis of influenza virus-infected cells. Biochem Biophys

Res Comm 1995; 217:1128-1136.
30. Lim B-L, Holmskov U. Expression of the carbohydrate recognition domain of bovine conglutinin and demonstration of its binding to iC3b and yeast mannan. Biochem Biophys Res Comm 1996; 218:260-266.
31. Vivier E, Sorrell J, Ackerly M et al. Developmental regulation of a mucin-like glycoprotein selectively expressed on natural killer cells. J Exp Med 1993; 178:2023-2033.
32. Ogata S, Maimonis PJ, Itzkowitz SH. Mucins bearing the cancer-associated sialosyl-Tn antigen mediate inhibition of natural killer cell toxicity. Cancer Res 1992; 52:4741-4746.
33. El Ouagari K, Teissié J, Benoist H. Glycophorin A protects K562 cells from natural killer cell attack. Role of oligosaccharides. J Biol Chem 1995; 270:26970-26975.
34. Bezouska K, Yuen CT, O'Brien J et al. Oligosaccharide ligands for NKR-P1 protein activate NK cells and cytotoxicity. Nature 1994; 372:150-157.
35. Ahrens PB. Role of target cell glycoproteins in sensitivity to natural killer cell lysis. J Biol Chem 1993; 268:385-391.
36. Voshol H, Dullens H, Otter W et al. Cell surface glycoconjugates as possible target structures for human natural killer cells: evidence against the involvement of glycolipids and N-linked carbohydrate chains. Glycobiology 1993; 3:69-76.
37. Mehta B, Collard H, Negrin R. The role of N-linked carbohydrate residues in lymphokine-activated killer cell-mediated cytolysis. Cell Immunol 1994; 155:95-110.
38. Arkwright P, Rademacher T, Boutignon F et al. Suppression of allogeneic reactivity in vitro by the syncytiotrophoblast membrane glycocalyx of the human term placenta is carbohydrate dependent. Glycobiology 1994; 4:39-47.
39. Kuhns W, Bramson S. Variable behavior of blood group H on HeLa cell population synchronized with thymidine. Nature 1968; 219:938-939.
40. Thomas D. Cyclic expression of blood group determinants in murine cells and their relationship to growth control. Nature 1971; 233:317-321.
41. Feizi T. Demonstration by monoclonal antibodies that carbohydrate structures of glycoproteins and glycolipids are onco-developmental antigens. Nature 1985; 314:15-17.
42. Reid M. Associations of red blood cell membrane abnormalities with blood group phenotype. In: Garratty G. Immunobiology of Transfusion Medicine. New York: Marcel Dekker.1994; 257-271.
43. Telen M. Erythrocyte blood group antigens: not so simple after all. Blood 1995; 85:299-306.
44. Hakomori S. New directions in cancer therapy based on aberrant expression of glycosphingolipids:anti-adhesion and ortho-signalling therapy. Cancer Cells 1991; 3:461-470.
45. Moulds J. Association of blood group antigens with immunologically important proteins. In: Garratty G. Immunobiology of Transfusion Medicine. New York: Marcel Dekker 1994; 273-297.
46. King M. Blood group antigens on human erythrocytes-distribution, structure and possible functions. Biochim Biophys Acta 1994; 1197:15-44.
47. Ugorski B, Blackall D, Påhlsson P et al. Recombinant Miltenberger I and II human blood group antigens:the role of glycosylation in cell surface expression and antigenicity of glycophorin A. Blood 1993; 82:1913-1920.
48. Ridgwell K, Eyers S, Mawby W et al. Studies on the glycoprotein associated with Rh (Rhesus) blood group antigen expression in the human red blood cell membrane. J Biol Chem 1994; 269:6410-6416.
49. Tanner M, Jenkins R, Anstee D et al. Abnormal carbohydrate composition of the major penetrating membrane protein of En(a-) human erythrocytes. Biochem J 1976; 155:701-703.
50. Yang Z, Bergström J, Karlsson K. Glycoproteins with Galα4Gal are absent from human erythrocyte membranes, indicating that glycolipids are the sole carriers of blood group P activities. J Biol Chem 1994; 269:14620-14624.
51. Karlsson K. Animal glycosphingolipids as membrane attachment sites for bacteria. Ann Rev Biochem 1989; 58:309-350.
52. Brown KE, Anderson SM, Young NS.

Erythrocyte P antigen:cellular receptor for B19 parvovirus. Science 1993; 262:114-117.
53. Hardisty R. Disorders of platelets. II Functional abnormalities. In: Lilleyman J, Hann I, eds. Pediatric Hematology. Edinburgh: Churchill Livingstone, 1992; 167-199.
54. Phillips D, Charo I, Parise L et al. The platelet membrane glycoprotein IIb-IIIa complex. Blood 1988; 71:831-843.
55. Da Silva M, Tamuri T, McBroom T et al. Tyrosine derivatization and preparative purification of the sialyl and asialyl N-linked oligosaccharides from porcine fibrinogen. Arch Biochem Biophys 1994; 312:151-157.
56. Gilman P. The role of the carbohydrate moiety in the biological properties of fibrinogen. J Biol Chem 1984; 259:3248-3253.
57. Martinez J, Palascak JE, Kwasniak D. Abnormal sialic acid content of the dysfibrinogenemia associated with liver disease. J Clin Invest 1978; 61:535-538.
58. Beacham DA, Cruz MA, Handin RI. Glycoprotein Ib can mediate endothelial cell attachment to a von Willebrand factor substratum. Thromb Hemostasis 1995; 73:309-317.
59. Clemetson K. Platelet activation:signal transduction via membrane receptors. Thromb Hemostas 1995; 74:111-116.
60. Handin R, Wagner D. Molecular and cellular biology of von Willebrand factor. Prog Hemostasis Thromb 1989; 9:233-259.
61. De Marco L, Mazzucato M, Masotti A et al. Localization and characterization of an α-thrombin binding site on platelet glycoprotein Ibα. J Biol Chem 1994; 269:6478-6484.
62. Gralnick H, Williams S, McKeown L et al. High-affinity α-thrombin binding to platelet glycoprotein Ibα:identification of two binding domains. Proc Natl Acad Sci USA 1994; 91:6334-6338.
63. Korrel S, Clemetson K, van Halbeek H et al. Identification of a tetrasialylated monofucosylated tetra-antennary N-linked carbohydrate chain in human platelet glycocalicin. FEBS Lett 1988; 228:321-326.
64. Tsuji T, Tsunchisa S, Watanabe Y et al. The carbohydrate moiety of human platelet glycocalicin. J Biol Chem 1983; 258:6335-6339.
65. Michelson A, Loscalzo J, Melnick B et al. Partial characterization of a binding site for von Willebrand factor on glycocalicin. Blood 1986; 67:19-26.
66. Ruggeri Z. The platelet glycoprotein Ib-IX complex. Prog Hemostasis Thromb 1991; 10:35-68.
67. Cruz M, Handin R, Wise R. The interaction of the von Willebrand factor-A1 domain with platelet glycoprotein Ib/IX. J Biol Chem 1993; 268:21238-21245.
68. Lynch D, Williams R, Zimmerman T et al. Biosynthesis of the subunits of factor VIIIR by bovine aortic endothelial cells. Proc Natl Acad Sci USA 1983; 80:2738-2742.
69. Tartakoff A. The combined function model of the Golgi complex:center for ordered processing of biosynthetic products of the rough ER. Int Rev Cytol 1983; 85:221.
70. Hironaka T, Furukawa K, Esmon P et al. Structural study of the sugar chains of porcine factor VIII- tissue and species specific glycosylation of Factor VIII. Arch Biochem Biophys 1993; 307:316-330.
71. Pirie-Sheperd S, Jett E, Andon N et al. Sialic acid content of plasminogen 2 glycoforms as a regulator of fibrinolytic activity. J Biol Chem 1995; 270:5877-5881.
72. Harris R, Leonard C, Guzzetta A et al. Tissue plasminogen activator has an O-linked fucose attached to threonine-61 in the epidermal growth factor domain. Biochemistry 1991; 30:2311-2314.
73. Hajjar K, Reynolds C. α-Fucose mediated binding and degradation of tissue type plasminogen activator by HepG2 cells. J Clin Invest 1994; 93:703-710.
74. Noorman F, Braat E, Rijken D. Degradation of tissue-type plasminogen activator by human monocyte derived macrophages is mediated by the mannose receptor and by the low density lipoprotein receptor-related protein. Blood 1995; 86:3421-3427.
75. Björk I, Ylinenjärvi K, Olsen ST et al. Decreased affinity of recombinant antithrombin for heparin due to increased glycosylation. Biochem J 1992; 286:793-800.

CHAPTER 11

Growth- and Hormone-Related Functions of Glycoproteins and Cell Surface Receptors

11. A. GROWTH, DIFFERENTIATION AND DEVELOPMENT

Development assures cellular diversity and continuation of life from one generation to the next. This is accomplished by differentiation into specialized cell types via asymmetric cell cycle events. For example, α2-Fuc-transferase, blood group B α3-Gal-transferase and blood group A α3-GalNAc-transferase do not appear active on human ova and sperm judged by the absence of ABO blood group reactivity; however, their products appear in early embryogenesis when differentiated epithelium can be demonstrated.[1,2] In self-renewing epithelium, alterations in glycoconjugates and glycosyltransferase levels occur as cell cycle-related events.[3,4]

Cell cycle-associated regulation of glycoconjugate production has been demonstrated in transgenic (I) mice expressing α3/4-Fuc-transferase in the small intestinal epithelium. Le^a and Le^b antigens have been localized in proliferating crypt cells and in post-mitotic non-cycling villus cells; in contrast, Le^x is restricted to crypt cells. When bi-transgenic mice are created by crossing I strain with mice expressing SV 40 tumor antigen (II), I-II offspring mice are able to expand the range of Le^x expression to include villus cells that have reentered the cell cycle. Thus, genetically engineered animals may have altered endogenous glycan metabolism.[5]

Continuous growth related changes of cellular interactions may be mediated by differentiation antigens on glycolipids and glycoproteins.[6,7] Regenerating rat liver shows an increased proportion of complex bi-antennary N-glycans with Fuc attached to the core, in comparison with controls.[8]

N-glycans of transferrin secreted by slowly growing HepG2 cells contain a greater proportion of bi-antennary complex-type species than rapidly dividing cells, as well as decreased activity of GlcNAc-transferase V, reflecting a growth associated property of this enzyme.[9]

11.A.1. Development and Embryogenesis

Changing glycosylation patterns have been observed during embryogenesis. Thus, the i antigen on fetal erythrocytes of several species is replaced by the I antigen in the adult. Some antigens that are transiently present during ontogeny can reappear later in tumors and are therefore termed onco-fetal antigens, for example Le^x(SSEA-1) and carcinoembryonic antigen (CEA).[10,11] Lectin staining studies of pancreatic zymogen granule membrane protein (gp300) have shown marked changes in sugar patterns during mouse embryogenesis. Fucosylation demonstrated by UEA-1 lectin binding is absent in mouse embryos, but abundant postnatally.[12]

Studies of glycoproteins in invertebrate and vertebrate species show progressive increases in glycosyltransferase activities and in glycan complexity during development, and a requirement for N-glycans for normal maturation.[13] In chick embryo development age-related carbohydrate changes occur at 8-16 days; the early glycans are N-linked but change to more complex O- and N-glycans at 16 days; studies of embryo cell adhesion show that this can be altered following tunicamycin treatment.[14,15]

GlcNAc-transferase I of the N-glycan pathway is required for normal embryogenesis in the mouse. Deletion of both GlcNAc-transferase I alleles in mice results in multiple structural defects, particularly in the nervous system, and death at about the tenth day of gestation (Fig. 5) (see chapter 16).[16]

GlcNAc-transferase V and core 2 GlcNAc-transferase expression were studied in post-implantation mouse embryos by in situ hybridization. mRNA of core 2 β6-GlcNAc-transferase is widely expressed at embryonic day 7, but later becomes restricted to a subset of mucin and cartilage producing tissues. GlcNAc-transferase V transcripts are not demonstrated at embryonic day 7, are widely expressed throughout 9.5 day embryos, then become restricted to the central nervous system, and specialized epithelia in a number of organs. The capacity of cells to migrate is thought to be mediated by glycans which are products of GlcNAc-transferase V.[17]

A number of organ systems possess a high content of self-renewing cells exhibiting cell cycle- and maturation-dependent glycosyltransferases and glycoproteins. Developing lungs of human embryos express i and Le^x antigens whereas fully differentiated epithelial cells express I but not Le^x; a re-expression of the fetal type occurs in lung cancers.[18] Reciprocal shifts of sialyl- and Fuc-transferases occur in mucosal fractions of rat small intestine during maturation and exhibit higher fucosylation as maturation progresses.[19] Sialic acid alterations have been noted in neonatal versus postnatal rat livers[20] and selective losses of sialic acid from small intestinal epithelium can be demonstrated postnatally.[21] This is consistent with the postnatal decrease of both α6-sialyltransferase (N) mRNA expression as well as activity.[22,23] Differentiation of human hepatoma HepG2 cells using sodium butyrate causes a significant reduction in α6-sialyltransferase activity, apparently through post-transcriptional reduction of sialyltransferase mRNA.[24]

11.A.2. Differentiation

When cancer cells are differentiated in culture, significant glycosylation changes take place. Undifferentiated HT-29 colon cancer cells have been shown to accumulate high mannose $Man_{9\text{-}8}$-$GlcNAc_2$ species in contrast to differentiated cells in which increased processing of high mannose structures has been demonstrated.[25] Human colon carcinoma Caco-2 cells contain polyfucosylated poly N-acetyl-lactosaminoglycan chains mainly attached to the lysosomal membrane protein, lamp-1.[26] Differentiation of these cells in culture decreases the

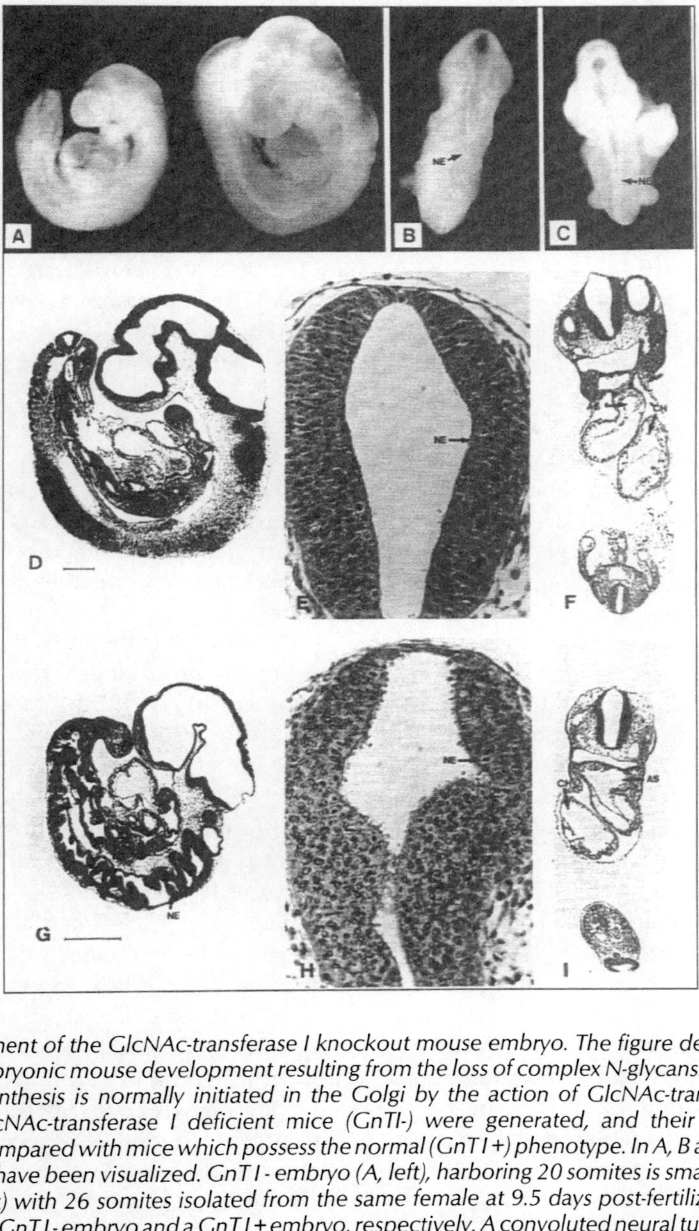

Fig. 5. Development of the GlcNAc-transferase I knockout mouse embryo. The figure depicts alterations of structures in embryonic mouse development resulting from the loss of complex N-glycans. Complex N-linked carbohydrate synthesis is normally initiated in the Golgi by the action of GlcNAc-transferase I. In these experiments GlcNAc-transferase I deficient mice (GnTI-) were generated, and their earlier embryonic development compared with mice which possess the normal (GnT I +) phenotype. In A, B and C photographs, whole embryos have been visualized. GnT I - embryo (A, left), harboring 20 somites is smaller than the GnTI+ embryo (A, right) with 26 somites isolated from the same female at 9.5 days post-fertilization. B and C are dorsal views of a GnT I - embryo and a GnT I + embryo, respectively. A convoluted neural tube with a staggered suture line is present in the GnTI - embryo. In D - I sections of embryos were stained with hematoxylin-eosin. Longitudinal sections D (GnT I - embryo) and G (control GnT I + embryo) show convoluted neural epithelium (NE, arrow). Transverse sections E (control GnT I +) and H (GnT I -) from the mid-trunk region show increased mitotic figures in neural epithelial cells of the GnT I - embryo (H). Transverse sections F (control GnT I +) and I (GnT I -) reveal a situs inversus phenotype in the GnT I - embryo with inverted tail position and heart structure. NE, neural epithelium; CH, common ventricular chamber of the heart; AS, aortic sac. Solid bar = 0.1 mm. In the absence of complex N-glycans, embryos have one-half to two-thirds of the size of wild-type or heterozygous embryos. Development proceeds until day 9, and embryos become non-viable shortly thereafter (about day 10.5). Neural system development is most affected in the mutants, as demonstrated by convoluted neural folds, but cardiovascular defects can also be demonstrated. Studies of tissue sections and of cell-type specific proteins indicate that differentiation into multiple cell types can proceed in animals with the GnT I - phenotype. Reprinted from Metzler et al (1994) with permission.

proportion of these poly-N-acetyllactosaminoglycan chains accompanied by the acquisition of an enterocytic phenotype. The loss of these glycans is not due to a decrease in the β4-Gal-transferase and i β3-GlcNAc-transferase involved in their synthesis, but GlcNAc-transferase II to V are increased, which appears to be in contrast to the structural findings. These enzymes synthesize additional branches on N-glycans, respectively, and thus provide more substrate for poly-n-acetyllactosamine addition. A possible explanation for the decrease in fucosylated poly-N-acetyllactosaminoglycans is the inhibitory action of GlcNAc-transferase III which may turn off further branching and elongation of the branches[27] with subsequent decreases of the substrate for poly-N-acetyl-lactosamine additions. Enzymes of the O-glycan pathway also change during Caco-2 cell differentiation. Core 1 β3-Gal-transferase synthesizing O-glycan core 1 is less active in the more differentiated cells while core 2 β6-GlcNAc-transferase is increased. The general tendency to increase branching may also be viewed as a compensatory mechanism for the decrease in poly N-acetyllactosaminoglycan chains.

The expression of O-glycan core 2 β6-GlcNAc-transferase appears to be generally regulated during growth and differentiation. The activity increases during activation of lymphocytes,[28] in lymphocytes and platelets from children with the Wiscott-Aldrich immunodeficiency syndrome[29] and in leukemias which exhibit a block of differentiation and an expression pattern of oligosaccharides normally present on immature cells.[30,31]

Differentiation of the promyelocytic cell line HL60 is associated with a decrease in GlcNAc-transferase III if differentiation is triggered with retinoic acid or vitamin D_3 along the myeloid or monocytic lineages, respectively.[32] GlcNAc-transferase IV increases only during differentiation induced by vitamin D_3 and not with a number of other differentiating agents. Differentiation of HL60 cells into granulocytes induced by dimethyl sulfoxide results in increases of complex poly-N-acetyllactosamines on N-glycans attached to lysosomal membrane proteins lamp-1 and lamp-2, specifically due to enhanced activity of i β3-GlcNAc-transferase.[33]

In the mouse, retinoic acid in teratocarcinoma F9 cells induces differentiation to primitive endoderm. Poly-N-acetyllactosamine structures on cell surfaces, as well as GlcNAc-transferase V and core 2 β6-GlcNAc-transferase activities providing substrate for poly-N-acetyllactosamine addition, increase during differentiation.

11.A.3. Cell Motility and Angiogenesis

During angiogenesis, tumor metastasis and invasion, cells migrate. Cell motility is controlled by factors affecting cell adhesion such as components of the extracellular matrix.

Oligosaccharides are instrumental in the process of angiogenesis. When N-glycan processing is inhibited by glucosidase I inhibitor castanospermine in nude mice, tumor growth and angiogenesis are significantly reduced. In vitro, the morphological differentiation of endothelial cells is prevented by castanospermine with no alteration of cell proliferation.[35]

The gene encoding β4-Gal-transferase appears to be a member of the cell-cycle dependent genes whose expression is growth related.[36] In addition to its role in fertilization related events, β4-Gal-transferase has been implicated in cell spreading and migration, neurite outgrowth and collagen attachment. Studies of β4-Gal-transferase in stably transfected mouse fibroblast cell lines indicate a direct involvement of the enzyme in the cell proliferation rate, possibly by regulating transduction via the EGF receptor.[37] It has been suggested that, even in the absence of UDP-Gal, cell surface β4-Gal-transferase may function catalytically during cell migration, either to dissociate from its oligosaccharide ligand and/or to modify the extracellular matrix.[38]

Gal-binding proteins have the ability to activate cells and stimulate growth. β-Galactoside specific lectins such as peanut lectin and bovine lung galaptin stimulate the growth of cultured vascular muscle or endothelial cells.[39] EC turn over very slowly but upon angiogenic stimulation, capillary endothelial cells proliferate and migrate and also provide signals for underlying smooth muscle cells. Ovarian angiogenesis can be measured in the bovine corpus luteum model which exhibits rapid vessel growth and regression.[40] These changes are accompanied by changes in glycosylation as seen by the different lectin staining patterns and upregulation of α6-sialyltransferase and α3-Gal-transferase mRNA expression. When bovine aorta EC migrate, binding sites for lectins such as ConA, WGA and peanut agglutinin after neuraminidase treatment increase compared to contact inhibited cells.[40,41] Novel glycoproteins are expressed in migrating EC similarly to those in subconfluent cells, but these were not found in confluent cells. These studies suggest a switch in the pathways of glycoprotein synthesis and the involvement of N- and O-glycans in the process of angiogenesis. Evidence for the role of glycans in the highly complex system of cell motility comes also from studies with anti-carbohydrate (anti-H) antibodies that inhibit cell motility as well as metastasis of human tumor cells.[42]

11.B. APOPTOSIS

Cells have the intrinsic capacity to self-destruct by activation of a cell suicide program (apoptosis) with accompanying morphologic and biochemical changes. Several studies suggest that cellular glycosylation is linked to apoptosis. The sensitivity of cultured U937 histiocytic lymphoma cells to apoptosis induced by tumor necrosis factor (TNF) is enhanced by the glycosylation inhibitor 2-deoxy-D-glucose, which suggests that this agent may be useful to enhance the anti-tumor potency of TNF.[43]

Protein N- and O-glycosylation may be involved in apoptotic events. Inhibition of N-glycosylation induced by tunicamycin in HL60 cells is associated with early apoptosis as determined by reductions in protein synthesis and typical DNA fragmentation.[44] The expression of Le^y antigen correlates closely with apoptosis as observed on stained tissue sections of normal and tumor tissue. However, this structure has not been correlated with cell proliferation or necrosis as deduced from DNA fragmentation patterns.[45] In skin cells, Le^y antigen expression has also been shown to be involved in cell death as a very early event.[46]

Apoptosis of activated human T cells appears to be mediated by galectin-1. Galectin-1 is found in human thymus and lymph nodes; during T-cell development, galectin-1 is associated with apoptotic cells.[47] Resting T cells do not undergo apoptosis although they also bind galectin1. The process is reduced when the conversion of N-glycans to complex chains is blocked by swainsonine. Conversely, GalNAcα–benzyl inhibition of O-glycan elongation potentiates apoptosis induced by galectin-1.

11.C. HORMONES AND CYTOKINES

11.C.1. Role of Hormone Glycosylation

Glycosylation of hormones provides a means by which their biological activities and interactions with cellular receptors can be regulated. In addition, glycosylation may serve to protect hormones from rapid degradation and thereby aid in maintaining function.[48] Both N-glycans and O-glycans may be carried by a variety of hormones.[49]

Studies concerning the role of glycosylation in erythropoietin (EPO) have been carried out with recombinant EPO produced in a variety of host cells. O-glycans were found in several studies to be unecessary for biological activity. Modification of N-glycosylation, however, has variable effects on activity.

Prevention of N-glycosylation of EPO by site-specific mutagenesis increases its in vitro activity, but decreases its in vivo activity. This shows the importance of N-glycans for the biosynthesis and possibly the secretion of, as well as the control of metabolic clearance of EPO.[50] Mutants of recombinant human EPO were constructed to demonstrate the effects of single site glycosylation deletion upon secretion, stability and in vitro biological activity. The mutants represent different glycoforms, and these differ by their expression levels and secretion rates. Deletions of the N-glycosylation sites at positions 24 and 38 result in increased reticulocyte proliferative activity of purified mutant proteins in mice.[51]

Deglycosylation of human EPO also results in decreased in vivo hormone activity while the in vitro activity increases.[52,53] The activity correlates with the number of sialic acid residues. Thus it appears that glycosylation is not necessary for activity, although in vivo the N-glycans and especially sialic acid residues may have a specific function.

The nature of the complex N-glycans may be of importance since EPO containing tetra-antennary N-glycans shows much higher specific activity in vivo than its bi-antennary counterpart.[54,55] The structure of N-glycans has been shown to influence the renal handling of recombinant EPO. Labelled EPO carrying bi-antennary N-glycans, injected into rats, disappears from the plasma more than three times as fast as its tetra-antennary counterpart.[56]

The pituitary glycoprotein hormones such as TSH, LH, FSH and hCG are glycosylated on both the common α– and the hormone-specific β-subunits, and N-glycosylation appears to be important for signaling events. Site directed mutagenesis of N-glycosylation sequences of human TSH indicates that ablation of N-glycans on TSH expressed in CHO cells increases its in vitro activity. However, the mutagenized hormone has a faster serum disappearance rate and a decreased effect on T_4 production in mice.[57] In a related study, deletion of either of the two N-glycosylation sites of the α-subunit increases the biological activity of human TSH expressed in CHO cells by 30%. Deletion of N-glycans from both sites significantly reduces cAMP formation as well as T_3 secretion by human thyroid follicles in comparison with wild-type TSH.[58] Complex glycans of secreted hormones may thus influence bioactivities.[59-61]

High mannose precursor oligosaccharides are thought to protect subunits from intracellular proteolysis and aggregation, and facilitate α–β–subunit combination. Terminal residues of α-subunit oligosaccharides have a major influence on TSH intrinsic potency.[62] Sequential removal of sialic acid, Gal or GlcNAc from recombinant human TSH by glycosidase treatment causes a greater than 10-fold increase in the in vitro activity while the in vivo activity decreases. Hormone half life, which is determined by its metabolic clearance, depends on the exposure of Gal and sialic acid. Asialo-TSH clears faster than agalacto-TSH and the deglycosylated hormone is removed faster than intact hormone. These effects seem to be modulated mainly by terminal sialylation since resialylation of terminal galactose reverses the effects noted above.[63]

Thyroid hormone is synthesized from thyroglobulin which carries N- and O-glycans. The synthesis appears to be modulated by two N-glycans[64] in the N-terminal domain. De-N-glycosylated N-terminal domain of thyroglobulin is inactive in hormone synthesis in contrast to fully N-glycosylated or desialylated protein. High mannose structures enhance hormone synthesis. This suggests a specific role for N-glycan structures in hormone synthesis.

Regulation of TSH by thyrotropin release hormone (TRH) results in changed N-glycan sulfation and sialylation of TSH when compared with the glycosylation of basal TSH.[65] TSHα, TSHβ and free TSHα subunits incorporate sialic acid and sulfate. Relative activities and clearance rates of sulfated and sialylated oligosaccharides of TSH α– and β-subunits were analyzed using combinations of dimers derived from complete and asialo pituitary and thyroid recombinant TSH subunits. N-glycans of the β-subunit have a more predominant role in clearance rates and in vivo bioactivity than those of α-subunits.

TSH, LH and FSH carry terminal sialic acid and 4-sulfated GalNAc residues variably distributed on complex N-glycans.[66] The β4-GalNAc-transferase involved in the synthesis of these structures is believed to recognize hormone-specific peptide,[67,68] specifically a cluster of basic amino acids,[69] whereas the sulfotransferase is oligosaccharide-specific and may therefore act on terminal GalNAc-residues of glycoproteins other than hormones. Pro-opiomelanocortin also carries terminal 4-sulfated GalNAc-structures and has the sequon Pro-X-Arg/Lys, a putative recognition signal for the β4-GalNAc-transferase. However, the GalNAcβ1-4 GlcNAc linkage occurs also on non-hormone glycoproteins. Neither sialic acid nor sulfate are required for hormone activity or receptor binding, but appear to function in the modulation of clearance of hormones from the blood. Sialic acid may decrease binding to the hepatic asialoglycoprotein receptor and may influence hormone potency.[70] The sulfate group attached to GalNAc might lead to efficient hormone clearance by a specific hepatic endothelial and Kupffer cell receptor recognizing sulfated sugars.[71]

Human chorionic gonadotropin (hCG) contains a mixture of N- and O-glycans, but apparently lacks sulfate residues. Its biological responsiveness can be abrogated by deglycosylation. Deglycosylated hCG fails to elicit cAMP, although its receptor binding affinity is maintained. A recombinant analog of hCG lacking O-glycans maintains its in vitro immunological and biological properties, indicating that O-glycoproteins are not involved in these functions.[72]

The effects upon hCG biological activity and receptor binding have been studied following enzymatic remodelling of the N- and O-linked oligosaccharide chains. Desialylated hCG has been found to exhibit reduced hormone activity assessed by cAMP production in MA-10 Leydig tumor cells; the reduction in biological activity is greater with the desialylated hormone from invasive mole and choriocarcinoma than with normal hCG.[73] Enzymatic removal and reglycosylation studies show that the Manα1-3 arm but not the Manα1-6 arm of N-glycans of hCG is essential for signal transduction. The Manα1-6 branch can potentially interfere with receptor binding. Sialic acid attached to the Manα1-3 branch appears to be responsible for complete expression of bioactivity.[74]

Lectin specificity studies of hCG processed in urine samples collected from the 7th-32nd weeks of pregnancy suggest the occurrence of more highly branched and fucosylated oligosaccharides as gestation progresses.[75] In choriocarcinoma, different isoforms of tri-antennary and unusual bi-antennary structures occur while complex N-glycans of normal hCG are of bi-antennary structures.[76] These studies suggest that in disease the hormone may have a modified function due to atypical glycosylation.

The significance of carbohydrate chains at individual glycosylation sites has been studied in corticosteroid-binding globulin following the construction of mutants, each containing only one of six possible N-glycosylation sites. The processing of Asn-linked glycans appears to be site-specific.[77] Earlier studies by these authors demonstrate that N-glycosylation at Asn 238 is essential for the production of steroid-binding activity. However, when the mutant containing only one glycan at this position is deglycosylated its binding activity is unaltered, suggesting that glycan-peptide interaction may be needed only for proper folding and creation of the steroid-binding site during protein synthesis.

11.C.2. Cytokines and Glycosylation

Cytokines are soluble non-immunoglobulin compounds which through their interactions with receptors serve as regulators of cell function. They may stimulate hormone production, and hormones in turn may modulate cytokine activities. A particular role of cytokines is to control modelling or remodelling of tissues, whether programmed or subsequent to inflammation or injury.[78,79] In addition, cytokines may have a profound effect on the cellular glycosylation machinery. The role of carbohydrates in cytokine activities may be comparable to their hormone-related function as protective residues and modulators of clearance velocity and biological activity.[80,81]

The inflammatory cytokines tumor necrosis factor-α and interleukin-1 have been shown to be effective in inducing three to four-fold increased expression of α6-sialyltransferase (N) acting on N-glycans in cultured human endothelial cells, along with endothelial glycoproteins bearing N-glycans with increased numbers of α2-6-linked sialic acid residues.[82] Interleukin-1 production can be potentiated by endogenous Gal binding protein galectin-3, present in human monocytic cells.[83]

Cytokines have been shown to modulate glycosyltransferase activities. Incubation of a human myeloma cell line with interleukin-6, while augmenting cell proliferation, results in a decrease of GlcNAc-transferase III activity and an increase of GlcNAc-transferase IV and V activities acting on N-glycans. The resulting increases of asialo-tri- and tetra-antennary N-glycans are suggested on the basis of increased reactivity with *Datura stramonium* lectin.[84]

The GLUT-1 gene product is an integral membrane glucose transport protein. In Swiss 3T3 fibroblasts, it is a 55 kDa glycoprotein with a single but heterogeneous N-glycan chain. Transforming growth factor β1 alters the N-glycan structures of GLUT-1 of Swiss 3T3 cells and larger molecules are produced with more branched and/or repeated poly-N-acetyllactosamines. In addition, the affinity to 2-deoxy-glucose is significantly increased. Treated cells reveal a two-fold increase in V_{max} for 2-deoxyglucose transport, suggesting that increased amounts of GLUT-1 protein are produced upon cytokine treatment in contrast to untreated control cells.[85]

11.D. CELL SURFACE RECEPTOR FUNCTIONS

Cell surface receptors, which are often glycoproteins, must be properly assembled, transported and anchored. The maintenance of the stability of newly formed functional receptors may require receptor-linked O- or N-glycans. In other cases, glycans are required for either ligand binding or biological activity.

11.D.1. Role of Glycosylation in Receptor Stability and Transport

The major antigen envelope protein of Epstein-Barr virus, human interleukin-2 receptor, low density lipoprotein receptor,[86] recombinant decay accelerating factor and other glycoproteins[87] are all stabilized by O-glycans as integral structural components.[88]

The transferrin receptor in human erythroleukemia K562 cells acquires sialic acid during recycling from the cell surface through the Golgi.[89] Although glycosylation is not essential for the binding of transferrin to the receptor, it influences the binding affinity.[90] Deletion of one N-glycosylation site carrying one of the three high mannose chains in the recombinant human receptor expressed in CHO cells leads to lack of cell surface expression and retention in the ER, probably due to failure to form dimers by disulfide bonds.[91,92]

Prevention of glycosylation by inhibition of glycosyltransferases or by site-directed mutagenesis results in decreased expression of β_2 adrenergic receptors or rhodopsin on the cell surface and also a change and appearance of receptor bands on polyacrylamide gels. However, loss of glycosylation does not have a marked effect on biological function.[93] The expression of recombinant platelet activating factor in COS cells has been studied after mutagenesis of the receptor to change N-glycosylation sites. The membrane expression of the receptor is facilitated by N-glycosylation, although there is no change in binding affinity or signal transduction.[94]

A human interleukin-4 receptor derived from CHO cells is highly glycosylated, containing N-linked complex oligosaccharides with bi-, tri-, and tetra-antennary structures. De-N-glycosylation increases non-specific aggregation and reduces solubility of the receptor but does not affect its interleukin-4 binding activity. The major role of glycosylation in this system, and in others, is believed to be that of insuring proper folding, translocation and insertion into the membrane.[95-98]

11.D.2. Role of Glycosylation in Receptor-Ligand Binding and Function

The lipoprotein receptor as well as the interferon receptor requires N-glycans for ligand binding.[99,100] Prevention of N-glycosylation of the human interferon-γ receptor blocks interferon binding but allows its cell surface expression. However, in vitro, the receptor can bind to the ligand suggesting that the N-glycans are essential in vivo for proper conformational stability of the receptor in the cell membrane.[99] Tunicamycin treatment of human skin fibroblasts results in about 50% lower affinity of the LDL receptor for LDL compared to the fully N-glycosylated receptor.[100]

The bovine low molecular weight Man-6-P receptors occur in different glycoforms. One of the receptor species has sialylated poly-n-acetyl-lactosamine sequences that demonstrate less ligand binding[101] than another species which lacks a poly-n-acetyl-lactosamine sequence with a majority of the outer branches of complex chains terminating in galactose residues. N-glycans appear to be required for the stability of the high affinity ligand binding conformation of this receptor.[102]

The GM-CSF receptor has two subunits, α and β, which have multiple N-glycosylation sites. COS cells expressing the α subunit and treated with tunicamycin express the unglycosylated α subunit and transport it to the cell surface. The unglycosylated receptor is not capable of binding GM-CSF,[103] suggesting that N-glycans are essential for ligand binding and signal transduction of the GM-CSF receptor.

The oligosaccharides of hormone receptors may provide a means by which binding to their ligands is regulated. Two CHO cell lines with gene defects in their glycosylation pathways were used to compare the binding properties of insulin and insulin-like growth factor. Mutant B4-2-1 cells are deficient in dolichol-Man-phosphate synthase activity and transfer a $Glc_3Man_5GlcNAc_2$ oligosaccharide to protein from the dolichol derivative and therefore cannot synthesize the normal high mannose N-glycans; Lec 1 cells lack GlcNAc-transferase I and therefore cannot process N-glycans to the complex-type structures. Insulin binding to B4-2-1 cells is increased, whereas binding to Lec 1 cells is decreased in comparison to its behavior on control CHO cells. This suggests that preferential binding occurred in the presence of complex N-glycans but not in their absence. In contrast, insulin-like growth factor binds equally well to all three cell types.[104]

References

1. Hughes RC. Membrane Glycoproteins. London:Butterworth & Son, 1976:119.
2. Vedtofte P, Dabelsteen E, Hakomori S et al. Changes in the expression of blood-group carbohydrates during oral mucosal development in human fetuses. Differentiation 1984; 27:221-228.
3. Kuhns W, Pann C. Differentiation of HeLa cells with respect to blood group H antigen. Nature New Biology 1972; 96:22-24.
4. Aoi Y. Biosynthesis of glycoprotein-glycosyl transferases during the cell cycle. J Exp Med 1978; 124:139-144.
5. Bry L, Falk PG, Gordon JI. Genetic engineering of carbohydrate biosynthetic pathways in transgenic mice demonstrates cell cycle-associated regulation of glycoconjugate production in small intestinal epithelial cells. Proc Natl Acad Sci USA 1996; 93:1161-1166.
6. Hakomori S. Aberrant glycosylation in cancer cell membranes as focused on glycolipids:overviews and perspectives. Cancer Res 1985; 45:2405-2414.
7. Feizi T. Demonstration by monoclonal antibodies that carbohydrate structures of gly-

coproteins and glycolipids are onco-developmental antigens. Nature 1985; 314:15-17.

8. Kato S, Akamatsu N. Alterations in fucosyl oligosaccharides of glycoproteins during rat liver regeneration. Biochem J 1985; 229:521-528.
9. Hahn T, Goochee C. Growth-associated glycosylation of transferrin secreted by HepG2 cells. J Biol Chem 1992; 267:23982-23987.
10. Feizi T. Antigenicity of mucins - their relevance to tumour associated and stage specific embryonic antigens. In:Chantler EN, Elder JB, Elstein M eds. Mucus in Health and Disease - II. New York:Plenum Press 1982:29-39.
11. Gold P, Freedman S. Specific carcinoembryonic antigens of the digestive system. J Exp Med 1965; 122:467-480.
12. De Lisle R, Isom K. Expression of sulfated gp300 and changes in glycosylation during pancreatic development. J Histochem Cytochem 1996; 44:57-66.
13. Welply J, Lau J, Lennarz W. Developmental regulation of glycosyltransferases involved in synthesis of N-linked glycoproteins in sea urchin embryos. Develop Biol 1985; 107:252-258.
14. Codogno P, Bernard B, Font J et al. Changes in protein glycosylation during chick embryo development. Biochim Biophys Acta 1983; 763:265-275.
15. Berjonneau C, Aubery M, Vernay M et al. Correlation between changes in cell adhesion and the ratio of N- to O-linked glycopeptides during chick embryo development. Biol Cell 1984; 52:21-26.
16. Metzler M, Gertz A, Sarkar M et al. Complex asparagine-linked oligosaccharides are required for morphogenic events during post-implantantion development. EMBO J 1994; 13:2056-2065.
17. Granovsky M, Fode C, Warren CE et al. GlcNAc-transferase V and core 2 GlcNAc-transferase expression in the developing mouse embryo. Glycobiol 1995; 5:797-806.
18. Itai S, Nishikata J, Takahashi N et al. Differentiation-dependent expression of I and sialyl I antigens in the developing lung of human embryos and in lung cancers. Cancer Res 1990; 50:7603-7611.
19. Chu S, Walker W. Developmental changes in the activities of sialyl-and fucosyltransferases in rat small intestine. Biochim Biophys Acta 1986; 883:496-500.
20. Kato S, Oda-Tamai S, Akamatsu N. Post natal changes in N-linked oligosaccharides of glycoproteins in rat liver. Biochem J 1988; 253:59-66.
21. Taatjes D, Roth J. Selective loss of sialic acid from rat small intestinal epithelial cells during post-natal development. Eur J Cell Biol 1990; 53:255-266.
22. Vertino-Bell A, Ren J, Black J et al. Developmental regulation of β-galactoside α2-6 sialyltransferase in small intestine epithelium. Develop Biol 1994; 165:126-136.
23. Dall'Olio F, Malagolini N, Di Stefano G et al. Postnatal development of rat colon epithelial cells is associated with changes in the expression of the β1-4 N-acetylgalactosaminyltransferase involved in the synthesis of Sd^a antigen and of α2-6 sialyltransferase activity towards N-acetyllactosamine. Biochem J 1990; 270:519-524.
24. Shah S, Lance P, Smith T et al. n-Butyrate reduces the expression of β-galactoside α2-6-sialyltransferase in HepG2 cells. J Biol Chem 1992; 267:10652-10658.
25. Denis E, Codogno P, Chantret I et al. The processing of asparagine linked oligosaccharides in HT-29 cells is a function of their state of enterocytic differentiation. J Biol Chem 1988; 263:6031-6037.
26. Youakim A, Romero P, Yee K et al. Decrease in polylactosaminoglycans associated with lysosomal membrane glycoproteins during differentiation of CaCo-2 human colonic adenocarcinoma cells. Cancer Res 1989; 49:6889-6895.
27. Brockhausen I, Romero PA, Herscovics A. Glycosyltransferase changes upon differentiation of CaCo-2 human colonic adenocarcinoma cells. Cancer Res 1991; 51:3136-3142.
28. Piller F, Piller V, Fox R et al. Human T lymphocyte activation is associated with changes in O-glycan biosynthesis. J Biol Chem 1988; 263:15146-15150.
29. Higgins E, Siminovitch K, Zhuang D et al. Aberrant O-linked oligosaccharide biosyn-

thesis in lymphocytes and platelets from patients with the Wiskott-Aldrich syndrome. J Biol Chem 1991; 266:6280-6290.
30. Saitoh O, Piller F, Fox R et al. T-lymphocytic leukemia expresses complex, branched O-linked oligosaccharides on a major sialoglycoprotein, leukosialin. Blood 1991; 77:1491-1499.
31. Brockhausen I, Kuhns W, Schachter H et al. Biosynthesis of O-glycans in leukocytes from normal donors and from patients with leukemia:increase in O-glycan core 2 UDP-GlcNAc:Galβ1-3GalNAcα-R (GlcNAc to GalNAc) β(1-6)-N-acetylglucosaminyltransferase in leukemic cells. Cancer Res 1991; 51:1257-1263.
32. Koenderman A, Wijermans P, van den Eijnden D. Changes in the expression of N-acetylglucosaminyltransferase III, IV, V associated with the differentiation of HL60 cells. FEBS Lett 1987; 222:42-46.
33. Lee N, Wang W, Fukuda M. Granulocytic differentiation of HL60 cells is associated with increase of poly-N-acetyllactosamine in Asn-linked oligosaccharides attached to human lysosomal membrane glycoproteins. J Biol Chem 1990; 265:20476-20487.
34. Heffernan, M, Lotan R, Amos B et al. Branding β1-6-N-acetylglucosamine transferases and polylactosamine expression in mouse F9 terabocarcinoma cells and differentiated counterparts. J Biol Chem 1993; 268:1242-1251.
35. Pili R, Chang J, Partis RA et al. The α-glucosidase I inhibitor castanospermine alters endothelial cell glycosylation, prevents angiogenesis, and inhibits tumor growth. Cancer Res 1995; 55:2920-2926.
36. Masibay A, Damewood G, Boeggeman E et al. Expression of β1-4 galactosyltransferase gene during 3T3 cell growth. Biochim Biophys Acta 1991; 1090:230-234.
37. Hinton D, Evans S, Shur B. Altering the expression of cell surface β1-4 galactosyltransferase modulates cell growth. Exper Cell Res 1995; 219:640-649.
38. Begovac PC, Shi YX, Mansfield D et al. Evidence that cell surface β1,4-galactosyltransferase spontaneously galactosylates an underlying laminin substrate during fibroblast migration. J Biol Chem 1994; 269:31793-31799.
39. Sanford G, Harris-Hooker S. Stimulation of vascular cell proliferation by β-galactoside specific lectins. FASEB J 1990; 4:2912-2918.
40. Augustin HG, Kozian DH, Johnson RC. Differentiation of endothelial cells:Analysis of the constitutive and activated endothelial cell phenotypes. BioEssays 1994; 16:901-906.
41. Augustin-Voss HG, Pauli BU. Migrating endothelial cells are distinctly hyperglycosylated and express specific migration-associated cell surface glycoproteins. J Cell Biology 1992; 119:483-491.
42. Miyake M, Hakomori S. A specific cell surface glycoconjugate controlling cell motility:evidence by functional monoclonal antobodies that inhibit cell motility and tumor cell metastasis. Biochemistry 1991; 30:3328-3334.
43. Halicka HD, Ardelt B, Li X et al. 2-deoxy-D-glucose enhances sensitivity of human histiocytic lymphoma U937 cells to apoptosis induced by tumor necrosis factor. Cancer Res 1995; 55:444-449.
44. Pérez-Sala D, Mollideno F. Inhibition of N-linked glycosylation induces early apoptosis in human promyelocytic HL60 cells. J Cell Physiol 1995; 163:523-531.
45. Hiraishi K, Suzuki K, Hakomori S et al. Le^y antigen expression is correlated with apoptosis (programmed cell death). Glycobiology 1993; 3:381-390.
46. Minamide S, Naora H, Adachi M et al. Apoptosis as a mechanism of skin renewal: Le^y-antigen expression is involved in an early event of a cell's commitment to apoptosis. Histochem 1995; 103:339-343.
47. Perillo NL, Pace KE, Seilhamer JJ et al. Apoptosis of T cells mediated by galectin-1. Nature 1995; 378:736-739.
48. Rademacher T, Parekh R, Dwek R. Glycobiology. Ann Rev Biochem 1988; 57:792-794.
49. Kobata A. Structural function and transformational changes of the sugar chains of glycohormones. J Cell Biochem 1988; 37:79-90.
50. Yamaguchi K, Akai K, Kawanishi G et al. Effects of site-directed removal of N-glyco-

sylation sites in human erythropoietin on its production and biological properties. J Biol Chem 1991; 266:20434-20439.
51. Fibi M, Hermentin P, Pauly J et al. N- and O-Glycosylation muteins of recombinant human erythropoietin secreted from BHK21 cells. Blood 1995; 85:1229-1236.
52. Delorme E, Lorenzini T, Giffin J et al. Role of glycosylation on the secretion and biological activity of erythropoietin. Biochemistry 1992; 31:9871-98776.
53. Higuchi M, Oh-eda M, Kuboniwa H et al. Role of sugar chains in the expression of the biological activity of human erythropoietin. J Biol Chem 1992; 267:7703-7709.
54. Wasley L, Timony G, Murtha P et al. The importance of N- and O-linked oligosaccharides for the biosynthesis and in vitro and in vivo biologic activities of erythropoietin. Blood 1991; 77:2624-2632.
55. Takeuchi M, Inoue N, Strickland T et al. Relationship between sugar chain structure and biological activity of recombinant human erythropoietin produced in Chinese Hamster Ovary cells. Proc Natl Acad Sci USA 1989; 86:7819-7822.
56. Misaizu T, Matsuki S, Strickland TW et al. Role of antennary structure of N-linked sugar chains in renal handling of recombinant human erythropoietin. Blood 1995; 86:4097-4104.
57. Grossmann M, Szkudlinski MW, Tropea JE et al. Expression of human thyrotropin in cell lines with different glycosylation patterns combined with mutagenesis of specific glycosylation sites. Characterization of a novel role for the oligosaccharides in the in vitro and in vivo bioactivity. J Biol Chem 1995; 270:29378-29385.
58. Fares FA, Gruener N, Kraiem Z. The role of the asparagine-linked oligosaccharides of the α-subunit in human thyrotropin bioactivity. Endocrinology 1996; 137:555-560.
59. Weintraub B, Stannard B, Meyers L. Glycosylation of thyroid-stimulating hormone in pituitary tumor cells:influence of high-mannose oligosacccharide units on subunit aggregation, combination and degradation. Endocrinology 1983; 112:1331-1345.
60. Joshi L, Weintraub B. Naturally occurring forms of thyrotropin with low bioactivity and altered carbohydrate content act as competetive antagonists to more bioactive forms. Endocrinology 1983; 113:2145-2154.
61. Morell A, Gregoriadis G, Scheinberg I et al. The role of sialic acid in determining the survival of glycoproteins in the circulation. J Biol Chem 1971; 246:1461-1467.
62. Szkudlinski M, Thotakura N, Weintraub B. Subunit-specific functions of N-linked oligosaccharides in human thyrotropin:role of terminal residues of α- and β-subunit oligosaccharides in metabolic clearance and bioactivity. Proc Natl Acad Sci USA 1995; 92:9062-9066.
63. Thotakura N, Szkudlinski M, Weintraub B. Structure-function studies of oligosaccharides of recombinant human thyrotropin by sequential deglycosylation and resialylation. Glycobiology 1994; 4:525-533.
64. Mallet B, Lejeune P-J, Baudry N et al. N-Glycans modulate in vivo and in vitro thyroid hormone synthesis. Study at the N-terminal domain of thyroglobulin. J Biol Chem 1995; 270:29881-29888.
65. Gesundheit N, Magner J, Chen T et al. Differential sulfation and sialylation of secreted mouse thyrotropin (TSH) subunits: regulation by TSH-releasing hormone. Endocrinology 1986; 119:455-463.
66. Baenziger J, Green E. Pituary glycoprotein hormone oligosaccharides:structure, synthesis and function of the asparagine-linked oligosaccharides on lutropin, follitropin and thyrotropin. Biochim Biophys Acta 1988; 947:287-306.
67. Smith P, Baenziger J. A pituitary N-acetylgalactosamine transferase that specifically recognizes glycoprotein hormones. Science 1988; 242:930-933.
68. Smith P, Baenziger J. Molecular basis of recognition by the hormone-specific N-acetylgalactosamine-transferase. Proc Natl Acad Sci USA 1992; 89:329-333.
69. Mengeling B, Manzella S, Baenziger J. A cluster of basic amino acids within an α-helix is essential for α-subunit recognition by the glycoprotein hormone N-acetylgalactosaminyltransferase. Proc Natl Acad Sci USA 1995; 92:502-506.
70. Baenziger J, Kumar S, Brodreck R et al.

Circulatory half-life but not interaction with the lutropin/chorionic gonadotropin receptor is modulated by sulfation of bovine lutropin oligosaccharides. Proc Natl Acad Sci USA 1992; 89:334-338.

71. Flete D, Srivastava V, Hindsgaul O et al. A hepatic reticuloendothelial cell receptor specific for SO_4-GalNAcβ GlcNAcβ1,2Manα that mediates rapid clearance of lutropin. Cell 1991; 67:1103-1110.
72. Chen W, Bahl O. Recombinant carbohydrate variant of human choriogonadotropin β-subunit descarboxyl terminus(115-145). J Biol Chem 1991; 266:6246-6251.
73. Amano J, Nishimura R, Sato S et al. Altered glycosylation of human chorionic gonadotropin decreases its hormonal activity as determined by cyclic adenosine 3,5'-monophosphate production in MA-10 cells. Glycobiology 1990; 1:45-50.
74. Nemansky M, De Leeuw R, Wijnands R et al. Enzymic remodelling of the N- and O-linked carbohydrate chains of human chorionic gonadotropin. Effects on biological activity and receptor binding. Eur J Biochem 1995; 227:880-888.
75. Skarulis M, Wehmann R, Nisula B et al. Glycosylation changes in human chorionic gonadotropin and free alpha subunit as gestation progresses. J Clin Endocrinol Metab 1992; 75:91-96.
76. Endo T, Nishimura R, Kawano T et al. Structural differences found in the asparagine-linked sugar chains of human chorionic gonadotropins purified from the urine of patients with invasive mole and with choriocarcinoma. Cancer Res 1987; 47:5242-5245.
77. Akkakumov G, Hammond G. Glycosylation of human corticosteroid-binding globulin:differential processing and significance of carbohydrate chains at individual sites. Biochemistry 1994; 33:5759-5765.
78. Nathan C, Sporn M. Cytokines in context. J Cell Biol 1991; 113:981-986.
79. Smith E. Hormonal activities of cytokines. Chem Immunol 1992; 52:154-169.
80. Fukushima K, Watanabe H, Takeo K et al. N-linked sugar chain structure of recombinant human lymphotoxin produced by CHO cells:the functional role of carbohydrate as to its lectin-like character and clearance velocity. Arch Biochem Biophys 1993; 304:144-153.
81. Ziltener H, Clark-Lewis I, Jones A et al. Carbohydrate does not modulate the in vivo effects of injected interleukin-3. Exp Hematol 1994; 22:1070-1075.
82. Hanasaki K, Varki A, Stamenkovic I et al. Cytokine-induced β-galactoside α2-6-sialyltransferase in human endothelial cells mediates α2-6-sialylation of adhesion molecules and CD22 ligands. J Biol Chem 1994; 269:10637-10643.
83. Jeng K, Frigeri L, Lui F. An endogenous lectin, galectin-3, potentiates IL-1 production by human monocytes. Immun Lett 1994; 42:113-116.
84. Nakao H, Nishikawa A, Karasuno T et al. Modulation of N-acetylglucosaminyltransferase III, IV and V activities and alteration of the surface oligosaccharide structure of a myeloma cell line by interleukin-6. Biochem Biophys Res Comm 1990; 172:1260-1266.
85. Masumi A, Akamatsu Y, Kitagawa T. Alteration by transforming growth factor-β1 of asparagine-linked sugar chains in glucose transporter protein in Swiss 3T3 cells. Biochim Biophys Acta 1994; 1221:330-338.
86. Kingsley DM, Kozarsky KF, Hobbie L et al. Reversible defects in O-linked glycosylation and LDL receptor expression in a UDP-Gal/UDP-GalNAc 4-epimerase deficient mutant. Cell 1986; 44:749-759.
87. Reddy P, Caras I, Krieger M. Effects of O-linked glycosylation on the cell surface expression and stability of decay accelerating factor, a glycophospholipid anchored membrane protein. J Biol Chem 1989; 264:17329-17336.
88. Schachter H, Brockhausen I. The biosynthesis of serine/threonine N-acetylgalactosamine linked carbohydrate moieties. In:Allen H, Kisailus E Eds. Glycoconjugates, Composition, Structure and Function. New York:Marcel Dekker, 1992; 263-332.
89. Snider M, Rogers O. Intracellular movement of cell surface receptors after endocytosis: resialylation of asialo-transferrin receptor in human erythroleukemia

cells. J Cell Biol 1985; 100:826-834.
90. Hunt RC, Riegler R, Davis AA. Changes in glycosylation alter the affinity of the human transferrin receptor for its ligand. J Biol Chem 1989; 264:9643-9648.
91. Williams A, Enns C. A mutated transferrin receptor lacking asparagine-linked glycosylation sites shows reduced functionality and an association with binding immunoglobulin protein. J Biol Chem 1991; 266:17648-17654.
92. Hoe M, Hunt R. Loss of one asparagine-linked oligosaccharide from human transferrin receptors results in specific cleavage and association with the endoplasmic reticulum. J Biol Chem 1992; 267:4916-4923.
93. Strader C, Fong T, Tota M et al. Structure and function of G protein-coupled receptors. Ann Rev Biochem 1994; 63:101-132.
94. Rodríguez CG, Cundell DR, Tuomanen EI et al. The role of N-glycosylation for functional expression of the human platelet-activating factor receptor. Glycosylation is required for efficient membrane trafficking. J Biol Chem 1995; 270:25178-25184.
95. Rajan N, Tsarbopoulos A, Kumarasamy R et al. Characterization of recombinant human interleukin 4 receptor from CHO cells:role of N-linked oligosaccharides. Biochem Biophys Res Comm 1995; 206:694-702.
96. Schülke N,Schmidt F. Effects of glycosylation on the mechanism of renaturation of invertase from yeast. J Biol Chem 1988; 263:8832-8837.
97. Gibson R, Schlesinger S, Kornfeld S. The nonglycosylated glycoprotein of vesicular stomatitis virus is temperature-sensitive and undergoes intracellular aggregation at elevated temperatures. J Biol Chem 1979; 254:3600-3607.
98. Gibson R, Kornfeld S, Schlesinger S. The effect of oligosaccharide chains of different sizes on the maturation and physical properties of the G protein of vesicular stomatitis virus. J Biol Chem 1981; 256:456-462.
99. Fischer T, Thoma B, Scheurich P et al. Glycosylation of the human interferon-γ receptor. N-linked carbohydrates contribute to structural heterogeneity and are required for ligand binding. J Biol Chem 1990; 265:1710-1717.
100. Filipovic I. Effect of inhibiting N-glycosylation on the stability and binding activity of the low density lipoprotein receptor. J Biol Chem 1989; 264:8815-8820.
101. Li M, Jourdian G. Isolation and characterization of the two glycosylation isoforms of low molecular weight mannose-6-phosphate receptor from bovine testis. J Biol Chem 1991; 266:17621-17630.
102. Wendland M, Waheed A, Schmidt B et al. Glycosylation of the M_r 46,000 mannose-6-phosphate receptor. Effect on ligand binding, stability and conformation. J Biol Chem 1991; 266:4598-4604.
103. Ding DX-H, Vera JC, Heaney ML et al. N-glycosylation of the human granulocyte-macrophage colony-stimulating factor receptor α subunit is essential for ligand binding and signal transduction. J Biol Chem 1995; 270:24580-24584.
104. Podskalny J, Rouiller D, Grunberger G et al. Glycosylation defects alter insulin but not insulin-like growth factor I binding to Chinese Hamster Ovary cells. J Biol Chem 1986; 261:14076-14081.

CHAPTER 12

General Glycoprotein Functions

12.A. ROLE OF MUCINS

Mucins are the major glycoprotein components of mucous secretions; they may also occur on cell membranes; the peptide backbone typically has highly O-glycosylated tandem repeat sequences and Cys rich amino and carboxyl-terminal ends that may polymerize to form long thread-like molecules of several million molecular weight. Mucins are produced and secreted by goblet cells, specialized epithelial cells which can store and condense mucin in secretory vesicles. Human MUC mucin genes encode mucin core proteins which are variably expressed and glycosylated in different tissues.[1-4]

About 50-80% of the dry weight of mucins is carbohydrate. The majority of glycan chains are O-linked, but a relatively minor population of N-glycans may be important for efficient oligomerization, folding and quality control of mucin precursors in the rough endoplasmic reticulum[5,6] and for mucin secretion.[7] The O-glycans of mucins are responsible for the chemical, physical and biological properties and protective effect of mucins. NMR and other biophysical studies indicate that the carbohydrate moieties greatly influence the structure of mucins and produce highly extended molecules with significant interactions between the peptide and GalNAc.[8] Mucin O-glycans can function as major antigens; terminal structures may possess blood group activity. Antigenic activity may be masked by terminal structures such as sulfate groups.[9]

Protection against microbial colonization or invasion normally occurs over the mucosal surfaces of the body. Epithelial cells, in addition to their soluble secretions, possess membrane-bound mucins which can bind various microorganisms via a mosaic of carbohydrate ligands representing potential bacterial binding sites. They incude O-glycan core 1, 2, 3, and 4 structures, I-i antigens with single or repeating N-acetyllactosamine units, Lewis antigens and terminal structures.[10] The rapid turnover of epithelial cells and removal of the mucus gel assures the relatively swift removal of trapped pathogens by non-immune methods. Mucin type O-glycans are involved in cell adhesion phenomena mediated by selectins.[11]

Mucins carrying sialyl-Tn antigens appear to strongly inhibit NK cell cytotoxicity in the presence of ammonium ions[12] whereas mucins with more complex O-glycans (human lung and breast cancer cell mucins) do not show this effect. The immunomodulatory role of mucins, at least with respect to NK cell function, may therefore depend on their carbohydrate structures. Elimination of tumor cells by components of the immune system may be rendered difficult because of the protective and immune modulatory effect of their mucin coat.

The effect of alcohol or drug intake and certain bacteria may degrade the protective mucin gel or decrease mucus glycoprotein sulfation. Glycosidases and sulfatases are produced by many bacterial colonies in the colon and cause mucin degradation for nutrient needs, but this process is normally balanced by replacement from new mucosal cells.[13-16]

Epithelial cancers and cultured cancer cells produce various types and amounts of mucins. Hypersecretion of mucins occurs in diseases such as cystic fibrosis and bronchiectasis. Mucin undersecretion and dryness of mucosal surfaces are common occurrences in senescence. Aberrant expression of mucin oligosaccharides occurs in diseases such as colon cancer and ulcerative colitis. Mucins may be involved in signal transduction events. A signal transduction component containing a glycoprotein complex has been identified on rat adenocarcinoma cell surface microvilli. It is the product of a single gene and is comprised of a transmembrane subunit and a sialomucin subunit. The transmembrane portion has two EGF-like domains and is capable of activating the proto-oncogene receptor kinase p185^{neu}. The mRNA for the sialomucin complex is expressed in rat brain and lung, but not in several other organs. Antibodies produced against the complex can demonstrate its presence by immunocytochemistry in some human breast tumors. The results suggest that this membrane glycoprotein is regulated in a tissue-specific manner, and is overexpressed in some tumors.[17,18]

12.B. FUNCTIONS OF SULFATED OLIGOSACCHARIDES

Sulfated glycoconjugates occur in a broad spectrum of organisms ranging from simple halobacteria,[19] sponges[20] and slime molds[21] to mammals.[22,23] Sulfated sugars have been isolated from, or synthesized by a variety of tissues and secretions,[24] and there are tissue-specific differences in structure and function.[25] The context in which the high negative charge of sulfated oligosaccharides is presented determines their functional roles. The diversity of structures permits interactions with many forms of microoganisms. In addition, sulfation may be altered in certain diseases; for example, increased O-glycan sulfation in cystic fibrosis is associated with colonization by *Pseudomonas aeruginosa*.[26]

Sulfate linked to the 6-position of Gal of Le^x and Le^a structures appears to favor adhesion of these ligands to P- and L- selectins, but not to E-selectins.[27] Sulfated oligosaccharides, including those of the Le^a and Le^x series, can support L-selectin adhesion, and thus serve as a link between selectin and integrin-mediated adhesins.[28] Chlorate, a metabolic inhibitor of sulfation, blocks sulfate incorporation into GlyCAM-1, an endothelial ligand for L-selectin; the reduction in sulfated macromolecules is accompanied by altered selectin-lymphocyte binding.[29a,b] The major recognition group for selectins of GlyCAM-1 is believed to possess the structure sialylα2-3 (6-sulfate-) Galβ1-4 (Fucα1-3) GlcNAc.[30,31]

The turnover and clearance of some glycoprotein hormones such as lutropin is mediated by 4-sulfated-GalNAc signals.[32] Regulation of the expression of the GalNAc- and sulfotransferases responsible for synthesis of sulfated oligosaccharides on LH units is in response to specific modulation by estrogen.[33] In earlier studies sulfation of O-glycans was reported to be regulated by female sex hormones as evidenced by an increased activity of sulfotransferase in microsomal fractions of uterine endometrium of rabbits treated with estrogen, whereas progesterone exerts a suppressive effect.[34]

Podocalyxin is a major sialoglycoprotein of rat glomerulus. Its high negative charge ensures that the glomerular slits are kept open and renal glomerular filtration proceeds normally. The presence of considerable amounts of sulfated N-and O-linked chains suggests that sulfate as well as sialic acid are important in maintaining this function.[35]

Leukosialin (CD43), the cell surface mucin of T lymphoid cells and other leukocytes, has been found to be intensely sulfated in the murine T lymphoma line RDM-4. Ten percent of the total sulfate is incorporated into a CD43-like molecule which is present on activated but not resting splenic T cells. The sulfate label can be released by β-elimination but not by endoglycosidase F, indicating that sulfate is attached to O- but not N-glycans.[36]

Sulfated mucins are of interest because of their prevalence in respiratory secretions, particularly in cystic fibrosis (see also chapter 14). The physicochemical properties of intestinal sulfomucins appear to be critical for the establishment of certain intestinal helminths (chapter 17.D).[37-39]

Certain proteoglycans are intensely sulfated. It is yet unknown if N- and O-sulfotransferases involved in proteoglycan synthesis also can act on glycoproteins. One important function of sulfated proteoglycans is their anticoagulant activity and the presence of sulfate on heparin/heparans is closely linked to this role.[40] Studies on heparan in a COS cell mutant defective in N-sulfotransferase, indicated that fibroblast growth factor binding is regulated by this enzyme.[41,42] Fibroblast growth factor-induced endothelial cell growth and chemotaxis can be inhibited by a bacteria-derived sulfated polysaccharide (D-gluco-D-galactan sulfate).[43]

Functions of sulfated compounds may be elucidated through model studies in lower organisms. Sperm-egg binding in the sea urchin is mediated by egg surface sulfated fucan polysaccharides which recognize bindin, a sperm adhesion protein.[44] Soluble sulfated polyanions interfere with binding whereas other polyanionic compounds are ineffective. In addition, fucoidan, a complex sulfated polysaccharide, is of interest as an inhibitor of several biological processes; it is capable of blocking human sperm-egg binding and cell-cell binding mediated by P- or L- selectin recognition of Le^a and Le^x ligands, and inhibits HIV reverse transcriptase activity.[45]

12.C. ION CHANNELS

Translocation of solutes across membranes occurs via specific protein transporters, either as a passive or an active, energy-dependent, process. Studies on the topology of ion channels by mutagenesis and specific antibodies suggest a role for N-glycans in the appropriate folding and three-dimensional orientation of channel proteins. Examples are the cGMP-gated channel from bovine rod photoreceptor[46] and the erythrocyte Glut 1 glucose transporter.[47] Carbohydrates may also assist in internalizing and targeting transporters, as proposed for lamp-1 glycans which are responsive to phosphate. Under conditions of phosphate excess, there is a decrease in Na^+-dependent phosphate transport associated with a decrease in lamp-1 N-glycans; the association has prompted the belief that lamp-1 could enable the targeting of phosphate transporter to lysosomes.[48]

A few studies have served to demonstrate the dynamic effects of glycosylation at the molecular level. N-glycans have been shown to be involved in the renal handling of K^+ and Na^+ ions. Inward rectifier potassium channels are of importance for potassium exchange in the kidney. ROMK1, an ion channel of the inward rectifier potassium channel family, is well suited for glycosylation studies because it has a single N-glycosylation consensus sequence.[49] Treatment of cells with tunicamycin or removal of the consensus sequence by mutation shows a marked decrease in potassium channel currents compared with normally glycosylated controls. The results suggest that the N-glycan stabilizes the open state of ROMK1.[50]

Water channel forming integral protein (Aquaporin, CHIP) is present in the membranes of red cells and renal proximal tubules and contains two potential N-glycosylation sites. Studies of mutant CHIP molecules expressed in *Xenopus* oocytes show that dysfunctional mutants are abnormally glycosylated with high mannose residues, presumably due to their retention in the ER.[51]

12.D. CHAPERONES

Molecular chaperones are proteins which bind to and stabilize non-native conformations of other proteins; through cycles of chaperone binding and release chaperones may promote competency for peptide folding or membrane translocation.[6,52] Chaperones are thus thought of as devices which assure quality control of newly synthesized proteins. Calnexin and calreticulin are two proteins which fulfill this function. Calnexin, recently shown to be a lectin,[53] associates with substrates containing monoglucosylated N-linked oligosaccharides (Fig. 6). They arise in the ER following removal of two of the three glucose residues (by glucosidases I and II) on the core oligosaccharide.[54] The monoglucosylated intermediates are maintained by a de- and re-glucosylation cycle until glycoproteins have acquired their correctly folded structures.[6] Glycan trimming and protein folding are thus close correlates in the complex glycosylation and trimming activities that occur in the ER. This process is typified by the assembly of MHC class I molecules in the ER. A complex of heavy chain-β_2m-transport peptide is required for stability and transport and this arises from a sequence of events involving calnexin-class 1 molecule heavy chain binding,[55] heavy chain-β_2m binding, altered conformation of heavy chain and then binding to the transporter.[52,56] The requirements for glucose trimming are variable, however, and MHC class I molecules may be expressed in the absence of glucose trimming in glucosidase-deficient thymoma cells.[57]

Chaperones appear to be involved in chloride channel defects observed in cystic fibrosis. The transmembrane conductance regulator (CFTR) is altered in CF due to mutations such as a deletion of phenylalanine at position 508. Misfolded mutant protein is retained in the ER rather than becoming inserted into the surface plasma membrane.[58] Calnexin helps to retain this misfolded protein and thus contributes to its mislocalization and subsequent degradation.[59]

Calreticulin has been shown to have sequence similarities with calnexin. It interacts transiently with newly synthesized human immunodeficiency virus (HIV) type 1 envelope protein gp160. Non-glycosylated envelope protein is not associated with calreticulin, indicating a requirement for N-glycan as reported for calnexin. The in vivo binding kinetics of calnexin and calreticulin to gp160 are similar.[60] The proper folding of hepatitis C virus glycoprotein appears to require a relatively long association with calnexin; clarifying the steps in this pathway may yield a better understanding of virus assembly, receptor binding and entry, and hopefully lead to improved vaccines.[61]

12.E. ENZYMES

Many enzymes are known to be glycosylated.[62] The presence of glycans on enzymes may be related to enzyme activity or stability and secretion, and the effects of altering glycosylation are variable.[63] A few examples will be described where various effects of enzyme glycosylation have been observed.

Glycosylation has been found to be necessary for the catalytic activity of human acid β-glucosidase. This lysosomal enzyme has five potential N-glycosylation sites, four of which are occupied; complete

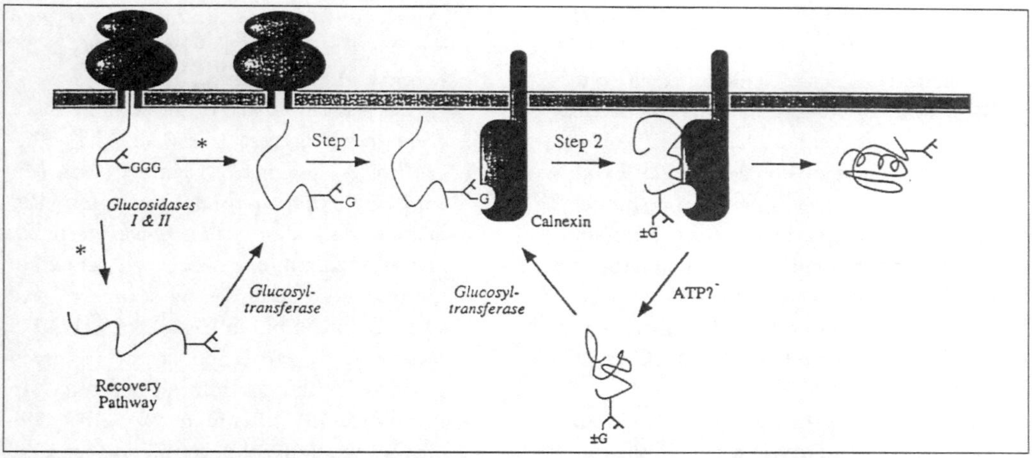

Fig. 6. Chaperone binding to glycoproteins in the endoplasmic reticulum. Chaperones are ER associated proteins which can "edit" newly formed proteins to assure their normal folding and secretion. Calnexin is a molecular chaperone which fulfills this function. Thus, secretion-incompetent variants remain bound to calnexin; examples of such defective variants include mutants of cystic fibrosis transmembrane conductance regulator and α-antitrypsin. This schematic drawing depicts a two-step model for the binding of calnexin to nascent unfolded protein. Subsequent to the removal of two glucose residues from the nascent glycoprotein molecule, contact by the protein is made by way of its monoglucosylated oligosaccharide chains (Step I). Following the initial contact, the unfolded polypeptide then associates more closely with calnexin by forming additional contacts (Step 2). Reglucosylation by glucosyltransferase may be of importance in recovering proteins that have lost all three glucose residues prior to contact with calnexin or in facilitating rebinding to calnexin during cycles of refolding. The asterisks designate steps blocked by castanospermine and 1-deoxynojirimycin. Reprinted from Williams (1995) with permission.

deglycosylation results in loss of activity. Glycan occupancy of at least one site is required for the formation and maintenance of active glucosidase.[64]

Activation of lipoprotein lipase requires trimming of glucose residues in the ER, but further processing or translocation to the cis-Golgi is not needed for full expression of lipolytic activity.[65-67]

The secretion of functional papain precursor (propapain) from insect cells depends upon N-glycosylation.[68] Similarly, yeast carboxypeptidase Y requires N-glycosylation for efficient intracellular transport, especially at higher temperatures, although not for sorting, stability or activity.[69] In some instances enzyme stability can be influenced by removal of glycan; thus the susceptibility of acid phosphatase toward proteolysis is markedly increased by deglycosylation.[70]

The secretion and activity of the soluble portion of recombinant core 2 β6-GlcNAc-transferase from insect cells also depends on the occupancy of its two N-glycosylation sites.[71] This has been shown by tunicamycin treatment and site-directed mutagenesis which resulted in total loss of the activity when both N-glycosylation sites were deleted and partial loss when only the amino-terminal localized site was eliminated.

Both the secretion and enzyme activity of *Mucor pusillus* rennin, an aspartic proteinase, are significantly decreased by mutation of two of three possible N-glycosylation sites.[72] The effect of N-glycosylation upon the folding of nascent yeast invertase chains and their maturation is protective in that solubility is improved and irreversible aggregation is suppressed.[73]

References

1. Gendler SJ, Spicer AP, Lalani E-N et al. Structure and biology of a carcinoma-associated mucin, MUC1. Am Rev Respir Dis

1991; 144:S42-S47.

2. Braga VMM, Pemberton LF, Duhig T et al. Spatial and temporal expression of an epithelial mucin, MUC-1, during mouse development. Development 1992; 115:427-437.
3. Audie JP, Janin A, Porchet N et al. Expression of human mucin genes in respiratory, digestive, and reproductive tracts ascertained by in situ hybridization. J Histochem Cytochem 1993; 41:1479-1485.
4. Verma M, Davidson E. Mucin gene: structure, expression and regulation. Glycoconj J 1994; 11:172-179.
5. Dekker J, Strous G. Covalent oligomerization of rat gastric mucin occurs in the rough endoplasmic reticulum, is N-glycosylation dependent, and precedes initial O-glycosylation. J Biol Chem 1990; 265:18116-18122.
6. Helenius A. How N-linked oligosaccharides affect glycoprotein folding in the endoplasmic reticulum. Molec Biol Cell 1994; 5:253-265.
7. McCool D, Forstner J, Forstner G. Synthesis and secretion of mucin by the human colonic tumor cell line LS180. Biochem J 1994; 302:111-118.
8. Gerken TA, Butenhof KJ, Shogren R. Effects of glycosylation on the conformation and dynamics of O-linked glycoproteins: carbon-13 NMR studies of ovine submaxillary mucin. Biochem 1989; 28:5536-5543.
9. Feizi T, Childs R. Carbohydrates as antigenic determinants of proteins. Biochem J 1987; 245:1-11.
10. Klein A, Carnoy C, Wieruszeski J et al. The broad diversity of neutral and sialylated oligosaccharides derived from human salivary mucins. Biochemistry 1992; 31:6152-6165.
11. Dowbenko D, Andalibi A, Young PE et al. Structure and chromosomal localization of the murine gene encoding GLYCAM 1. A mucin-like endothelial ligand for L selectin. J Biol Chem 1993; 268:4525-4529.
12. Ogata S, Maimonis PJ, Itzkowitz SH. Mucins bearing the cancer-associated sialosyl-Tn antigen mediate inhibition of natural killer cell toxicity. Cancer Res 1992; 52:4741-4746.
13. Slomiany BL, Murty VLN, Piotrowski J et al. Glycosulfatase activity of Helicobacter pylori toward gastric mucin. Biochem Biophys Res Comm 1992; 183:506-513.
14. Roberton A, McKenzie C, Sharfe N et al. A glycosulfatase that removes sulfate from mucus glycoprotein. Biochem J 1993; 293:683-689.
15. Corfield A, Wagner S, Clamp J et al. Mucin degradation in the human colon. Production of sialidase, sialate O-acetylesterase, N-acetylneuraminate lyase, arylesterase and glycosulfatase activities by strains of fecal bacteria. Infec Immun 1992; 60:3971-3978.
16. Hoskins L. Degradation of mucus glycoproteins in the gastrointestinal tract. In: Horowitz M, Pigman W eds. The Glycoconjugates Vol 2. New York:Academic Press, 1978:235-253.
17. Wu K, Salas P, Yee L et al. Tissue and tumor expression of a cell surface glycoprotein complex containing an integral membrane glycoprotein activator of p185^{neu}. Oncogene 1994; 9:3139-3147.
18. Li Y, Carraway K, Carraway C. A glycoprotein complex serves as a core for a p185 containing, microfilament-associated signal transduction particle in ascites mammary tumor cell microvilli. Proc Amer Assoc Cancer Res 1995; 36:53.
19. Lechner J, Wieland F, Sumper M. Biosynthesis of sulfated saccharides N-glycosidically linked to the protein via glucose. J Biol Chem 1985; 260:860-866.
20. Spillmann D, Thomas-Oates J, van Kuik J et al. Characterization of a novel sulfated carbohydrate unit implicated in the carbohydrate-carbohydrate-mediated cell aggregation of the marine sponge Microciona prolifera. J Biol Chem 1995; 270:5089-5097.
21. Freeze H, Yeh R, Miller A et al. Structural analysis of the asparagine-linked oligosaccharides from three lysosomal enzymes of dictyostelium discoideum. J Biol Chem 1983; 258:14874-14879.
22. Heifetz A, Watson C, Johnson A et al. Sulfated glycoproteins secreted by human vascular endothelial cells. J Biol Chem 1982; 257:13581-13586.
23. Green E, Gruenebaum J, Bielinska M et al. Sulfation of lutropin oligosaccharides with a cell-free system. Proc Natl Acad Sci USA 1984; 81:5320-5324.
24. Mulder G, Jakoby W. Sulfation Chapt 5.

In:Mulder G. ed. Conjugation reactions in drug metabolism. London:Taylor and Francis Ltd, 1990:114-121.

25. Kato M, Wang H, Bernfield M et al. Cell surface syndecan-1 on distinct cell types differs in fine structure and ligand binding of its heparan sulfate chains. J Biol Chem 1994; 269:18881-18890.
26. Ramphal R, Carnoy C, Fiebre S et al. Pseudomonas aeruginosa recognizes carbohydrate chains containing type 1(Galβ1-3GlNAc) or type 2(Galβ1-4GlNAc) disaccharide units. Infect Immun 1991; 59:700-704.
27. Green P, Tamatani T, Watanabe T et al. High affinity binding of the leukocyte adhesion molecule L-selectin to 3' sulfated Le a and Le x oligosaccharides and the predominance of sulfate in this interaction demonstrated by binding studies with a series of lipid-linked oligosaccharides. Biochem Biophys Res Comm 1992; 188:244-251.
28. Green P, Yuen C, Childs R et al. Further studies of the binding specificity of the leukocyte adhesion molecule, L-selectin, towards sulfated oligosaccharides-suggestion of a link between the selectin- and the integrin-mediated lymphocyte adhesion systems. Glycobiology 1995; 5:29-38.

29a. Imai Y, Rosen S. Direct demonstration of heterogeneous sulfated O-linked carbohydrate chains on an endothelial ligand for L-selectin. Glycoconj J 1993; 10:34-39.

29b. Imai Y, Lasky L, Rosen S. Sulfation requirement for GlyCAM-1 an endothelial ligand for L-selectin. Nature 1993; 361:555-557.

30. Hemmerich S, Rosen SD. 6'-sulfated sialyl Lewis x is a major capping group of GlyCAM-1. Biochemistry 1994; 33:4830-4835.
31. Hemmerich S, Bertozzi CR, Leffler H. Identification of the sulfated monosaccharides of GlyCAM-1, an endothelial derived ligand for L-selectin. Biochemistry 1994; 33:4820-4829.
32. Drickamer K. Clearing up glycoprotein hormones. Cell 1991; 67:1029-1032.
33. Dharmesh S, Baenziger J. Estrogen modulates expression of the glycosyltransferases that synthesize sulfated oligosaccharides on lutropin. Proc Natl Acad Sci USA 1993; 90:11127-11131.
34. Munakata H, Isemura M, Yosizawa Z. Enzymatic sulfation of exogenous high molecular weight glycopeptides by microsomal fraction of the rabbit uterine endometrium. J Biol Chem 1985; 260:6851-6856.
35. Dekan G, Gabel C, Farquhar M. Sulfate contributes to the negative charge of podocalyxin, the major sialoglycoprotein of the glomerular filtration slits. Proc Natl Acad Sci USA 1991; 88:5398-5402.
36. Wilson A, Rider C. Evidence that leukosialin, CD43, is intensely sulfated in the murine T lymphoma line RDM-4. J Immun 1992; 148:1777-1783.
37. Shi B-B, Ishikawa N, Itoh H et al. Goblet cell mucins of four genera of the subfamily cricetinae with reference to the protective activity against stronglyoides venezuelensis. Parasite Immunol 1994; 16:553-559.
38. Forstner J, Maxwell B, Roomi N. Intestinal secretion of mucus in chronically reserpine treated rats. Am J Physiol 1981; 241:G443.
39. Ishikawa N, Shi B-B, Khan AI et al. Reserpine-induced sulphomucin production by goblet cells in the jejunum of rats and its significance in the establishment of intestinal helminths. Parasite Immunol 1995; 17:581-586.
40. Ofosu F, Modi G, Blajchman M et al. Increased sulfation improves the anticoagulant activities of heparan sulfate and dermatan sulfate. Biochem J 1987; 248:889-896.
41. Ishihara M, Guo Y, Swiedler S. Selective impairment of the synthesis of basic fibroblast growth factor binding domains of heparan sulfate in a COS cell mutant defective in N-sulfotransferase. Glycobiology 1993; 3:83-88.
42. Ishihara M, GuoY, Wei Z et al. Regulation of biosynthesis of the basic fibroblast growth factor binding domains of heparin sulfate by heparin sulfate N-deacetylase/N-sulfotransferase expression. J Biol Chem 1993; 268:20091-20095.
43. Nakayama Y, Iwahana M, Sakamoto N et al. Inhibitory effects of a bacteria-derived sulfated polysaccharide against basic fibroblast growth factor-induced endothelial cell growth and chemotaxis. J Cell Physiol 1993; 154:1-6.

44. DeAngelis P, Glabe C. Specific recognition of sulfate esters by bindin, a sperm adhesion protein from sea urchins. Biochim Biophys Acta 1990; 1037:100-105.
45. Patankar M, Oehninger S, Barnett T et al. A revised structure for fucoidan may explain some of its biological activities. J Biol Chem 1993; 268:21770-21776.
46. Wohlfart P, Haase W, Molday R et al. Antibodies against synthetic peptides used to determine the topology and site of glycosylation of the cGMP gated channel from bovine rod photo receptors. J Biol Chem 1992; 267:644-648.
47. Hresko R, Kruse M, Strube M et al. Topology of the Glut-1 glucose transporter deduced from glycosylation scanning mutagenesis. J Biol Chem 1994; 269:20482-20488.
48. Helps CR, McGivan JD. Regulation of glycosylation of Lamp-1 in the bovine epithelial cell line NBL-1 by changes in the concentration of extracellular phosphate. Biochem J 1994; 303:613-618.
49. Ho K, Nichols CG, Lederer WJ et al. Cloning and expression of an inwardly rectifying ATP-regulated potassium channel. Nature 1993; 362:31-38.
50. Schwalbe R, Wang Z, Wible B et al. Potassium channel structure and function as reported by a single glycosylation sequon. J Biol Chem 1995; 270:15336-15340.
51. Smith B, Preston G, Spring F et al. Human red cell aquaporin CHIP. J Clin Invest 1994; 94:1043-1049.
52. Williams D. Calnexin: a molecular chaperone with a taste for carbohydrate. Biochem Cell Biol 1995; 73:123-132.
53. Ware F, Vassilakos A, Peterson P et al. The molecular chaperone calnexin binds $Glc_1Man_9GlcNAc_2$ oligosaccharide as an initial step in recognizing unfolded proteins. J Biol Chem 1995; 270:4697-4604.
54. Kornfeld R, Kornfeld S. Assembly of asparagine-linked oligosaccharides. Ann Rev Biochem 1985; 54:631-664.
55. Tector M, Salter R. Calnexin influences folding of human class I histocompatability proteins but not their assembly with β_2 microglobulin. J Biol Chem 1995; 270:19638-19642.
56. Williams D, Watts T. Molecular chaperones in antigen presentation. Curr Biol 1995; 7:77-84.
57. Balow JP, Weissman JD, Kearse KP. Unique expression of major histocompatibility complex class I proteins in the absence of glucose trimming and calnexin association. J Biol Chem 1995; 270:29025-29029.
58. Lukacs G, Mohamed A, Kartner N et al. Conformational maturation of CFTR but not its mutant counterpart (δ508 CFTR) occurs in the endoplasmic reticulum and requires ATP. EMBO J 1994; 13:6076-6086.
59. Pind S, Riordan J, Williams D. Participation of the endoplasmic reticulum chaperone calnexin (p88, IP90) in the biogenesis of the cystic fibrosis transmembrane conductance regulator. J Biol Chem 1994; 269:12784-12788.
60. Otteken A, Moss B. Calreticulin interacts with newly synthesized human immunodeficiency virus type 1 envelope glycoprotein, suggesting a chaperone function similar to that of calnexin. J Biol Chem 1996; 271:97-103.
61. Dubuisson J, Rice CM. Hepatitis C virus glycoprotein folding:disulfide bond formation and association with Calnexin. J Virol 1996; 70:778-786.
62. Varki A. Biological roles of oligosaccharides: all of the theories are correct. Glycobiology 1993; 3:97-130.
63. Fujita H, Kawamoto K, Hanasaki K et al. Glycosylation-dependent binding of pancreatic type I phospholipase A_2 to its specific receptor. Biochem Biophys Res Comm 1995; 209:293-299.
64. Grace M, Grabowski G. Human acid β-glucosidase:glycosylation is required for catalytic activity. Biochem Biophys Res Commun 1990; 168:771-777.
65. Carroll R, Ben-Zeev O, Doolittle M et al. Activation of lipoprotein lipase in cardiac myocytes by glycosylation requires trimming of glucose residues in the endoplasmic reticulum. Biochem J 1992; 285:69-696.
66. Ben-Zeev O, Doolittle M, Davis R et al. Maturation of lipoprotein lipase. J Biol Chem 1992; 267:6219-6227.

67. Semenkovich C, Luo C, Nakanishi M et al. In vitro expression and site-specific mutagenesis of the cloned lipoprotein lipase gene. J Biol Chem 1990; 265:5429-5433.
68. Vernet T, Tessier D, Richardson C et al. Secretion of functional papain precursor from insect cells. J Biol Chem 1990; 265:16661-16666.
69. Winther J, Stevens T, Kielland-Brandt M. Yeast carboxypeptidase Y requires glycosylation for efficient intracellular transport, but not for vacuolar sorting, in vivo stability, or activity. Eur J Biochem 1991; 197:681-689.
70. Barbaric S, Mrsa V, Ries B et al. Role of the carbohydrate part of yeast acid phosphatase. Arch Biochem Biophys 1984; 234:567-575.
71. Toki D, Sarkar M, Yip B et al. Role of N-linked glycosylation of recombinant human core 2 β6-N-acetylglucosaminyltransferase expressed in Sf9 insect cells. Submitted.
72. Aikawa J, Yamashita T, Nishiyama M et al. Effects of glycosylation on the secretion and enzyme activity of Mucor rennin, an aspartic proteinase of Mucor pusillus, produced by recombinant yeast. J Biol Chem 1990; 265:13955-13959.
73. Schülke N, Schmid F. Effect of glycosylation on the mechanism of renaturation of invertase from yeast. J Biol Chem 1988; 263:8832-8837.

CHAPTER 13

Glycosylation in Leukemia and Blood-Related Disorders

13.A. LEUKEMIA

Since cellular glycosylation appears to be a function of growth and differentiation, it is not surprising that in disease there are profound alterations of carbohydrate structures and the enzymes involved in their synthesis (Table 4). These complex alterations are difficult to classify, but based on many studies in this field we begin to understand emerging patterns.

Leukemias are malignant neoplasms of hematopoietic tissues which exhibit altered cell surface properties associated with premature release of cells from the bone marrow, an arrest at an early stage of differentiation, abnormal growth and decreased cell turnover in the blood (Table 5). These neoplasms may share common cell surface glycoproteins. For example, leukosialin may show fine structural differences of carbohydrates, depending upon the cell type. Altered expression of blood group structures is noted in leukemia but not in a consistent fashion; the most common alterations involve the ABH and Ii systems.[1,2]

13.A.1. Cell Surface Glycosylation Changes in Leukemia Cells

Granulocytes from patients with chronic myelogenous leukemia (CML) have different cell surface properties compared to normal granulocytes. Functional abnormalities of CML cells may be explained by the presence of hypersialylated membrane glycoproteins.[3,4] These cells are hypersialylated and adhere poorly to peanut lectin compared to normal granulocytes. The abnormal cell surface properties and functions of CML cells, including chemotaxis, are restored following removal of sialic acid with sialidase, and during chemotherapy.[5,6] Oligosaccharide chains in CML exhibit a characteristically high amount of sialyl-Le^x and they are more

Glycoproteins and Human Disease, by Inka Brockhausen and William Kuhns. © 1997 R.G. Landes Company.

Table 4. Glycosylation changes in human disease

Disease	Carbohydrate Abnormality	Enzyme Abnormality	Chapter/ Reference
Alzheimer's disease	↑ O-GlcNAc		18/14
	abnormal glycosylation		18/16
Batten's disease	↑ Dol-P-sugars		18/6
Carbohydrate deficiency disease CDGS Type I		↓ phosphomannomutase	16/20
	truncated or abnormal N-glycans		16/18,19
CDGS Type II	defective N-glycans	↓GlcNAc-T II	16/21
Cystic fibrosis	↑ O-glycosylation,		14/14,24
	↑ mucin sulfation		14/10-15
	↑ Fucα6 in fibroblasts		14/28
	Le^x in fibroblasts	↑ fibroblast Gal-T	14/29
		↓ SA-T	14/32,33
		↑Gal-T, Fuc-T, Man-T	14/37
Diabetes	advanced protein glycation		18/1
HEMPAS	truncated N-glycans	↓ mannosidase II	16/10
	↑ polyglucosylceramide		
		↓ GlcNAc-T II	16/11
Inflammatory disease	sialyl-Le^x		15/1
	Gal-deficient IgG		15/13,17
Intestinal inflammatory disease	abnormal mucin		15/20,22, 27,28,30
	↓ sulfation		15/23,26
	↑SAα6, ↓ SAα3		15/32,34
	↓O-acetyl SA		
Leukocyte adhesion deficiency Type II	Bombay (hh) blood group	↓ GDP-Fuc	16/7
	↓ sialyl-Le^x		
Liver disease	↑ Fucα6		21/16
		↑ serum Gal-T	15/37
Osteoarthritis		↑Gal-T, Fuc-T	
		↑SA-T, Man-T	18/11
		↑Glc-T, GlcNAc-T	
	↑ chondroitin-sulfate		18/10
Rheumatoid arthritis	↓ Gal on IgG	abnormal β4Gal-T	13/53
			15/13
		↓ Gal-T	15/14
Thrombosis	↑ cell surface SA-T		19/93
Xenograft rejection	antigenic Galα1-3Gal		18/19,20, 22

Abbreviations: ↓ decreased; Dol-P, dolichol-phosphate; Fuc, fucose; HEMPAS, hereditary erythroblastic multinuclearity-associated anemia with positive acidified serum test; ↑ increased; Gal, galactose; Glc, glucose; GalNAc, N-acetylgalactosamine; GlcNAc, N-acetylglucosamine; Man, mannose; SA, sialic acid; -T, transferase.

Table 5. Glycosylation changes in leukemia and hematopoietic disease

Disease Abnormality	Carbohydrate Abnormality	Glycosyltransferase	Chapter/ Reference
Leukemia, general	unique O-glycans		13/10
		Fuc-T isoenzymes	13/26
		various α6SA-T	13/24
Acute lymphocytic leukemia		↑ core 2 β6GlcNAc-T	13/11
	↑branched disialyl core 2 leukosialin		
Acute myeloid leukemia	↑ SA-Le^{x}		13/7,8
	↑ SA		
	↓ α 3Fuc		
	shorter chains, ↓ i	↑α3SA-T(O-glycan)	13/4,12,13
		↑ core 2 β6GlcNAc-T	13/14
		↑ i β3GlcNAc-T	
		↑ β4Gal-T	
		↑ core 1 β3Gal-T	
		loss of elongation β3GlcNAc-T	
	blood groups	↓ α2Fuc-T	13/17
		↑ Fuc-T	13/26
		↑ GlcNAc-T V	13/22
Chronic granulocytic leukemia		↑ α2Fuc-T	13/18
Chronic lymphocytic leukemia	↓ i		13/2
	↑ disialyl core 1		13/11
	Tn on leukosialin from T-cells		
Chronic myelogenous leukemia	↑ sialyl-Le^{x}		13/7, 8
	↑ SA		
	↑ SA-core 1		
	↓ α 3Fuc		
	shorter chains		
	↑ SA		13/3
		↑ α3SA-T (O-glycan)	13/4,12,13
		↑ core 2 β6GlcNAc-T	13/14
		↓ elongation β3GlcNAc-T	
		↑ Fuc-T	13/26
	bisected N-glycans	↑ GlcNAc-T III	13/20
Permanent mixed field agglutinability	↓ SA, ↓ Gal on rbc	↓ core 1 β3Gal-T	13/51,52
Myeloma	abnormal N-glycans		13/53
		↑ SA-T	13/55
		↑ Fuc-T	
	↑ SAα6Gal on Bence-Jones protein		13/56
Wiscott Aldrich syndrome	↑ core 2 on leukosialin	↑ core 2 β6GlcNAc-T	13/69,70

Abbreviations:↓ decreased; Fuc, fucose; HEMPAS, hereditary erythroblastic multinuclearity-associated anemia with positive acidified serum test; ↑ increased; Gal, galactose; GlcNAc, N-acetylglucosamine; Man, mannose; SA, sialic acid; T, transferase;

sialylated, although chains are shorter and contain less α1-3 fucosylated residues than normal mature granulocytes.[7]

Cells from patients with acute myeloid leukemia (AML), poorly differentiated myeloid cells, contain primarily short O-linked oligosaccharides with core 1 and 2 structures and α2-6 and α2-3 linked sialic acid residues.[8] Antibody binding studies indicate that AML myeloblasts carry sialyl-Le^x antigens.[9] Leukosialin isolated from erythroid, myeloid and T-lymphoid cell lines expresses unique sets of O-glycans which differ in core structures as well as in sialic acid content.[10]

The i antigen is detected on lymphocytes; it is less common in T-cell lymphocytic leukemia (CLL)[2] and in AML.[8] However, blast cells from AML patients show an increase in the blood group i β3-GlcNAc- and β4-Gal-transferases involved in the synthesis of the i antigen (Table 5).

Oligosaccharides derived from T-cell acute lymphocytic leukemia (ALL) contain elevated levels of disialyl core 2 hexasaccharides on leukosialin which are also expressed in activated normal T-lymphocytes but are absent in resting normal T-lymphocytes. An associated increase of core 2 β6-GlcNAc-transferase activity may be responsible for the predominance of core 2 structures.[11] CLL cells exhibit O-glycans with mainly disialylated core 1 structures in addition to a smaller amount of core 2 hexasaccharides; the latter structure is also found on thymocytes.

13.A.2. Altered Glycoprotein Biosynthesis in Leukemia Cells

The structural changes observed in leukemia cells can be explained in part on the basis of their altered glycosyltransferase activities. A significant activation of core 2 β6 GlcNAc-transferase and α3-sialyltransferase (O) has been observed in CML and AML, possibly due to the less differentiated status of cells (Table 5).[4,12-14] The increased core 2 branching of leukemia cells which may provide more substrate for sialyltransferases may contribute to the hypersialylation phenomenon. Synthetic core 1 derivatives, lacking the active hydroxyl groups that are required for core 2 β6-GlcNAc-transferase and α3-sialyltransferase (O) action (Fig. 1), reduce the abnormal activities in leukemic cell homogenates.[15] Exposure of autologous bone marrow cells to sialidase may possibly result in clinical remission following transfusion. This has been demonstrated in one AML patient[16] but the clinical effectiveness of this procedure has to be examined in further detail.

Other enzymes in the O-glycan biosynthetic pathway are significantly altered in AML and CML; the elongation β3-GlcNAc-transferase that elongates core 1 and 2 is undetectable in leukemic cells but present at very low levels in normal granulocytes. Core 1 β3 Gal-transferase, β4 Gal-transferase, and blood group i β3 GlcNAc-transferase are present at elevated levels in AML cells but the increase is less pronounced in CML.[14]

α6-Sialyltransferase I acting on O-glycans or sialyltransferases acting on N-glycans appear to be unaltered.[4,12] The β6-GlcNAc-transferase activity synthesizing the I antigen has not been detected in leukemia cells.[14]

The α2-Fuc-transferase that synthesizes the blood group H determinant is deficient in untreated AML but high in chronic granulocytic leukemia.[17,18] This H determinant has been shown to vary with cell cycle changes in normal and transformed cell lines.[19] Striking changes in the levels of serum α2 Fuc-transferase have been observed during the course of AML, and these correlate well with patterns of clinical remission and relapse. The original interpretation of that observation is that serum α2-Fuc-transferase is derived from fluctuations in the enzyme content of cell debris from normal granulocytes or AML cells. A more recent study suggests that fluctuating platelet levels correlate with serum α2-Fuc-transferase;[18] contributions from other fragmented cell sources might be difficult to assess.

N-linked glycan processing is also altered in leukemia. Elevated expression of GlcNAc-transferase III activity synthesizing bisected N-glycans (Fig. 3), and mRNA levels are found in CML blast crisis and multiple myeloma compared to chronic phase CML cells or normal granulocytes.[20] The expression of bisected N-glycans has been detected by E-PHA binding to CD45 cell surface glycoprotein. Leukocytes from patients with other types of leukemia, including AML, contain much less elevated levels. Additionally, the ratio of GlcNAc-transferase III activity to the level of a single mRNA transcript differs in the various leukemias, which indicates that the stability or activity of the enzyme may be regulated by post-translational processing.[21] GlcNAc-transferase III would be expected to downregulate branching and extension of branches and therefore reduce sialylation.

GlcNAc-transferase V activity also appears to be elevated in AML cells.[22] This increase may contribute to the cell surface hypersialylation of leukemia cells. Changes in glycosyltransferase activities in vivo utilizing UDP-GlcNAc as the donor substrate may be attributed to an imbalance in nucleotide-sugar pools, as previously observed in CML and HL60 cells.[23]

Some leukemic cells, as well as normal T- and B-lymphocytes, express the α6-sialyltransferase (N) at various levels but not the α3-sialyltransferase acting on N-glycans. The activities and linkage specificities were determined by sialyltransferase assays combined with methylation analysis.[24]

13.A.3. Cell Lines Derived from Leukemia Cells

Rossowsky and Srivastava[25] assayed Gal, Fuc- and sialyltransferases in a variety of leukemia cell lines, and in normal and leukemic cells using desialylated and degalactosylated fetuin as substrates. The results suggest that these glycosyltransferase activities are characteristic of certain leukemic cell types and cell lines and may be useful for the characterization of leukemias. The profiles of Fuc-transferases observed upon isoelectric focusing differ between leukemic and normal cells and can distinguish myeloid and lymphoid leukemias.[26] It should be recognized that Fuc-transferases are glycoproteins and may be affected by aberrant glycosylation of leukemia cells.[27,28]

Leukosialin from several human leukemia cell lines contains characteristic O-glycans with sialylated core 1 and core 2 structures, whereas promyeloid HL60 cell lines contain mainly sialylated core 2 oligosaccharides. In contrast, the erythroleukemia cell line K562 possesses primarily sialylated core 1 oligosaccharides. Differentiation of HL60 leads to a redistribution of core structures, possibly resulting from alterations in sialyltransferase activities. α3-Sialyltransferase activities are low in two differentiation-deficient HL60 cell variants whereas α6-sialyltransferase levels are similar to those in wild-type cells. α3-Sialylation may relate to abnormal growth and differentiation patterns. Leukemia derived human myeloblasts and myeloblastic cell lines all contain appreciable α3-sialyl-transferase (O) activity at different levels.[13]

The Jurkat T-cell leukemia cell line contains largely truncated O-glycans with GalNAc-Ser/Thr (Tn-antigen) structures.[29,30] This is attributed to the low activity of core 1 β3-Gal-transferase in these cells although β4-Gal-transferase is present. The presence of polypeptide GalNAc-transferase and several sialyl-transferase activities have also been demonstrated in Jurkat cells. The core 1 β3-Gal-transferase activity is not reactivated by differentiating cells with sodium butyrate[31] suggesting that this cell line has a selective defect in the enzyme.

Myeloid differentiation of HL60 cells induced by dimethylsulfoxide is associated with increases in poly-N-acetyl-lactosamine structures attached to lysosomal membrane glycoproteins and with high levels of i β3-GlcNAc-transferase but not β4-Gal-transferase.[32] GlcNAc-transferase I and V activity levels (of the N-glycan pathways) are unchanged although the proportion of biantennary N-glycans decreases with differentiation compared to more complex type chains.

Retinoic acid differentiated HL60 cells exhibit decreased GlcNAc-transferase III levels compared to the undifferentiated counterpart. Iα.25-dihydroxyvitamin D treated cells also show decreased GlcNAc-transferase III activity and an increase in GlcNAc-transferase IV.[33] This demonstrates altered regulation of N-glycan branching GlcNAc-transferases in leukemia cells, possibly due to variations in cellular differentiation status.

Sialidase and other glycosidases have been found in increased levels in HL60 cells upon differentiation by retinoic acid or dimethylsulfoxide when compared with undifferentiated HL60 cells.[34] These results suggest that hypersialylation found in leukemic cells could result from decreased sialidase activity.

Robinson et al[35] surveyed a number of cell lines by an antibody staining technique. Myeloid cell lines and the B cell line RPMI 8226 express Le^x, sialyl-Le and the VIM2 epitope (resembling sialyl-Le^x but containing an internal α3-linked Fuc residue). In contrast, two T-cell lines express only Le^x and VIM2. Of the enzymes synthesizing the Le^x antigen, β4-Gal-transferase is expressed by all cell types as well as α3-Fuc-transferase IV, based on specificity and inhibition studies as well as mRNA analysis. Several types of α3-Fuc-transferases are present in myeloid cell lines and especially in the B cell line, but only a fraction of lymphocytic lines contains this transferase activity.

Much higher levels of α3-Fuc-transferase activity are found in blast cells than in neutrophils using non-sialylated substrates. In neutrophils, the enzyme reacts with sialylated and non-sialylated acceptors; in contrast, the blast cell enzyme showed a marked preference for non-sialylated substrate.

The α6-sialyltransferase (acting on N-glycans) and α3-Fuc-transferase (towards neutral substrate) are expressed strongly in blasts in contrast to neutrophils, but α3-sialyltransferase (acting on N-glycans) is low in both cell types. Blasts, however, express lower levels of sialyl-Le^x and Le^x antigens. Dimethylsulfoxide differentiated HL60 cells contain reduced levels of α6-sialyltransferase (N) and α3-Fuc-transferase although increased levels of sialyl Le^x are expressed.[36] These observations suggest that α6-sialyltransferase (N) activity may prevent sialyl-Le^x synthesis.

Lymphomas comprise a heterogenous group of tumors of the reticuloendothelial and lymphatic systems. Murine lymphoma cell leukosialin may exhibit excessive sulfate in comparison with normal cells.[37] Sulfated chains could be released by beta-elimination but not by endoglycosidase F, indicating that mainly O-glycans and not N-glycans carry sulfate. Heterotypic adherence between murine leukemia/lymphoma cells and marrow stromal cells was reported to involve a lectin recognizing Gal and Man residues.[38]

13.A.4. Functions of Cell Surface Receptors and Growth Factors in Leukemia Cells

Hematopoietic growth factors and their receptors may be glycosylated; disease-induced alterations of glycosylation patterns may affect their biological functions. Colony stimulating factors (CSF) and their receptors are of special importance since they initiate clonal proliferation, cell division and differentiation of progenitor cells.[39] CSF and other cytokines are glycoproteins;[40] the O-glycan chains of granulocyte-CSF enhance the conformational stability of the growth factor and this directly relates to its biological activity.[41]

The more highly glycosylated form of the granulocyte-macrophage-colony stimulating factor (GM-CSF), a potent cytokine, shows lower affinity for its receptor than the less or non-glycosylated species.[41-43] The GM-CSF receptor (GM-CSF R) is a membrane glycoprotein composed of glycosylated α and β chains where the β-chain is a subunit common to other cytokine receptors.[44] The α– and β- subunits of the cloned human receptor have eleven and three potential N-glycosylation sites, respectively, and may contain an undetermined number of O-glycosylation sites.[45,46] Although the structures have not yet been determined, processing of N-glycans appears to be required for effective receptor expression on the cell surface of TF-1 erythroleukemia cells. Wheat germ agglutinin is highly effective in inhibiting binding of (^{125}I)-labeled GM-CSF to TF-1 cells. However, other GlcNAc-binding lectins and peanut lectin are not effective inhibitors of growth factor binding. These findings suggest a role for highly sialylated N- or O-glycans. Using glycoprotein synthesis inhibitors it has been shown that N-glycan processing of the α–chain, especially glucose removal by glucosidase I and II, is required for its cell surface expression. In contrast, Ding et al showed that inhibition of N-glycosylation by tunicamycin does not affect the cell surface expression of the α chain in HL60 and COS cells; however, N-glycosylation is apparently required for ligand binding and the resulting signal transduction, i.e. Tyr phosphorylation.[47] The β-chain expression is little affected by N-glycosylation inhibitors.[44]

A functional role for highly sialylated N- or O-glycans of leukemic cell surfaces is suggested by studies which demonstrate that desialylation increases binding of GM-CSF to CML granulocytes. In addition, binding of the synthetic chemotactic peptide formyl-Met-Leu-Phe to CML granulocytes is deficient but can be restored following sialidase treatment. Undifferentiated HL60 cells exhibit greater binding of GM-CSF after sialidase treatment, compared to HL60 cells differentiated with dibutyryl-cAMP.[48] The hypersialylation of leukemic cells may therefore be related to their less differentiated status, and may bear directly on their impaired functions.

The sialomucin CD34 is expressed in stem cells as well as in the vascular endothelium in mice. The functional role of CD34 in hematopoietic development has been studied by producing embryonic stem cells and knockout mice that do not express this mucin. The deficiency of CD34 causes delayed erythroid and myeloid differentiation and an impairment of colony forming ability and response to hematopoietic growth factors. These deficiencies are reversed by transfection of cells with complete or cytoplasmic domain containing CD34 constructs. However, mice develop normally and mature null mice appear to be comparable to wild-type animals, and they possess typical hematopoietic profiles.[49] The experiments suggest that CD34 plays a role in cellular maturation, but probably does so cooperatively with other cell adhesion molecules. CD34 has structural similarity to CD43.[50]

13.B. BLOOD-RELATED AND VASCULAR DISEASES

The erythrocytes of patients with permanent mixed-field polyagglutinability are characterized by undersialylated and undergalactosylated glycophorin. Erythrocytes from these patients react increasingly with anti-Tn antibodies. The cause for this abnormality is the deficiency of core 1 β3-Gal-transferase.[51,52] The deficiency results in abnormal exposure of the Tn antigen. This is probably due to persistent repression since the enzyme can be reactivated by induction of differentiation in the patient's T cells.[31] The enzyme appears to be essential for the maintenance of normal red cell membrane function.

It has been suggested that various B cell clones differ in their glycosylation potential and that selected clones proliferate in myeloma. Normal human IgG is a glycoprotein containing 2 to 3 bi-antennary N-glycans per molecule. IgG produced in certain myeloma patients contain different bi-antennary chains and there is great variability in the proportion of bisected chains.[53] Normal B but not T cells contain GlcNAc-transferase III activity[54] but this activity may vary between different types of B cells. Sialyltransferase and Fuc-transferase activities are significantly higher in the sera of multiple myeloma patients[55] but transferase levels decrease in response to treatment. Bence-Jones proteins also contain abnormal oligosaccharide structures, and in contrast to IgG, these highly sialylated chains contain sialylα2-6Gal-linkages.[56]

The availability of animal models of heart failure and subtractive hybridization technology permits screening of mRNAs for comparisons of gene expression. In a rat model, cDNA from controls and heart failure subjects was amplified by PCR, cloned and sequenced. A cDNA was found to be 86% homologous to murine β1-4Gal-transferase. The expression of Gal-transferase transcript as shown by RNA blot analysis is increased in failing hearts relative to non-failing heart controls. In addition, biochemical assays demonstrated increased Gal-transferase activity in failing hearts.[57]

Antibodies (cold autoagglutinins) against the I antigen (or other oligosaccharides) on lipids or glycoproteins may cause hemolytic anemia,[58] and may be present in immunoglobulin fractions produced by B cell neoplasms.[59]

A hemorrhagic condition has been found to be associated with a deficiency of platelet cell surface sialyltransferase acting on asialo-fetuin.[60] The total and membrane bound platelet sialic acid content appears to be normal, suggesting that a specific sialyltransferase responsible for normal platelet release may be lacking in this condition.

13.B.1. Blood Clotting Disorders

Many diseases of the cardiovascular system involve changes in the types and levels of O- or N-glycans. The major features of diseases such as Bernard Soulier syndrome, von Willebrand disease and Glanzmann's thrombasthenia are the lack of expression of glycoproteins functional in blood coagulation and platelet aggregation. The biochemical background of most diseases is not well defined (for example, Dowton thrombocytopenia) and may also involve abnormal glycosylation. Glycosylation is known to be necessary for the maintenance of healthy membranes. It is therefore anticipated that abnormal glycosylation may contribute to faulty enzyme expression, binding to ligands, cell adhesion and instability of glycoproteins and membranes leading to blood coagulation problems. Many of the blood coagulation factors are synthesized in the liver. Liver disease in which glycosyltransferases are abnormally expressed may adversely affect the functions of glycoproteins.

Inherited disorders of platelet function frequently possess defects in glycoprotein synthesis or structure.[61] Glanzmann's thrombasthenia is a rare autosomal recessive disease caused by a defect of the GPIIb/IIIa platelet membrane glycoprotein complex which results in failure of platelets to bind fibrinogen–essential to the blood clotting process.[62] The Bernard-Soulier syndrome is characterized by a deficient GPIb/IX glycoprotein complex, and a reduced sialic acid content, which is thought to contribute to a shortened platelet life span and thrombocytopenia in these patients.[63,64] Some patients with Bernard-Soulier syndrome are deficient in GPIb, the platelet receptor for vWF, and exhibit multiple bands of GPIb on PAGE. Multiple bands could arise from different glycoforms, and abnormal glycosylation may promote instability of this membrane protein. In addition, platelets exhibit reduced electrophoretic mobility, possibly due to abnormal cell surface properties. A similar reduction can be achieved by removal of sialic acid from the cell surface. Since the sialic acid content of platelets is significantly decreased, it is assumed that a deficiency of sialic acid causes the membrane defect of platelets in these patients.[63]

Platelet-type von Willebrands disease is associated with a deficiency of the higher molecular weight forms of vWF and thus impairment of binding properties; vWF is noncovalently bound to factor VIII and deficiencies of the former are reflected in reduced factor VIII levels.[65]

Acquired glycoprotein alterations in platelets may relate to a basic underlying disease state. For example, diabetes mellitus is associated with abnormalities in platelet adhesiveness, aggregation and metabolism. Histochemical and lectin staining methods have been used to compare the carbohydrate content and distribution in diabetic and normal platelets, and have revealed that the distribution of anionic sites on the glycocalyx is clustered in diabetes but even in normals. In diabetic platelets WGA (sialic acid) binding sites are 1.5-fold lower and RCA (Gal) sites are 2.5-fold higher as related to normal platelets.[66] Non-enzymatic glycation of plasma proteins and of platelets and endothelium is one of the most important molecular alterations in the pathogenesis of diabetic retinopathy.[67]

A dysfibrinogenemia characterized by an abnormality of fibrin monomer polymerization has been described in patients with liver disease. The sialic acid content of patient fibrinogen samples is considerably higher than that found in controls. A progressive increase in blood clotting time was directly correlated with increasing sialic acid content of fibrinogen. Removal of sialic acid from patient platelets normalizes blood clotting times, suggesting that the hypersialylation is the cause of this functional defect in patients' fibrinogen.[68]

13.B.2. Wiscott Aldrich Syndrome

Blood-related abnormalities are manifested in immunodeficiency diseases. Wiscott-Aldrich syndrome (WAS) is an X-linked recessive disorder characterized by immunodeficiency, which leads to death at an early age. Reduced platelet counts are associated with abnormal functions. Lymphocytes from patients with WAS are abnormal in their leukosialin expression and apparently lack the normal maturation mechanism that is accompanied by O-glycan changes.[69,70] O-glycans of sialylated core 2 structure found on T-lymphocytes from WAS patients are similar to those normally found on activated lymphocytes but are present in increased amounts.[71] These structural changes are accompanied by an increase in core 2 β6-GlcNAc-transferase in lymphocytes and platelets (Fig. 1). When WAS lymphocytes are subjected to further activation, core 2 β6-GlcNAc-transferase activity decreases. When lymphocytes are activated with anti CD3 antibody and interleukin-2, the level of enzyme activity decreases in patients but increases in normal lymphocytes.[71] This suggests that the enzyme is inappropriately activated in Wiscott Aldrich patients and may possibly serve as a diagnostic marker for the disease (Table 5).[72] These results are consistent with a model of cyclic core 2 β6-GlcNAc-transferase activity that exhibits a peak at a certain stage of lymphocyte differentiation.

References

1. Kuhns W, Primus F. Alteration of blood groups and blood group precursors in cancer. Prog Clin Biochem Med 1985; 2:49-98.
2. Shumak K, Beldotti L, Rachkewich R. Diagnosis of hematological disease using anti-i. Brit J Hematol 1979; 41:399-405.
3. Baker MA, Taub RN, Whelton CH et al. Aberrant sialylation of granulocyte membranes in chronic myelogenous leukemia. Blood 1984; 63:1194-1197.
4. Baker MA, Taub RN, Kanani A et al. Increased activity of a specific sialyltransferase in chronic myelogenous leukemia. Blood 1985; 66:1068-1071.
5. Taub R, Baker M, Madyastha K. Masking of neutrophil surface lectin binding sites in chronic myelogenous leukemia (CML). Blood 1980; 55:294-303.
6. Baker MA, Kanani A, Hindenberg A et al. Changes in the granulocyte membrane following chemotherapy for chronic myelogenous leukemia. Br J Hematol; 1986; 62:431-438.
7. Fukuda M, Bothner B, Ramsamoo P et al. Structures of sialylated fucosyl polylactosaminoglycans isolated from chronic myelogenous leukemia cells. J Biol Chem 1985; 260:12957-12967.
8. Fukuda M, Carlsson S, Klock J et al. Structures of O-linked oligosaccharides isolated from normal granulocytes, chronic myelogenous leukemia cells, and acute myelogenous leukemia cells. J Biol Chem 1986; 261:12796-12806.
9. Muroi K, Suda T, Nojiri H et al. Reactivity profiles of leukemic myeloblasts with monoclonal antibodies directed to sialosyl Le x and other lacto series type 2 chain antigens:absence of reactivity with normal hematopoietic progenitor cells. Blood 1992; 79:713-719.
10. Carlsson S, Sasaki H, Fukuda M. Structural variations of O-linked oligosaccharides

present in leukosialin isolated from erythroid, myeloid, and T-lymphoid cell lines. J Biol Chem 1986; 261:12787-12795.

11. Saitoh O, Piller F, Fox R et al. T-lymphocytic leukemia expresses complex, branched O-linked oligosaccharides on a major sialoglycoprotein, leukosialin. Blood 1991; 77:1491-1499.
12. Baker MA, Kanani A, Brockhausen I et al. Presence of cytidine 5'-monophospho-N-acetylneuraminic acid:Galβ1-3GalNAc-R α(2-3)-sialyltransferase in normal human leukocytes and increased activity of this enzyme in granulocytes from chronic myelogenous leukemia patients. Cancer Res 1987; 47:2763-2766.
13. Kanani A, Sutherland DR, Fibach E et al. Human leukemic myeloblasts and myeloblastoid cells contain the enzyme cytidine 5'-monophosphate-N-acetylneuraminic acid:Galβ1-3GalNAc α(2-3)-sialyltransferase. Cancer Res 1990; 50:5003-5007.
14. Brockhausen I, Kuhns W, Schachter H et al. Biosynthesis of O-glycans in leukocytes from normal donors and from patients with leukemia:increase in O-glycan core 2 UDP-GlcNAc:Galβ1-3GalNAcα–R (GlcNAc to GalNAc) β(1-6)-N-acetylglucosaminyltransferase in leukemic cells. Cancer Res 1991; 51:1257-1263.
15. Kuhns W, Rutz V, Paulsen H et al. Processing O-glycan core 1, Galβ1-3GalNAcα-R. Specificities of core 2, UDP-GlcNAc: Galβ1-3GalNAc-R(GlcNAc to GalNAc) β-6-N acetylglucosaminyltransferase and CMP-sialic acid:Galβ1-3GalNAc-R α3-sialyltransferase. Glycoconj J 1993; 10:381-394.
16. Ball ED, Vredenburgh JJ, Mills LE et al. Autologous bone marrow transplantation for acute myeloid leukemia following in vitro treatment with neuraminidase and monoclonal antibodies. Bone Marrow Transplant 1990; 6:277-284.
17. Kuhns W, Oliver R, Watkins W et al. Leukemia-induced alterations of serum glycosyltransferase enzymes. Cancer Res 1980; 40:268-275.
18. Skacel P, Watkins W. Significance of altered α-2-L-fucosyltransferase levels in serum of leukemic patients. Cancer Res 1988; 48:3998-4001.
19. Kuhns W, Pann C. Differentiation of HeLa cells with respect to blood group H antigen. Nature New Biology 1972; 96:22-24.
20. Yoshimura M, Nishikawa A, Ihara Y et al. High expression of UDP-N-acetylglucosamine:β-D-mannoside β-1,4-N-acetylglucosaminyltransferase III (GnT III) in chronic myelogenous leukemia in blastic crisis. Int J Cancer 1995; 60:443-449.
21. Yoshimura M, Ihara Y, Taniguchi N. Changes of β-1,4-N-acetyl-glucosaminyltransferase III (GnT-III) in patients with leukaemia. Glycoconj J 1995; 12:234-240.
22. Brockhausen I, Reck F, Kuhns W et al. Substrate specificity and inhibition of UDP-GlcNAc:GlcNAcβ1-2Manα1-6R β1,6-N-acetylglucosaminyl-transferase V using synthetic substrate analogues. Glycoconj J 1995; 12:371-379.
23. de Korte D, Haverkort WA, de Boer M et al. Imbalance in the nucleotide pools of myeloid leukemia cells and HL-60 cells: correlation with cell-cyle phase, proliferation, differentiation, and transformation. Cancer Res 1987; 47:1841-1847.
24. Augener W, Brittinger G, Schiphorst WECM et al. Activities and specificities of N-acetylneuraminyltransferases in leukemic cells. In:Neth R, Gallo RC, Greaves MF, Moore MAS, Winkler K eds. Hematology and Blood Transfusion. Vol. 28. Modern Trends in human leukemia V. Springer Verlag 1983:135-138.
25. Rossowsky W, Srivastava B. Glycosyl transferase activities in leukemia cells from patients and human leukemic cell lines. Eur J Cancer Clin Oncol 1983; 19:1431-1437.
26. Suda K, Sakamoto S, Hida K et al. Electrofocusing pattern of fucosyltransferase activity in human leukemic cells. Cancer Res 1987; 47:2782-2786.
27. Sarnesto A, Köhlin T, Hindsgaul O et al. Purification of the secretor-type β-galactoside α1-2 fucosyltransferase from human serum. J Biol Chem 1992; 267:2737-2744.
28. Sarnesto A, Köhlin T, Hindsgaul O et al. Purification of the β-N-acetylglucosiminide α1-3 fucosyltransferase from human serum. J Biol Chem 1992; 267:2745-2752.
29. Piller V, Piller F, Fukuda M. Biosynthesis of truncated O-glycans in the T cell line

Jurkat. J Biol Chem 1990; 265:9264-9271.
30. Inoue M, Nakada H, Tanaka N et al. Tn antigen is expressed on leukosialin from T-lymphoid cells. Cancer Res 1994; 54:85-88.
31. Thurnher M, Rusconi S, Berger EJ. Persistent repression of a functional allele can be responsible for galactosyltransferase deficiency in Tn syndrome. J Clin Invest 1993; 91:2103-2110.
32. Lee N, Wang W, Fukuda M. Granulocytic differentiation of HL60 cells is associated with increase of poly-N-acetyllactosamine in Asn-linked oligosaccharides attached to human lysosomal membrane glycoproteins. J Biol Chem 1990; 265:20476-20487.
33. Koenderman A, Wijermans P, van den Eijnden D. Changes in the expression of N-acetylglucosaminyltransferase III, IV, V associated with the differentiation of HL60 cells. FEBS Lett 1987; 222:42-46.
34. Nojiri H, Takaku F, Tetsuka T et al. Stimulation of sialidase activity during cell differentiation of human promyelocytic cell line HL60. Biochem Biophys Res Comm 1982; 104:1239-1246.
35. Robinson N, de Vries T, Davis R et al. Expression of fucosylated antigens and α1-3 fucosyltransferases in human leukemia cell lines. Glycobiology 1994; 4:317-326.
36. Skacel P, Edwards A, Harrison C et al. Enzymic control of the expression of the X determinant (CD15) in human myeloid cells during maturation:the regulatory role of 6'-sialyltransferase. Blood 1991; 78:1452-1460.
37. Wilson A, Rider C. Evidence that leukosialin, CD43, is intensely sulfated in the murine T lymphoma line RDM-4. J Immunol 1992; 148:1777-1783.
38. Juneja H, Schmalsteig F, Rajaraman S et al. Heterotypic adherence between murine leukemia/lymphoma cells and marrow stromal cells involves a recognition mechanism with galactosyl and mannosyl specificities. Exp Hematol 1992; 20:405-411.
39. Metcalf D. The granulocyte-macrophage colony stimulating factors. Cell 1985; 43:5-6.
40. Katlinsky A, Petrov R, Mikhailova A et al. Purification and properties of a 155 kDa bone marrow derived glycoprotein enhancing the activity of granulocyte-macrophage colony stimulating factor (GM-CSF). FEBS Lett 1993; 320:67-70.
41. Oh-eda M, Hasegawa M, Hattori K et al. O-linked sugar chain of human granulocyte colony stimulating factor protects it against polymerization and denaturation allowing it to retain its biological activity. J Biol Chem 1990; 265:11432-11435.
42. Cebon J, Nicola N, Ward M et al. Granulocyte-macrophage colony stimulating factor from human lymphocytes:effect of glycosylation on receptor binding and biological activity. J Biol Chem 1990; 265:4483-4491.
43. Moonen P, Mermod J, Ernst J et al. Increased biological activity of deglycosylated recombinant human granulocyte/macrophage colony-stimulating factor produced by yeast or animal cells. Proc Natl Acad Sci USA 1987; 84:4428-4431.
44. Shibuya K, Chiba S, Miyagawa K et al. Structural and functional analyzes of glycosylation on the distinct molecules of human GM-CSF receptors. Eur J Biochem 1991; 198:659-666.
45. Hayashida K, Kitamura T, Gorman D et al. Molecular cloning of a second subunit of the receptor for the human GM-CSF:Reconstitution of a high affinity GM-CSF receptor. Proc Natl Acad Sci USA 1990; 87:9655-9659.
46. Gearing D, King J, Gough N et al. Expression cloning of a receptor for human granulocyte/macrophage colony stimulating factor. EMBO J 1989; 8:3667-3676.
47. Ding DXH, Vera JC, ML Heaney et al. N-glycosylation of the human granulocyte-macrophage colony-stimulating factor receptor α subunit is essential for ligand binding and signal transduction. J Biol Chem 1995; 270:24580-24584.
48. Cyopick P, Culliton R, Brockhausen I et al. Role of aberrant sialylation of chronic myeloid leukemia granulocytes on binding and signal transduction by chemotactic peptides and colony stimulating factors. Leuk and Lymph 1993; 11:79-90.
49. Cheng J, Baumhueter S, Cacalano G et al. Hematopoietic defects in mice lacking the sialomucin CD34. Blood 1996; 87:479-490.
50. Simmons D, Satterthwaite A, Tenen D et

al. Molecular cloning of a cDNA encoding CD34, a sialomucin of hematopoietic stem cells. J Immunol 1992; 148:267-271.
51. Berger EG, Kozdrowski I. Permanent mixed-field polyagglutinable erythrocytes lack galactosyltransferase activity. FEBS Lett 1978; 93:105-108.
52. Cartron J, Andrev J, Cartron J et al. Demonstration of T-transferase deficiency in Tn-polyagglutinate blood samples. Eur J Biochem 1978; 92:111-119.
53. Kobata A. Function and pathology of the sugar chains of human immunoglobulin G. Glycobiology 1990; 1:5-8.
54. Narasimhan S, Lee J, Cheung R et al. β-1,4-mannosyl glycoprotein β-1,4-N-acetylglucosaminyl transferase III in human B and T lymphocyte lines and in tonsillar B and T lymphocytes. Biochem Cell Biol 1988; 66:889-900.
55. Frithz G, Ronquist G, Ericsson P. Serum sialyltransferase and fucosyltransferase activities in patients with multiple myeloma. Eur J Clin Oncol 1985; 21:913-917.
56. Okhura T, Isobe T, Yamashita K et al. Structures of the carbohydrate moieties of two monoclonal human λ-type immunoglobulin light chains. Biochemistry 1985; 24:503-508.
57. Humphries DE, Sirokman G, Bing OHL. Enhanced galactosyltransferase expression in the failing hearts of spontaneously hypertensive rats. Biochem Biophys Res Comm 1996; 218:320-324.
58. Pirofsky B. In:Rose N, Mackay I eds. The Autoimmune Diseases. Academic Press, Orlando 1985:469-491.
59. Crisp D, Pruzanski W. B-cell neoplasms with homogeneous cold-reacting antibodies (cold agglutinins). Am J Med 1982; 72:915-922.
60. Wu KK, Ku CSL, Chen Y-C. Reduced platelet sialyltransferase activity in patients with primary release disorder. Lancet 1980; 2:440-443.
61. Hardisty R. Disorders of platelets. II Functional abnormalities. In: Lilleyman J, Hann I. eds. Pediatric Hematology. Edinburgh: Churchill Livingstone, 1992:167-199.
62. Peerschke EI, Zucker MB, Grant RA et al. Correlation between fibrinogen binding to human platelets and platelet aggregability. Blood 1980; 55:841-847.
63. Gröttum KA, Solum NO. Congenital thrombocytopenia with giant platelets:a defect in the platelet membrane. Brit J Hemat 1969; 16:277-289.
64. Berndt MC, Gregory C, Chong BH et al. Additional glycoprotein defects in Bernard-Soulier's syndrome:confirmation of genetic basis by parental analysis. Blood 1983; 62:800-807.
65. Bithell T. Disorders of hemostasis and coagulation. In:Lee GR, Bithell TC, Foerster J, Athens JW, Lukens JN. eds. Wintrobe's Clinical Hematology 9th ed. Vol 2. Philadelphia:Lea and Febiger, 1993:1432-1433.
66. Lupu C, Rizescu M, Calb M. Altered distribution of some surface glycosaminoglycans and glycoconjugates on human platelets in diabetes mellitus. Platelets 1994; 5:201-208.
67. Sampietro T, Lenzi S, Cecchetti P et al. Nonenzymatic glycation of human platelet membrane proteins in vitro and in vivo. Clin Chem 1986; 32:1328-1331.
68. Martinez J, Palascak JE, Kwasniak D. Abnormal sialic acid content of the dysfibrinogenemia associated with liver disease. J Clin Invest 1978; 61:535-538.
69. Piller F, Le Deist F, Weinberg KI et al. Altered O-glycan synthesis in lymphocytes from patients with Wiscott-Aldrich syndrome. J Exp Med 1991; 173:1501-1510.
70. Higgins E, Siminovitch K, Zhuang D et al. Aberrant O-linked oligosaccharide biosynthesis in lymphocytes and platelets from patients with the Wiskott-Aldrich syndrome. J Biol Chem 1991; 266:6280-6290.
71. Piller F, Piller V, Fox R et al. Human T-lymphocyte activation is associated with changes in O-glycan biosynthesis. J Biol Chem 1988; 263:15146-15150.
72. Datti A, Orlaccio A, Siminovitch KA et al. Core 2 N-acetyl-glucosaminyltransferase activity:a diagnostic marker for Wiskott-Aldrich syndrome. Glycosyl Dis 1994; 1:127-135.

CHAPTER 14

Glycosylation in Cystic Fibrosis

Cystic fibrosis (CF) is characterized by increased viscosity of secreted mucus and clogging of exocrine gland ducts with mucosal debris.[1] There is an increased susceptibility to infections in the lungs, particularly to *Pseudomonas aeruginosa* and *Pseudomonas cepacia* strains, contributing to the high mortality.[2,3] Intestinal blockage by mucus, i.e. meconeum ileus, may occur in affected infants. The protein that is abnormal in CF is the cystic fibrosis transmembrane conductance regulator (CFTR), a chloride transport channel. The CFTR gene has been cloned and expressed.[4,5]

14.A. CFTR

A number of mutants of the CFTR protein have been reported in CF patients. The most common mutation is a replacement of phenylalanine at amino acid 508. The immature form of the mutant δF 508 CFTR protein associates with the molecular chaperone protein calnexin in the ER due to misfolding or incomplete processing and this may lead to a mislocalization of mutant CFTR and lack of function.[6,7] Thus, certain mutants may be inactive as transporter molecules and others may be active but do not reach the proper cell surface location.

It is likely that a deficiency of chloride transport is the basis of biochemical or functional mucin abnormalities noted in CF. Recent studies support a direct role for CFTR in mucin secretion by columnar epithelial cells. CFTR has been overexpressed in dog gallbladder cells using an appropriate vector containing CFTR cDNA. Mucin secretion was measured by the incorporation of N-(^{3}H)-acetyl-D-glucosamine into secreted glycoproteins. Secretion was found to be greatly elevated compared with controls in cells overexpressing CFTR. Inhibition of CFTR-mediated chloride transport was found to markedly diminish mucin secretion in the dog gallbladder model.[8] These results support the hypothesis that CF pathogenesis lies in a CFTR-related defect in the mechanism of secretion.

CFTR is an N-glycosylated protein. The glycosylation status of endogenous CFTR does not appear to affect membrane targeting or cAMP stimulated secretion of chloride in epithelial cells as suggested from experiments using glycosylation inhibitors.[9]

14.B. OLIGOSACCHARIDE STRUCTURES OF CYSTIC FIBROSIS MUCINS

A number of mucin abnormalities have been reported in CF. It is not known how glycosylation or sulfation abnormalities are related to the defective chloride transport in CF, and if these phenomena are secondary to the disease process. CF mucins secreted from various tissues are highly sulfated.[10-14] The structures of normal mucins, however, have not been sufficiently studied for comparison. The level of sulfation appears to increase with severity of the disease (Table 4).[15,16]

Increased membrane transport of sulfate in CF is thought to account for these changes and it is also possible that high mucin sulfation is directly related to the CFTR defect in CF.[17,18] This is supported by the finding that elevated cellular sulfate could be normalized in a pancreatic ductal adenocarcinoma cell line from a CF patient by retrovirus-mediated CFTR gene transfer.[19]

Sulfated mucin oligosaccharides are relatively abundant in CF respiratory mucins and have been shown to contain core 2 structures with blood group H and Le^x determinants, α2-3-linked sialic acid and sulfate attached to the 3-hydroxyl of Gal or the 6-hydroxyl of GlcNAc residues.[20] Other CF respiratory mucin oligosaccharides contain core 1, 2, 3 and 4 structures which are elongated and terminated similarly to those from bronchiectasis patients.[21-23] O-glycans of respiratory mucin from bronchiectasis patients also contain core 1 and 2 core structures with blood group H and Le^x determinants as well as α2-3-linked sialic acid.[24]

A human bronchial xenograft model has been helpful in evaluating pathophysiologic differences between CF and non-CF airway epithelium, in the absence of secondary effects such as goblet cell hyperplasia. The extent of sulfation within secreted mucus glycoproteins has been found to be significantly higher in CF than in non-CF derived tissue. Allelic variation within the CF genotype appears to be correlated with levels of sulfation.[14] In general, CF respiratory mucins compared to non-CF mucins contain more sugar residues per carbohydrate chain, which could be due to overproduction of carbohydrate chains.[14,25] The appearance of glycoprotein and mucin epitopes may originate from host inflammatory airway responses to bacterial colonization.[26] Le^a epitopes are highly expressed in CF patients with severe pulmonary disease and may serve to monitor the response to antibiotic therapy.

Salivary mucin oligosaccharide structures have been compared between one CF and one normal individual.[27] The major O-glycans are fucosylated and sialylated core 1 structures and minor structures with Fuc α1-2 and Fuc α1-3 and sialyl α2-3 residues attached to core 2 structures. The patterns of neutral and sialic acid containing oligosaccharides are apparently similar for normal and CF mucins.

CF fibroblasts exhibit increased fucosylation of membrane N-linked glycoproteins when compared with normal cells.[28,29] The Le^x determinant is present in glycopeptides from CF fibroblasts but not from normal fibroblasts. The increase in fucosylation is not a general phenomenon but is characteristic of specific glycoproteins which may be redistributed in CF. It is also possible that increased fucosylation results from enhanced Gal-transferase activity, thereby providing more substrate for fucosylation.[30]

Mouse mammary epithelial cells which express mutant CFTR (δF 508) contain significantly decreased amounts of sialylated glycoproteins and gangliosides on the cell surface, as measured by lectin binding and cholera toxin B subunit binding, respectively, whereas cells expressing normal CFTR exhibit normal sialylation. This suggests that δF 508 perturbs glycoprotein processing in the Golgi.[31]

Non-mucin glycoproteins may also be aberrantly glycosylated in CF. CF-related reduction in the sialylation and galactosylation of IgG may contribute to higher IgG levels due to reduced hepatic clearance and the likelihood of immune complex formation.[32]

14.C. BIOSYNTHESIS OF MUCINS IN CYSTIC FIBROSIS CELLS

Reduced sialyltransferase activity towards desialylated α2-macroglobulin has been reported in the plasma and fibroblasts of CF patients compared with normal controls.[33] Other sialyltransferase activities such as those towards asialo-fetuin and asialo-ovine submaxillary mucin are decreased in CF liver homogenates to about 40% of the average normal activity levels. This is apparently caused by generalized liver pathology.[34] Increases in fucosylation relative to sialylation of mucins secreted from various CF tissues and epithelial cells has been reported by several groups.[14,35-37] Hyperactive microsomal Gal-, Fuc- and Man-transferases have been observed in lungs from CF patients. These enhanced transferase activities may be secondary to hypersecretion characteristic of CF.[38]

14.D. BACTERIAL INFECTIONS IN CYSTIC FIBROSIS PATIENTS

CF mucins provide important substrates for microbial-host interactions. *Pseudomonas aeruginosa* and *P. cepacia* are two bacterial strains that cause virulent lung infections in CF and both strains adhere to human respiratory mucins. Asialo carbohydrate chains may function as specific receptors for *Pseudomonas aeruginosa* bacteria.[39] *P. aeruginosa* recognizes type I and II chain disaccharides, Galβ1-3/4GlcNAc, via a non-pilus adhesion protein.[40] *Pseudomonas cepacia* binding to buccal epithelial cells is inhibited by CF respiratory and intestinal normal mucin. Chemical deglycosylation experiments revealed that GlcNAc and GalNAc are important constituents of the receptors for *P. cepacia.* The *Pseudomonas* response to colonization involves the expression of its mucoid phenotype which produces copious quantities of alginate, an extracellular polysaccharide. This protects the bacteria from phagocytosis, reduces ciliary movements and obstructs lungs because of the large hydrated volume of the polysaccharide. In addition, alginate interacts with mucin glycoproteins in the presence of Ca^{2+} which leads to increased elasticity and viscosity of secretions.[41]

14.E. ANIMAL MODELS FOR STUDIES OF CYSTIC FIBROSIS

Previously, reserpinized rats were used as a model for CF. Treated rats show mucus abnormalities resembling those found in CF.[42] Reserpine administration causes an increase in adrenergic receptors and mucus production and secretion. Forstner et al[43] have shown that intestinal mucins from reserpinized rats are highly sulfated associated with higher intestinal sulfotransferase activities.

A CF mouse model ($Cftr^{-/-}$) has recently been developed based on gene targeting techniques.[44,45] Special diets are necessary to maintain such mice for periods over 100 days.[46] $Cftr^{-/-}$ mice develop abnormal intestinal histology including deep crypts, epithelial flattening and increased secretion from goblet cells. A decreased intensity of polysaccharide-specific staining suggests that mucins are chemically abnormal. No pathological lesions have been detected in the lungs. The $Cftr^{-/-}$mouse model may be of importance to study the relationship of CFTR dysfunction to mucin abnormalities, to determine if lung disease in CF is the result of intrinsic or environmental factors, and to assess potential therapeutic agents.

References

1. Wood R, Boat T, Doershuk C. Cystic fibrosis. Am Rev Resp Dis 1976; 113:833-878.
2. Sajjan U, Corey M, Karmali M et al. Binding of nonmucoid Pseudomonas aeruginosa to normal human intestinal mucin and respiratory mucin from patients with cystic fibrosis. J Clin Invest 1992; 89:657-665.
3. Sajjan U, Corey M, Karmali M et al. Binding of Pseudomonas cepacia to normal human intestinal mucin and respiratory mucin from patients with cystic fibrosis. J Clin Invest 1992; 89:648-656.

4. Rommens J, Iannuzzi M, Kerem B et al. Identification of the cystic fibrosis gene: chromosome walking and jumping. Science 1989; 245:1059-1065.
5. Quinton P. Cystic fibrosis: a disease in electrolyte transport. FASEB J 1990; 4:2709-2717.
6. Pind S, Riordan J, Williams D. Participation of the endoplasmic reticulum chaperone calnexin (p88, IP90) in the biogenesis of the cystic fibrosis transmembrane conductance regulator. J Biol Chem 1994; 269:12784-12788.
7. Demolombe S, Baro I, Laurent M et al. Abnormal subcellular localization of mutated CFTR protein in a cystic fibrosis epithelial cell line. Eur J Cell Biol 1994; 65:214-219.
8. Kuver R, Ramesh N, Savard C et al. Mucin secretion linked to CFTR expression in epithelial cells. Pediat Pulmonol 1994; 16:244.
9. Morris A, Cunningham S, Benos D et al. Glycosylation status of endogenous CFTR does not affect cAMP stimulated Cl secretion in epithelial cells. Am J Physiol 1993; 265:C688-694.
10. Cheng P, Boat T, Cranfill K et al. Increased sulfation of glycoconjugates by cultured nasal epithelial cells from patients with cystic fibrosis. J Clin Invest 1989; 84:68-72.
11. Mawhinney T, Adelstein E, Morris D et al. Structure determination of five sulfated oligosaccharides derived from tracheobronchial mucus glycoproteins. J Biol Chem 1987; 262:2994-3001.
12. Roussel P, Lamblin G, Degand P et al. Heterogeneity of the carbohydrate chains of sulfated bronchial glycoproteins isolated from a patient suffering from cystic fibrosis. J Biol Chem 1975; 250:2114-2122.
13. Lamblin G, Rahmoune H, Wieruszeski J-M et al. Structure of two sulfated oligosaccharides from respiratory mucins of a patient suffering from cystic fibrosis. Biochem J 1991; 275:199-206.
14. Wesley AW, Forstner JF, Forstner GG. Structure of intestinal-mucus glycoprotein from human post-mortem or surgical tissue:inferences from correlation analyzes of sugar and sulfate composition of individual mucins. Carbohydr Res 1983; 115:151-163.
15. Chace K, Leahy D, Martin R et al. Respiratory mucous secretions in patients with cystic fibrosis:relationship between levels of highly sulfated mucin component and severity of the disease. Clin Chim Acta 1983; 132:142-155.
16. Zhang Y, Doranz B, Yankaskas JR et al. Genotypic analysis of respiratory mucous sulfation defects in cystic fibrosis. J Clin Invest 1995; 96:2997-3004.
17. Cheng P, Boat T, Cranfill K et al. Increased sulfation of glycoconjugates by cultured nasal epithelial cells from patients with cystic fibrosis. J Clin Invest 1989; 84:68-72.
18. Elgavish A, Meezan E. Increased sulfate uptake in skin fibroblasts isolated from cystic fibrosis patients. Biochem Biophys Res Comm 1988; 152:99-106.
19. Elgavish A, Meezan E. Altered sulfate transport via anion exchange in CFPAC is corrected by retrovirus-mediated CFTR gene transfer. Am J Physiol 1992; 253:C176-C186.
20. Lo-Guidice J, Wieruszeski J, Lemoine J et al. Sialylation and sulfation of the carbohydrate chains in respiratory mucins from a patient with cystic fibrosis. J Biol Chem 1994; 269:18794-18813.
21. Breg J, Van Halbeek H, Vliegenthart JF et al. Primary structure of neutral oligosaccharides derived from respiratory-mucus glycoproteins of a patient suffering from bronchiectasis, determined by combination of 500-MHz 1H-NMR spectroscopy and quantitative sugar analysis. 2. Structure of 19 oligosaccharides having the GlcNAc beta (1-3)GalNAc-ol core (type 3) or the GlcNAc beta (1-3) [GlcNAc beta (1-6)]GalNAc-ol core (type 4). Eur J Biochem 1988; 171:643-654.
22. Lamblin G, Boersma A, Klein A et al. Primary structure determination of five sialylated oligosaccharides derived from bronchial mucus glycoproteins of patients suffering from cystic fibrosis. The occurrence of the NeuAcα(2 → 3)Galβ(1 → 4)[Fucα(1 → 3)]GlcNAcβ(1 → •) structural element revealed by 500-MHz ^{1}H NMR spectroscopy. J Biol Chem 1984; 259:9051-9058.
23. van Halbeek H, Dorland L, Vliegenthart J

et al. Primary-structure determination of fourteen neutral oligosaccharides derived from bronchial mucus glycoproteins of patients suffering from cystic fibrosis, employing 500 MHz H NMR spectrscopy. Eur J Biochem 1982; 127:7-20.

24. van Halbeek H, Breg J, Vliegenthart J et al. Isolation and structural characterization of low-molecular-mass monosialyl oligosaccharides derived from respiratory mucus glycoproteins of a patient suffering from bronchiectasis. Eur J Biochem 1988; 177:443-460.
25. Carnoy C, Ramphal R, Scharfman A et al. Altered carbohydrate composition of salivary mucins from patients with cystic fibrosis and the adhesion of Pseudomonas aeruginosa. Am J Respir Cell Molec Biol 1993; 9:323-334.
26. Kane R, Penny J, Walker K et al. Changes in the CA 19-9 antigen and Lewis blood group with pulmonary disease severity in cystic fibrosis. Pediat Pulmonol 1992; 12:221-226.
27. Reddy M, Levine M, Prakobphol A. Oligosaccharide structures of the low-molecular-weight salivary mucin from a normal individual and one with cystic fibrosis. J Dent Res 1985; 64:33-36.
28. Scanlin T, Wang Y, Glick M. Altered fucosylation of membrane glycoproteins from cystic fibrosis fibroblasts. Pediat Res 1985; 19:368-374.
29. Wang Y, Hare T, Won B et al. Additional fucosyl residues on membrane glycoproteins but not a secreted glycoprotein from cystic fibrosis fibroblasts. Clin Chim Acta 1990; 188:193-210.
30. Rao GJS, Spells G, Nadler HL. Enhanced UDP-galactose:glycoprotein galactosyl transferase activity in cultivated skin fibroblasts from patients with cystic fibrosis and its possible relationship to the pathogenesis of the disease. Pediat Res 1977; 11:981-985.
31. Dosanjh A, Lencer W, Brown D et al. Heterologous expression of delta F508 CFTR results in decreased sialylation of membrane glycoconjugates. Am J Physiol 1994; 266:C360-C366.
32. Margolies R, Boat T. The carbohydrate content of IgG from patients with cystic fibrosis. Pediat Res 1983; 17:931-935.
33. Ben-Yoseph Y, DeFranco CL, Nadler HL. The metabolism of sialic acid in cystic fibrosis. Pediat Res 1981; 15:839-842.
34. Alhadeff J, Cimino G. Cystic fibrosis liver sialyltransferase. Clin Genet 1978; 13:207-212.
35. Dische Z, Pallavicini C, Kavasaki H et al. Influence of the nature of the secretory stimulus on the composition of the carbohydrate moiety of glycoproteins of the submaxillary saliva. Arch Biochem 1962; 97:459-469.
36. Clamp JR, Gough M. Study of the oligosaccharide units from mucus glycoproteins of meconium from normal infants and from cases of cystic fibrosis with meconium ileus. Clin Sci 1979; 57:445-451.
37. Lazatin J, Glick M, Scanlin T. Fucosylation in cystic fibrosis airway epithelial cells. Glycosl Dis 1994; 1:263-270.
38. Louisot P, Levrat C. A new pathogenic hypothesis for cystic fibrosis: hyperactivity of glycosyltransferases at microsomic level. Clin Chim Acta 1973; 48:373-376.
39. Garber N, Guempel U, Belz A et al. On the specificity of the D-galactose-binding lectin (PA-1) of Pseudomonas aeruginosa and its strong binding to hydrophobic derivatives of D-galactose and thiogalactose. Biochem Biophys Acta 1992; 1116:331-333.
40. Ramphal R, Carnoy C, Fiebre S et al. Pseudomonas aeruginosa recognizes carbohydrate chains containing type 1(Glcβ1-3GlcNAc) or type 2(Galβ1-4GalNAc) disaccaride units. Infect Immun 1991; 59:700-704.
41. Zach M. Lung disease in cystic fibrosis-an updated concept. Pediat Pulmonol 1990; 8:188-202.
42. Forstner J, Maxwell B, Roomi N. Intestinal secretion of mucus in chronically reserpine treated rats. Am J Physiol 1981; 241:G443.
43. Forstner J, Roomi N, Khorasani R et al. Effect of reserpine on the histochemical and biochemical properties of rat intestinal mucin. Exp Mol Path 1991; 54:129-143.
44. Snouwaert J, Brigman K, Latour A et al. An animal model for cystic fibrosis made by gene targeting. Science 1992; 257:1083-1088.

45. Dorin J, Dickinson P, Alton E et al. Cystic fibrosis in the mouse by targeted insertional mutagenesis. Nature 1992; 359:211-215.

46. Kent G, Oliver M, Foskett K et al. Phenotypic abnormalities in long-term surviving cystic fibrosis mice. Pediat Res 1996; 40:1-9.

CHAPTER 15

Inflammatory Diseases

Inflammation may be acute or chronic, and may occur as a primary event or secondary to another disease process. Common to all forms of inflammation is endothelial injury followed by recruitment of phagocytes and immunocytes. Carbohydrates are key to the ensuing cellular response. Sialyl-Le^x has been recognized as a critical cell surface ligand in inflammatory reactions since it can bind to selectins that participate in early leukocyte-endothelial interactions[1] and may also participate in integrin-mediated interactions. E-selectin is induced on endothelial cells in a variety of acute inflammatory conditions such as tonsillitis, appendicitis and delayed hypersensitivity reactions and its appearance correlates with the extravasation of neutrophils. P-selectin is expressed in Grave's disease and rheumatoid arthritits.

In leukocyte adhesion deficiencies, sialyl-Le^x and $Lewis^x$ antigens are absent and recurrent infections are common.[2,3] Le^x is expressed on carcinoembryonic antigen-related glycoproteins which are crossreacting with anti CEA antibody.[4] The expression of Le^x is reduced in inflammation suggesting that glycosylation changes accompany this process. In contrast, $alpha_1$-acid glycoprotein changes glycosylation during acute inflammation after surgery or trauma and the N-glycans exhibit more sialyl-Le^x and higher branching (Table 4).[5] The availability of dolichol-derivatives as precursors for N-glycosylation increases during inflammation in rats and in dexamethasone-stimulated hepatocytes in culture.[6] It is not certain, however, how this relates to the completeness of N-glycosylation.

Sialylated oligosaccharides have anti-inflammatory effects in several conditions. In an IgG immune complex animal model of lung injury, which is E-selectin dependent, sialyl-Le^x preparations provide dose-dependent protection as assessed by diminished tissue neutrophils, changes in vascular permeability and hemmorhage. Protection by sialyl-Le^x compounds is not afforded against lung injury induced by IgA immune complexes which is not E-selectin dependent.[7] In a rat model of traumatic shock characterized by severe hypotension and circulatory collapse, intravenous sialyl-Le^x improved survival by reducing intestinal neutrophil accumulation and diminishing shock-induced tissue injury.[8]

The potential for inflammatory response is reduced by naturally occurring protectants including mucosal mucins, glycosaminoglycans, sulfated polysaccharides[9,10] and cell surface keratans. The effectiveness of these compounds as protectants may be based on their ability to bind to selectins. The protective effect of sulfation of glycoproteins or proteoglycans may be diminished by desulfation induced by drugs, chemicals and microbial enzymes.[11,12]

15.A. RHEUMATOID ARTHRITIS

Rheumatoid arthritis, one of the most common autoimmune inflammatory disorders, is characterized by a deficient galactosylation of IgG (Table 4), containing bisected as well as fucosylated N-glycans especially in the synovial fluid.[13] Degalactosylated IgG binds less to C1q and Fc-receptors, suggesting that important biological functions are associated with carbohydrate structures of IgG.[14] Increased binding of mannose-binding protein to IgG is observed when IgG is undergalactosylated and terminal GlcNAc residues are exposed.[15] The Man-binding is followed by complement activation. Undergalactosylation may be limited to IgG since other glycoproteins in rheumatoid arthritis patients, for example transferrin, possess fully sialylated bi-antennary complex N-glycans.[16] Tuberculosis, sarcoidosis, a form of leprosy (erythema nodosum leprosum) and other inflammatory diseases such as osteoarthritis and Crohn's disease also show Gal-deficient IgG.[17] It is possible that the Gal deficiency of IgG induces its antigenicity, leading to the binding of rheumatoid factor to the Fc portion of IgG, although this could not be confirmed in vitro.[14]

The enzymatic basis for undergalactosylation was studied in patients' and normal B cells. While core 1 β3-Gal-transferase levels were similar in both groups, Gal-transferase activity towards asialo agalacto IgG was much lower in the patients' B cells. Gal-transferase activity towards asialo-agalacto-transferrin, which has N-glycan chains similar to those of IgG, however, was unchanged. This suggests that two different Gal-transferases are involved in IgG and transferrin galactosylation. This was confirmed by the finding that the sensitivity of the two Gal-transferases towards α-lactalbumin differs from one another.[16] A similar Gal deficiency of IgG has been observed in the mouse model of type II collagen-induced arthritis.[17]

Agalactosyl IgG may also be pathogenic in the mouse model.[17] The GlcNAc content of IgG is much higher in collagen-induced mice compared to control mice.[18] Acute synovitis in mice is aggravated by passive transfer of undergalactosylated anti type II collagen IgG, although the transfer is ineffective when galactosylated antibodies are employed. The T cell cytokine, interleukin-6 (IL-6), may be one factor which regulates Ig glycosylation and it may be elevated in human rheumatoid arthritis.[19] The levels of agalactosyl IgG increase following injection of IL-6 into mice.

15.B. INFLAMMATORY DISEASE IN THE INTESTINES

The role of mucins in the protection of the gastric and intestinal epithelium is well established. Peptic ulcer disease is associated with erosion of the mucus gel by acid and possibly by bacterial infections. Mucin changes also accompany intestinal inflammatory disease such as ulcerative colitis and Crohn's colitis.[20-22]

Reduced radioactive sulfate incorporation into rectal mucins has been observed in patients with inflammatory bowel disease (Table 4).[23] Reduced sulfation of mucins also occurs in gastritis. Sulfatase produced by *Helicobacter pylori* is thought to be a causative factor in the latter case. Sulfatases acting on mucins have been isolated from a number of colonic bacteria.[11,24,25]

In ulcerative colitis, incorporation studies were carried out using ^{3}H-glucosamine and ^{35}S-sulfate on cultured biopsy specimens from patients and controls. The rate of glycosylation but not sulfation of mucins is elevated in active ulcerative colitis compared with normal controls, but returns to control values upon remission. Poorly sulfated glycoconjugates are found in active ulcerative colitis, but during remission, the extent of sulfation increases beyond control values.[26] Generally, an increased risk of neoplasia in ulcerative colitis is associated would not be identified in granulomatosus and ulcerative colitis patients.[29] Mucin polymerization also appears to be abnormal in ulcerative colitis.[30]

An increased rate of incorporation of the sialic acid precursor N-acetyl mannosamine into mucins has been demonstrated in cultured ulcerative colitis mucosal explants.[31] The changes of specific sugar linkages in the disease has been demonstrated by lectin binding studies of tissue sections. While the overall sialylation is increased in ulcerative colitis tissue, an increase of sialylα2-6 linkages is correlated with the severity of inflammation and a decrease of SAα2-3 linkages is observed in severely inflamed tissue.[32] Secretions from the colorectal areas contain high amounts of sialomucin. Anal glands secrete more sulfomucins and mucins containing O-acyl derivatives of sialic acid.[33] The cancer associated sialyl-Tn antigen is also expressed in ulcerative colitis. Although the occurrence of this antigen does not correlate with the degree of inflammation, it does precede the formation of neoplasia by several years and may therefore be used in early diagnosis of cancer.[34] O-acetylation is reduced in ulcerative colitis. Since O-acetylation confers resistance to bacterial sialidases, colitis mucin is more susceptible to degradation by sialidase.[35]

15.C. INFLAMMATION OF THE LIVER

Inflammation of the liver induces an acute phase response. In chronic viral, autoimmune and alcoholic liver diseases and biliary diseases, the expression of Le^y antigen is observed in membranous fractions of clustered hepatocytes.[36] Kim et al found that alcoholic liver disease as well as viral hepatitis is accompanied by an increase in serum Gal-transferase activities. The serum enzyme probably originates from cell damage in the liver.[37]

In a rat model, subcutaneous turpentine injections cause an acute phase response with stimulation of synthesis and secretion of glycoproteins from the liver. Membrane-bound proteins are released by the action of cathepsin D-like protease cleavage. α6-Sialyltransferase, but not β4-Gal-transferase, is released into the serum.[38,39] Liver homogenates derived from inflamed rats express increased activities of each enzyme.[40] Secretion of sialyltransferase may also be induced hormonally by glucocorticoids, cytokines and other factors.[41]

References

1. Welply JK, Keene JL, Schmuke JJ et al. Selectins as potential targets of therapeutic intervention in inflammatory diseases. Biochim Biophys Acta 1994; 1197:215-226.
2. Etzioni A, Frydman M, Pollack S et al. Brief report:recurrent severe infections caused by a novel leukocyte adhesion deficiency. New Eng J Med 1992; 327:1789-1792.
3. von Andrian UH, Berger EM, Ramezani L et al. In vivo behavior of neutrophils from two patients with distinct inherited leukocyte adhesion deficiency syndrome. J Clin Invest 1993; 91:2893-2897.
4. Stocks SC, Hopwood D, Sanders DSA et al. The expression of $Lewis^x$ on carcinoembryonic antigen (CEA)-related glycoproteins of normal and inflamed oesophageal squamous mucosa. Glycosyl Dis 1994; 1:279-286.
5. van Dijk W, Havenaar EC, Brinkman-van der Linden ECM. α_1-Acid glycoprotein (orosomucoid):pathophysiological changes in glycosylation in relation to its function. Glyconconj J 1995; 12:227-233.
6. Sarkar M, Moorkerjea S. Effect of dexamethasone on the synthesis of dolichol-linked saccharides and glycoproteins in hepatocytes prepared from control and inflamed rats. Biochem J 1985; 227:675-682.
7. Mulligan MS, Lowe JB, Larsen RD et al. Protected effects of sialylated oligosaccharides in immune complex-induced acute lung injury. J Exp Med 1993; 178:623-631.
8. Skurk C, Buerke M, Guo J-P et al. Sialyl $Lewis^x$-containing oligosaccharide exerts beneficial effects in murine traumatic shock. Am J Physiol 1994; 267:H2124-2131.
9. Bazzoni G, Nunez AB, Mascellani G et al. Effect of heparin, dermatan sulfate and related oligo-derivatives on human polymor-

phonuclear functions. J Lab Clin Med 1993; 121:268-275.
10. Arfors K, Ley K. Sulfated polysaccharides in inflammation. J Lab Clin Med 1993; 121:201-202.
11. Slomiany BL, Murty VLN, Piotrowski J et al. Glycosulfatase activity of Helicobacter pylori toward gastric mucin. Biochem Biophys Res Comm 1992; 183:506-513.
12. Slomiany BL, Murty VLN, Piotrowski J et al. Glycosulfatase activity of Porphyromonas gingivalis, a bacterium associated with peridontal disease. Biochem Mol Biol Int 1993; 29:973-980.
13. Parekh RB, Dwek RA, Sutton BJ et al. Association of rheumatoid arthritis and primary osteoarthritis with changes in the glycosylation pattern of total serum IgG. Nature 1985; 316:452-457.
14. Tsuchiya N, Endo T, Matsuta K et al. Effects of galactose depletion from oligosaccharide chains on immunological activities of human IgG. J Rheumatol 1989; 16:285-290.
15. Malhotra R, Wormald MR, Rudd PM et al. Glycosylation changes of IgG associated with rheumatoid arthritis can activate complement via the mannose-binding protein. Nature Med 1995; 1:237-243.
16. Furukawa K, Kobata A. IgG galactosylation - its biological significance and pathology. Mol Immun 1991; 28:1333-1340.
17. Rademacher TW, Williams P, Dwek RA. Agalactosyl glycoforms of IgG autoantibodies are pathogenic. Proc Natl Acad Sci USA 1994; 91:6123-6127.
18. Jones MG, Dilly SA, Bond A et al. Changes in the glycosylation of IgG in the collagen-induced model of arthritis. Glycosyl Dis 1994; 1:105-110.
19. Gleeson PA. Glycoconjugates in autoimmunity. Biochim Biophys Acta 1994; 1197:237-255.
20. Allen D, Connolly N, Biggart J. Mucin profiles in ulcerative colitis with dysplasia and carcinoma. Histopathology 1988; 13:413-424.
21. Jacobs LR, Huber PW. Regional distribution and alterations of lectin binding to colorectal mucin in mucosal biopsies from control and subjects with inflammatory bowel disease. J Clin Invest 1985; 75:112-118.
22. Clamp JR, Fraser G, Read AE. Study of the carbohydrate content of mucus glycoproteins from normal and diseased colons. Clin Sci 1981; 61:229-234.
23. Raouf AH, Tsai HH, Parker N et al. Sulphation of colonic and rectal mucin in inflammatory bowel disease:reduced sulphation of rectal mucus in ulcerative colitis. Clin Sci 1992; 83:623-626.
24. Roberton A, McKenzie C, Sharfe N et al. A glycosulfatase that removes sulfate from mucus glycoprotein. Biochem J 1993; 293:683-689.
25. Rhodes J, Black R, Gallimore R et al. Histochemical demonstration of desialation and desulfation of normal and inflammatory bowel disease rectal mucus by fecal extracts. Gut 1985; 26:1312-1318.
26. Morita H, Kettlewell MGW, Jewell DP et al. Glycosylation and sulphation of colonic mucus glycoproteins in patients with ulcerative colitis and in healthy subjects. Gut 1993; 34:926-932.
27. Boland CR, Lance P, Levin B et al. Abnormal goblet cell glycoconjugates in rectal biopsies associated with an increased risk of neoplasia in patients with ulcerative colitis:early results of a prospective study. Gut 1984; 25:1364-1371.
28. Kim YS, Byrd JC. Ulcerative colitis:a specific mucin defect? Gastroenterology 1984; 87:1193-1195.
29. Kim YS, Isaacs R. Glycoprotein metabolism in inflammatory and neoplastic diseases of the human colon. Cancer Res 1975; 35:2092-2097.
30. Rankin BJ, Srivastava ED, Record CO et al. Patients with ulcerative colitis have reduced mucin polymer content in the adherent colonic mucus gel. Biochem Soc Trans 1995; 23:104S.
31. Parker N, Tsai HH, Ryder SD et al. Increased rate of sialylation of colonic mucin by cultured ulcerative colitis mucosal explants. Digestion 1995; 56:52-56.
32. McMahon RFT, Jones CJP, Dutt S et al. Altered oligo-saccharide expression in ulcerative colitis with increasing grades of inflammation. Glycosyl Dis 1994;

1:235-245.

33. Fenger C, Filipe MI. Mucin histochemistry of the anal canal epithelium. Studies of normal anal mucosa and mucosa adjacent to carcinoma. Histochem J 1981; 13:921-930.
34. Itzkowitz SH, Young E, Dubois D et al. Sialosyl-Tn antigen is prevalent and precedes dysplasia in ulcerative colitis:a retrospective case-control study. Gastroenterol 1996; 110:694-704.
35. Reid PE, Culling CFA, Dunn WL et al. Chemical and histochemical study of normal and diseased human gastrointestinal tract. A comparison between histologically normal colon, colonic tumors, ulcerative colitis and diverticular disease of the colon. Histochem J 1984; 16:235-251.
36. Sasaki M, Kono N, Nakanuma Y. Membranous expression of Lewis Y antigen in clustered hepatocytes in chronic viral, autoimmune, and alcoholic liver diseases but not in biliary diseases. Mod Path 1994; 7:339-346.
37. Kim YS, Perdomo J, Whitehead JS et al. Glycosyltransferases in human blood. II. Study of serum galactosyltransferase and N-acetylgalactosaminyltransferase in patients with liver diseases. J Clin Invest 1972; 51:2033-2039.
38. Lammers G, Jamieson JC. The role of a cathepsin D-like activity in the release of Galβ1-4GlcNAc α2-6-sialyltransferase from rat liver Golgi membranes during the acute-phase response. Biochem J 1988; 256:623-631.
39. Lammers G, Jamieson JC. Cathepsin D-like activity in the release of Galbeta1-4GlcNAcα2-6sialyltransferase from mouse and guinea pig liver Golgi membranes during the acute phase response. J Comp Biochem Physiol 1990; 95:327-334.
40. Fraser IH, Coolbear T, Sakar M. Increase of sialytransferase activity in the serum and liver of inflamed rats. Biochim Biophys Acta 1984; 799:102-105.
41. Kaplan HA, Woloski BMRNJ, Hellman M et al. Studies on the effect of inflammation on rat liver and serum sialyltransferase. Evidence that inflammation causes release of Galβ1→4GlcNAc α2-6 sialyltransferase from liver. J Biol Chem 1983; 258:11505-11509.

CHAPTER 16

Carbohydrate Deficiency Diseases

Recent refinements in carbohydrate analytical methodology and the application of molecular cloning techniques have defined a whole new area of carbohydrate pathology. Several forms of carbohydrate deficiency diseases have been described. These relate to impairment of glycoprotein processing or biosynthesis.

The importance of N-glycan processing by GlcNAc-transferase I to complex type chains (Fig. 3) has been studied in the mouse. Mouse embryos were generated lacking a gene for GlcNAc-transferase I.[1] GlcNAc-transferase I activity is found at normal levels in wild-type mice, but is absent in these mutants; heterozygotes contain approximately one-half the normal level. Complex N-glycans are deficient in these mutants. Embryos develop into multiple cell types by day 9, but a number of morphogenic processes is impaired (Fig. 5). All mutant embryos die prior to birth. However, mutant GlcNAc-transferase I-deficient cell lines can grow normally in culture.[2] The results support the view that complex N-glycans are not essential for growth, at least in cultured cells, but have important roles in embryonal development and morphogenesis, particularly in the nervous system.

Familial goiter is associated with a thyroid sialyltransferase deficiency towards asialo-thyroglobulin and hyposialylated thyroglobulin.[3] These abnormalities may be the cause of defective thyroglobulin iodination and transport from the cell to the follicular lumen of the thyroid.

Another important deletion of glycan structures is seen in I-cell disease and pseudo-Hurler polydystrophy, which is characterized by the lack of UDP-GlcNAc-phospho-transferase in fibroblasts.[4] This defect leads to a general inability to synthesize the Man-6-phosphate signal involved in targeting lysosomal enzymes to the lysosomes.

16.A. Leukocyte Adhesion Deficiency

Leukocyte adhesion deficiency (LAD) is a syndrome characterized by mental retardation, severe recurrent infections and localized non-purulent cellulitis. Mutations affecting leukocyte integrin complexes have been described in LAD 1 patients.[5,6] Two cases of LAD type 2 have been described with a defect in GDP-fucose (Table 4). In each patient the blood group phenotype was hh (Bombay) and secretor status and Lewis antigens including glycans containing sialyl-Le^x were absent. In addition, marked defects in leukocyte motility as well as selectin-mediated adhesion were observed, which is explained by the absence of sialyl-Le^x.[7] Neutrophils from LAD type 2 patients are deficient in sialyl-Le^x and show no or very little binding to E- or P-selectin. It appears that a general defect in fucose metabolism may result in the lack of GDP-Fuc, the donor substrate for Fuc-transferase reactions. However, patients have normal levels of L-selectin and CD11b/CD18 integrin, and neutrophils adhere to and migrate across endothelium following CD11b/CD18 activation.[8]

16.B. HEMPAS

HEMPAS, hereditary erythroblastic multinuclearity-associated anemia with positive acidified serum lysis test, is a rare autosomal recessive anemia characterized by ineffective bone marrow erythropoiesis.[9] Red blood cells are abnormally sensitive to agglutination by anti i antibodies and are susceptible to lysis by anti i and anti I antibodies.

HEMPAS is thought to be due to heterogeneous genetic defects of glycosylation of erythrocyte membrane glycoproteins (Table 4), apparently confined to cells of erythrocytic lineages.[10] The alterations observed in HEMPAS could be due to several defects. There is a block of glycosylation resulting in underglycosylation of N-glycans and a decrease in the molecular weight of erythrocyte Band 3 and Band 4.5 glycoproteins. A corresponding shift of poly-n-acetyl-lactosamine units to glycolipid acceptors may occur. The N-glycan structures of Band 3 in HEMPAS patients exhibit varying amounts of truncated hybrid N-glycans containing unprocessed Man residues on the Manα1-6 arm and only one α6-sialylated poly-n-acetyl-lactosamine unit on the Manα1-3 arm. Complex type N-glycans are also present in various proportions relative to the truncated hybrid forms.

Lymphocytes from one HEMPAS patient were shown by mass spectrometry, methylation analysis and glycosidase digestion to exhibit N-glycans lacking extensions at the Manα1-6 arm. This patient exhibited a 90% decrease in GlcNAc-transferase II activity towards rhodopsin oligosaccharide substrate (lacking extension on the Manα1-6 arm) in lymphocytes compared with normal cells.[11] The reduction is probably not due to an endogenous inhibitor. Another patient had 30% of the normal level of lymphocytic GlcNAc-transferase II. The i β3-GlcNAc-transferase activity involved in the synthesis of poly N-acetyllactosamine chains is present at normal levels. Subsequently, a third patient with HEMPAS was found to possess significantly reduced α–mannosidase II activity and its mRNA expression, although genomic DNA showed normal restriction fragment patterns.[12] The lack of processing of the Manα1-3/1-6 residues of the Manα1-6 arm prevents GlcNAc-transferase II action and results in hybrid structures. Taken together, these studies indicate that normal N-glycosylation with poly-n-acetyllactosamine chains is required for the functions of hematopoietic membrane glycoproteins and that the poly N-acetyllactosamine chains of glycolipids cannot be appropriate substitutes.

16.C. CARBOHYDRATE DEFICIENT GLYCOPROTEIN SYNDROME

Carbohydrate deficient glycoprotein syndromes (CDGS) are genetic multisystemic diseases characterized by incomplete glycosylation of glycoproteins, including blood clotting factors, lysosomal enzymes and probably membrane glycoproteins (Table 4).[13] The syndrome is recognizable at birth or shortly thereafter; the nervous system is affected in all cases and most other organs are involved as well. The diagnosis is usually made by isoelectric focusing of serum transferrin which has two

N-glycosylation sites. The lack of sialic acid explains the significant electrophoretic shifts and high isoelectric point of glycoproteins. On this basis, three types of CDGS have been delineated. Patients with liver disease and recent chronic alcohol abuse may also demonstrate the CDGS-type of transferrin.

In CDGS type I, by far the most prevalent form, isoelectric focusing of serum transferrin shows an increase in asialo- and disialo-transferrin and decreased tetrasialo- and pentasialo-transferrin variants (Fig. 7). The oligosaccharide profiles are shown in Figure 8. The N-glycans exhibit normal structures, suggesting that the defect is not in the later stages of N-glycan processing. Lack of glycosylation of one or both of the two N-glycoslation sites is thought to account for the disease-related patterns.[14] Normal patterns of cotranslational glycosylation have been observed in CDGS fibroblasts.[15]

Fibroblasts derived from three CDGS patients exhibit dolichyl phosphate and GlcNAc-1-phosphate levels that are similar to those of normal controls.[16] The transfer of oligosaccharides from Dol-P-P-sugars to peptide by oligosaccharyltransferase from fibroblasts membranes of CDGS type I patients also appears to be normal.[17] In CDGS type I, a deficiency of (2-^{3}H)-mannose incorporation into lipid-linked oligosaccharide as well as glycoproteins has been demonstrated in fibroblasts derived from two affected sisters.[18] The oligosaccharides assembled on dolichol-pyrophosphate are partially truncated and oligosaccharides of alpha$_1$-antitrypsin show greater than normal heterogeneity. In addition, the high Man structures are probably abnormal since they cannot be released with endoglycosidase H.[19] Therefore CDGS may be due to decreased or deficient synthesis of dolichol-linked oligosaccharides, resulting in the transfer of abnormal sugar chains from dolichol to proteins. Van Schaftingen and Jaeken[20] reported a defect in phosphomannomutase in CDGS type I fibroblasts, liver and leukocytes. The enzyme synthesizes mannose-1-phosphate which is the precursor for the synthesis of GDP-Man and subsequently, dolichol-phospho-mannose.

Type II CDGS patients differ from type I in their clinical and biochemical features, for example, in their serum transferrin patterns. Serum disialo-transferrin studied from one type II CDGS patient possesses two moles of truncated monoantennary sialyl-Gal-GlcNAc-Manα1-3-(Manα1-6)Manβ1-4GlcNAcβ1-4GlcNAc-Asn per mole of transferrin.[21] Thus both N-glycosylation sites are occupied but lack normal processing. The patient's leukocytes and fibroblasts show a specific deficiency of GlcNAc-transferase II (Fig. 3).[22] Analysis of the patient's relatives suggests that the defect is transmitted in an autosomal recessive fashion. Fibroblasts from CDGS type II patients have normal phosphomannomutase levels.

Type II CDGS patients differ from those with HEMPAS in clinical presentation and the serology of erythrocytes. Bands 3 and 4.5 from type II CDGS have 50% of the normal amount of poly N-acetyllactosamine chains, and erythrocytes from HEMPAS patients almost completely lack these chains. Another patient has been shown to exhibit undergalactosylation of IgG molecules while Man and GlcNAc contents are normal.[22] A type III CDGS variant has been described; transferrin from these patients has shown approximately equal increase of tri-, di-, mono- and asialoferritin with appreciable amounts of normal isoforms.[23] The diversity of glycoprotein structures found in CDGS indicates that this disease comprises a number of different phenotypes.

Additional instances of genetic or familial types of glycosylation defects have been reported. They include galactosemia, a multisystemic disease resulting from a deficiency of Gal-1-phosphate uridyltransferase. Glycoprotein abnormalities similar to those of patients with CDGS are seen in this disease and are due to lack of Gal addition and therefore sialic acid.[24]

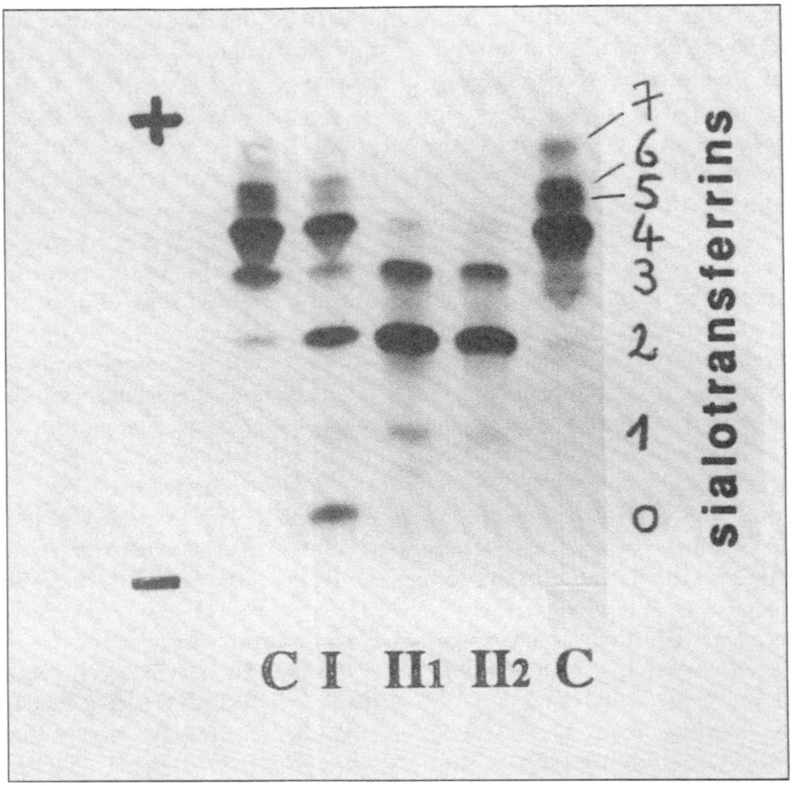

Fig. 7. Glycosylation of transferrin in carbohydrate deficient glycoprotein syndrome (CDGS). Isoelectric focusing pattern of serum transferrin in controls (C), CDGS type I (I), and in two patients with CDGS type II (II1 and II2), showing distinct abnormalities in CDGS type II patients. Compared with controls, tetrasialotransferrin (position 4) is nearly absent but there is a large increase of disialotransferrin (position 2). In addition, a band is present at the monosialotransferrin position (midway between 0 and 2) as well as a faint additional unidentified band between the positions of monosialotransferrin and asialotransferrin (position 0). The patient's parents show a normal pattern, presumably because they are heterozygous for the trait. A different pattern is displayed by type I patients, i.e. a less pronounced decrease of tetrasialoferrin, smaller increase of disialoferrin, no additional band, and a significant asialotransferrin fraction. Type III patients show approximately equal increases of tri-, di-, mono- and asialotransferrin with appreciable amounts of normal isoforms. Reprinted from Jaeken et al[21] with permission.

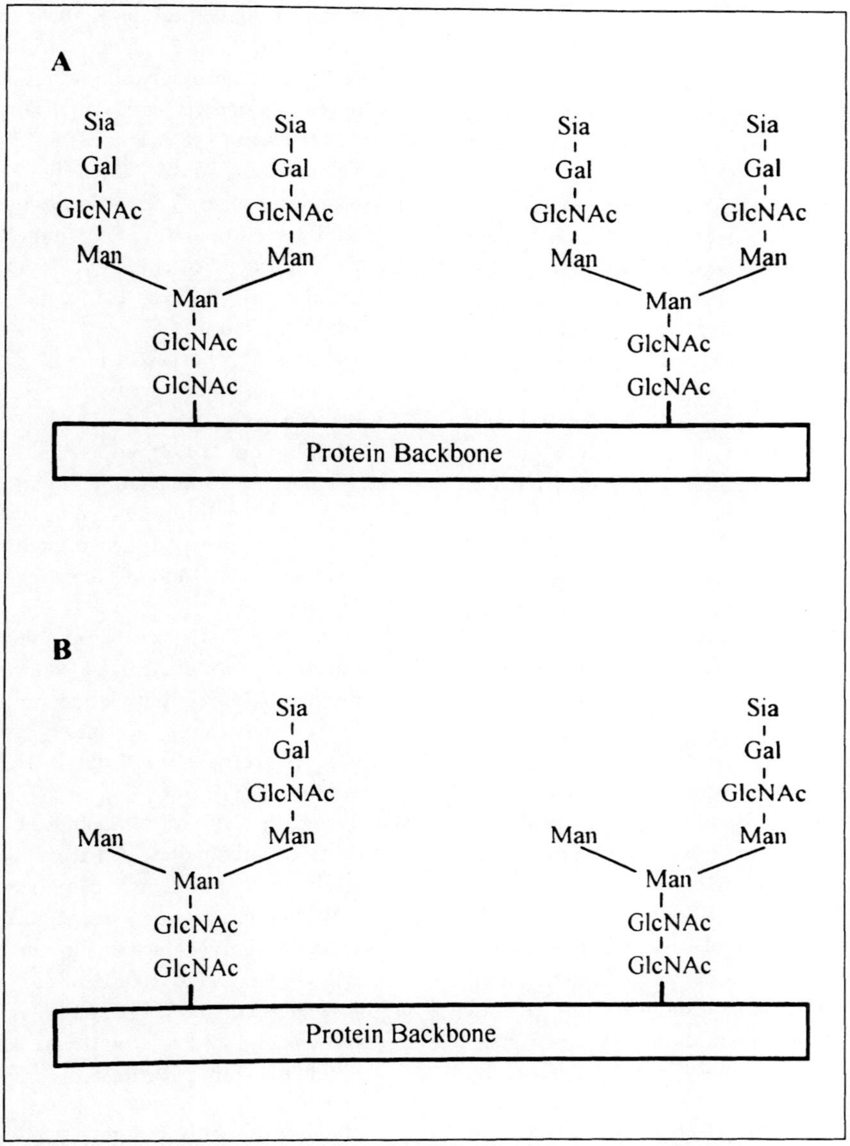

Fig. 8. Sialylated N-glycan structures of transferrin. Structures of two N-glycans in normal serum tetrasialotransferrin (A). Analysis of CDGS type II fractions by nuclear magnetic resonance and electrospray mass spectrometry demonstrates two truncated monoantennary, monosialylated N-glycans per mole of transferrin (B). The difference between the molecular weights of normal serum tetrasialotransferrin and patient's disialotransferrin (about 1.3 kDa) indicates that two moles of the trisaccharide chain SAα2-6 Galβ1-4 GlcNAc are missing from transferrin. Reprinted from Jaeken et al[21] with permission.

REFERENCES

1. Metzler M, Gertz A, Sarkar M et al. Complex asparagine-linked oligosaccharides are required for morphogenic events during post-implantantion development. EMBO J 1994; 13:2056-2065.
2. Stanley P, Narasimhan S, Siminovitch L et al. Chinese Hamster Ovary cells selected for resistance to the cytotoxicity of phytohemagglutinin are deficient in a UDP N-acetylglucosaminyl glycoprotein N-acetylglucosaminyltransferase activity. Proc Natl Acad Sci USA 1975; 72:3323-3327.
3. Grollman EF, Doi SQ, Weiss P et al. Hyposialylated thyroglobulin in a patient with congenital goiter and hypothyroidism. J Clin Endocrin Metab 1992; 74:43-48.
4. Reitman ML, Varki A, Kornfeld S. Fibroblasts from patients with I-cell disease and pseudo-Hurler polydystrophy are deficient in uridine 5'-diphosphate-N-acetylglucosamine:glycoprotein N-acetylglucosaminylphosphotransferase. J Clin Invest 1981; 67:1574-1579.
5. Sligh JE Jr, Hurwitz MY, Zhu C et al. An initiation codon mutation in CD18 in association with the moderate phenotype of leukocyte adhesion deficiency. J Biol Chem 1992; 267:714-718.
6. Back AL, Kwok WW, Hickstein DD. Identification of two molecular defects in a child with leukocyte adherence deficiency. J Biol Chem 1992; 267:5482-5487.
7. Etzioni A, Frydman M, Pollack S et al. Brief report:recurrent severe infections caused by a novel leukocyte adhesion deficiency. New Eng J Med 1992; 327:1789-1792.
8. Phillips ML, Schwartz BR, Etzioni A et al. Neutrophil adhesion in leukocyte adhesion deficiency syndrome type 2. J Clin Invest 1995; 96:2898-2906.
9. Crookston JH, Crookston MC, Burnie KL et al. Hereditary erythroblastic multinuclearity associated with a positive acidified-serum test:a type of congenital dyserythropoietic anemia. Brit J Hemat 1969; 17:11-26.
10. Fukuda MN. HEMPAS disease: genetic defect of glycosylation. Glycobiology 1990; 1:9-15.
11. Fukuda MN, Dell A, Scartezzini P. Primary defect of congenital dyserythropoietic anemia type II: failure in glycosylation of erythrocyte lactosaminoglycan proteins caused by lowered N-acetylglucosaminyltransferase II. J Biol Chem 1987; 262:7195-7206.
12. Fukuda MN, Masri KA, Dell A et al. Incomplete synthesis of N-glycans in congenital dyserythropoietic anemia type II caused by a defect in the gene encoding α-mannosidase II. Proc Natl Acad Sci USA 1990; 87:7443-7447.
13. Jaeken J, Carchon H, Stibler H. The carbohydrate-deficient glycoprotein syndromes: pre-Golgi and Golgi disorders? Glycobiology 1993; 3:423-428.
14. Winchester B, Clayton P, Mian N et al. The carbohydrate-deficient glycoprotein syndrome: an experiment of nature in glycosylation. Biochem Soc Proc 1995; 23:185-188.
15. Marquardt T, Ullrich K, Zimmer P et al. Carbohydrate-deficient glycoprotein syndrome (CDGS)-glycosylation, folding and intracellular transport of newly synthesized glycoproteins. Eur J Cell Biol 1995; 66:268-273.
16. Yasugi E, Nakasuji M, Dohi T et al. Major defect of carbohydrate-deficient-glycoprotein syndrome is not found in the synthesis of dolichyl phosphate or N-acetylglucosaminylpyrophosphoryl-dolichol. Biochem Biophys Res Comm 1994; 200:816-820.
17. Knauer R, Lehle L, Hanefeld F et al. Normal N-oligosaccharyltransferase activity in fibroblasts from patients with carbohydrate-deficient glycoprotein syndrome. J Inherit Metabol Dis 1994; 17:541-544.
18. Powell LD, Paneerselvam K, Vij R et al. Carbohydrate-deficient glycoprotein syndrome: not an N-linked oligosaccharide processing defect, but an abnormality in lipid-linked oligosaccharide biosynthesis? J Clin Invest 1994; 94:1901-1909.
19. Krasnewich DM, Holt GD, Brantly M et al. Abnormal synthesis of dolichol-linked oligosaccharides in carbohydrate-deficient glycoprotein syndrome. Glycobiol 1995; 5:503-510.
20. van Schaftingen E, Jaeken J. Phosphomannomutase deficiency is a cause of carbohydrate-deficient glycoprotein syn-

drome type I. FEBS Letts 1995; 377:318-320.

21. Jaeken J, Schachter H, Carchon H et al. Carbohydrate deficient glycoprotein syndrome type II:a deficiency in Golgi localised N-acetylglucosaminyltransferase II. Arch Dis Child 1994; 71:123-127.
22. Tan J, Dunn J, Jaeken J. Carbohydrate-deficient glycoprotein syndrome type II. An autosomal recessive disease due to point mutations in the coding region of the N-acetylglucosaminyltransferase III gene. Glycoconj J 1995; 12:478.
23. Gu J, Kondo A, Okamoto N et al. Oligosaccharide structures of immunoglobulin G from two patients with carbohydrate-deficient glycoprotein syndrome. Glycosyl Dis 1994; 1:247-252.
24. Stibler H, Westerberg B, Hanefeld F et al. Carbohydrate-deficient glycoprotein (CDG) syndrome—a new variant, type III. Neuropediatrics 1993; 24:51-52.
25. Jaeken J, Kint J, Spaapen L. Serum lysosomal enzyme abnormalities in galactosaemia. Lancet 1992; 340:1772-1773.

CHAPTER 17

MICROORGANISMS

Microorganisms may have cell surface carbohydrate binding proteins that participate in host interactions. These may involve microbial binding, colonization and infectivity, utilization of host glycoproteins as an energy source and interactions with the immune system. Microbial carbohydrates may be highly immunogenic for the host immune defense system.

17.A. BACTERIAL INFECTIONS

Bacterial cell walls contain complex polymers of saccharides, some of which possess unusual carbohydrate residues. Glycosylation differences have been reported between Gram negative and Gram positive bacteria. The binding of bacteria to cell membranes may be facilitated by adhesins which have affinities for specific oligosaccharide ligands such as Galα1-4Gal.[1] Bacteria adhere to glycoproteins, and in particular to mucins present on internal epithelial cells, for example, in the respiratory or the gastro-intestional tracts. Pathogenic bacteria often interact with glycolipids[2] and this process can be measured by thin layer chromatography.[3] Some of the glycolipid determinants, such as blood group antigens, may also be found on glycoproteins, for example.

Certain intestinal bacteria adhere to mucins and possess glycosidases acting on these mammalian glycoproteins. The interaction of intestinal mucins with bacteria may result in erosion of the mucus gel and subsequent ulcer formation.[4] In addition, bacteria are capable of synthesizing blood group-like structures capable of reacting with host antibodies.[5] It is therefore conceivable that bacteria influence the expression of antigens in the gut.

Gram negative *E. coli* strains bind preferentially to O-linked glycans of glycophorin A, whereas Gram-positive *Streptococcus suis* strains, that colonize pig tonsil or nose and are associated with meningitis in human, bind to α2-3-sialylated poly-N-acetyl-lactosamine glycans[6] shown by hemagglutinin assays using desialylated erythrocytes and to cells specifically resialylated with α3-sialyltransferase (N) acting on Galβ1-4GlcNAc residues. This effect is not seen when erythrocytes are resialylated with α6-sialyltransferase acting on N-glycans or α3-sialyltransferase acting on O-glycans. Another adhesin from *Streptococcus suis* recognizes the Galα1-4

Gal sequence present on blood group P_1 and P^k oligosaccharides.[7] A digalactoside receptor for these adhesins has been found on urinary tract epithelium.[8]

Membrane-associated glycolipids containing relatively short oligosaccharide chains function as receptors for many bacteria and other unicellular forms. The pili of *Pseudomonas aeruginosa* strains can bind to glycolipids with short carbohydrate sequences such as GalNAcβ1-4Gal found in asialo GM_1 and asialo GM_2 and on glycoproteins.[9] Similar adhesion receptors have been found in *Neisseria gonorrheae*.[10]

The carbohydrate binding specificity of *E. coli* verotoxin is dependent on terminal Galα1-4Gal linked to Gb_3 (globotriosylceramide); Gb_4 which contains an additional GalNAcβ1-3 residue is not recognized.[11] A single mutation of the binding subunit (Asp18 → Asn) alters the binding specificity of verotoxin and results in binding of Gb_4.[12]

Intestinal Shiga toxin receptors are determined by a developmentally regulated Gal-transferase acting on lactosyl-ceramide to form Gb_3, and an α–galactosidase.[13] A single cell membrane glycolipid carbohydrate sequence may bind different ligands depending on its fatty acid moiety; the lipid moiety can therefore regulate the availability of epitopes.[14]

Galactose binding lectins have been described in the *Pseudomonas* strain PA 1 while another strain (PA II) exhibits high affinity for L-fucose.[15,16] Mannose-specific adhesins of *E. coli* type I fimbriae bind strongly to high mannose structures located on granulocytes and macrophages (CD66) and resting B lymphocytes which then are stimulated to proliferate and to secrete immunoglobulins.[17,18]

As in the case of viruses, bacterial sialic acids play an important role in binding of host ligands. Excessive in vitro sialylation of lipopolysaccharides of virulent meningococcal C strains enables them to evade neutrophil killing, and this process might enhance the epidemic potential of certain strains.[19] The capsule of *E. coli* K1 containing multiple sialicα2-8 acid linkages is a neurovirulent determinant associated with neonatal meningitis.[20]

Human serum contains mannose binding proteins released from the liver during the acute phase response. These lectins may be important in the opsonization of bacteria and in complement activation. Some truncated forms of mannose-binding protein bind to mannose-rich *Salmonella* but fail to activate complement due to loss of a collagen-like region.[21]

17.B. VIRAL INFECTIONS

Enveloped DNA and RNA viruses are assembled by their host cells. Viruses bud from intracellular or plasma membrane regions and retain part of these, including membrane lipids and glycoproteins. Virus glycoproteins are glycosylated according to the glycosylation potential of the host cell[22] whose glycosyltransferase expression depends on the cell type and its growth and differentiation status. The role of viral carbohydrates may be to shield the virion from immune attack or from degradation, or to interact with components of the host cell or other viruses. Some viral surface proteins are lectins or agglutinins that recognize specific carbohydrate structures on host cell surfaces. These interactions may in part be responsible for viral infectivity. Viral glycoproteins may possess binding sites for host cell messenger molecules, and by this means may increase their infectivity and mediate viral entry and viral release.[23-25] Sialylated lipids, i.e. gangliosides, are present on human granulocytes to which *influenza A* and *Sendai* virus bind; it is thought that they mediate viral phagocytosis which could play a role during virus infection and recovery.[26,27]

Carbohydrate ligands mediate microbial adherence, infectivity and toxicity and influence the defense systems of the host. Biologically important roles of viral glycoproteins include immunoglobulin binding and inhibition of complement function.[28,29]

Viral glycoproteins are known to possess N-linked as well as O-linked glycans, for example in herpes simplex virus (HSV) glycoproteins. N-glycans are of particular importance in the expression of viral infectivity, since treatment of HSV-infected cells with tunicamycin blocks production of infectious progeny, although viral binding to cells is unaffected.[29]

Terminal sialic acid residues are important in ligand recognition by many viruses.[30,31] *Sendai* virus-mediated hemagglutination is abolished by removal of sialic acid from erythrocytes but can be restored by reassembling the sialylα2-3Gal-R moiety, indicating that the sialylα2-3Gal-structure is a virus receptor.[31] Rat α-macroglobulin and bovine submandibular mucin, containing the diacetyl sialic acid Neu5,9Ac_2, are inhibitors of hemagglutination. Influenza virus C binds to N-acetyl-9-O-acetyl-sialic acid on erythrocytes and destroys its receptor by hydrolyzing the 9-O-acetyl group.[32,33]

Neuraminidase of the active influenza WSN virus which is neurovirulent in mice lacks a glycosylation site at amino acid Asn 130, although this glycosylation site is conserved in other neuraminidases. However, mutant WSN virus containing neuraminidase into which an N-glycosylation site was introduced lacks replication potential in mouse brain.[34] This suggests that glycosylation of influenza WSN virus interferes with its neurovirulence.

Sialidase substrate specificity studies relative to glycosidic linkages of sialic acids have been carried out on six sialidases derived from viruses and bacteria. α2-3-Glycosidic linkages of small oligosaccharides were found to be hydrolyzed faster than α2-6-linkages, and sialidase activity was higher with Neu5Ac when compared with Neu5Gc.[35]

When the gene encoding GlcNAc-transferase III (adding the bisected GlcNAc into N-glycans) is introduced into a human hepatoma cell line HB611, the expression of hepatitis B virus (HBV) in these cells is reduced. The effect of introducing the bisected GlcNAc residue is seen in the increased binding of E-PHA lectin. Tunicamycin or swainsonine treatment causes an enhanced expression of HBV-related mRNA. Thus, overexpression of glycoproteins carrying the bisected GlcNAc residues appears to suppress HBV gene expression.[36]

17.C. AIDS

The human immunodeficiency virus (HIV) is responsible for the acquired immunodeficiency syndrome (AIDS).[37] Infection by this retrovirus leads to the destruction of CD4 lymphocytes which culminates in immunodeficiency and AIDS. The initial step in this process is the binding of a viral envelope glycoprotein GP120 to host lymphocyte surface membranes[38] on a domain of 113 amino acids with immunoglobulin-like properties.[39] GP120 is a proteolytic fragment of the larger glycoprotein GP160, which is cleaved into GP120 and the associated integral membrane glycoprotein GP41.[40] Increased GP120 and interferon levels are found in the sera of AIDS patients. GP120 has been reported to induce interferon α and the polyclonal activation of B cells.[41]

GP120 has 50% carbohydrate with 24 structurally heterogeneous N-glycans and 2-4 O-glycans.[42-44] All N-glycosylation sites appear to be occupied. The structures of carbohydrate chains of recombinant gp120 may vary between expression systems.[45,46] GalNAc and sialylα2-6GalNAc O-glycans were detected using anti Tn and anti sialyl-Tn antibodies and glycosidase treatment.[47] It has been shown that anti Tn and anti sialyl-Tn antibodies, as well as anti Le^y and anti A blood group antibodies, neutralize both HIV-1 and HIV-2 infections of T lymphocytes.[48] Since Tn and related antigens are expressed restrictively in healthy persons, their presence on HIV creates a potential target in immunotherapy using anti Tn antibodies.[49] It should be noted, however, that the syncytium score of HIV infected lymphocytes is not affected by treatment with *Helix pomatia*, an αGalNAc-binding lectin.

Viral carbohydrate may serve to mask peptide epitopes and to influence protein conformation. The peptide portion of GP120 is subject to variation due to errors that occur during reverse transcription in host cells. This variability may cause shifts in the degree and type of glycosylation of GP120 with resultant changes in epitopes. Patients may form antibodies to various glycoforms of GP120.[50]

Diverse N-glycans have been found in GP120 obtained from the human lymphoblastoid cell line H9. About two-thirds of the N-glycans are of the high mannose type containing five to nine mannose residues, a small percentage of hybrid types and bisected multiantennary N-glycan structures. The presence of bisected structures is surprising since human T-cells are low in GlcNAc-transferase III activity.[51] The remaining structures are bi-, tri-, or tetra-antennary complex N-glycans which may contain fucose or sialic acid.[52-54] Recombinant GP120 produced in insect cells is likely to contain more high mannose type N-glycans and is more potent than GP120 expressed in CHO cells.[55]

Approximately 14% of the oligosaccharides contain 6-sulfated-GlcNAc residues within N-acetyl-lactosamine chains.[56] Neoantigens such as the difucosylated Le^y and blood group A or other structures may be expressed de novo on GP120 subsequent to infection within the limits of the glycosyltransferases indigenous to the host cell.[45,46,57-58]

The major cell surface glycoprotein of HIV-infected T-cells is closely related or identical to leukosialin, CD43, and is expressed in high amounts in AIDS. CD43 from infected human acute lymphocytic leukemia CEM cells is sulfated and O-glycans are hyposialylated. The loss of homotypic cell aggregation induced by CD43 of infected cells can be restored by resialylation of O-glycans using α3-sialyltransferase (O).[59] These results suggest that cell binding and infection by HIV causes a shift in the expression pattern of glycans presumably related to glycosyltransferase gene activation or gene modification.[45,46]

The significance of GP120 glycosylation or of changes in host cell glycosylation in the pathogenesis or amelioration of AIDS is controversial. It has been shown that viral particles deglycosylated with N-glycanase F remained capable of binding and infecting human $CD4^+$ lymphoid cells. Enzymatic removal of carbohydrate from recombinant GP160 or GP120, or purified soluble GP120 or inhibition of N-glycosylation likewise results in minimal decreases in the affinity of GP120 for its receptor.[60,61] The question remains whether complete deglycosylation is achieved by this treatment since recombinant non-glycosylated GP120 made in tunicamycin treated cells or recombinant mutant GP120 does not bind CD4.[62]

Glycosylation influences the binding of antibodies to GP120 by masking underlying epitopes or influencing protein folding and conformation.[63-65] Thus glycosylation of GP120, which is cell type-specific and may change during the course of infection, may induce variability in the structure and antigenicity of GP120 and may influence the immune response. The carbohydrate itself may not be a major target for the immune response in HIV-infected patients since it is produced by the host cell and may be recognized as "self." Induction of a cytotoxic response to the V3 and other regions of GP120 in mice is enhanced by removal of N-glycans by N-glycosidase F.

When potential glycosylation sites are deleted by mutagenesis, marked influences on cell tropism and infectivity are found in some cases.[41,66-68] This suggests a role for N-glycans, either directly or perhaps due to an effect upon GP120 envelope protein folding.[69] Treatment of GP120 with inhibitors of glycan processing yields diverse effects, including reduction of infectivity, reduced syncytia formation, GP120 surface expression and virus replication. These effects suggest a potential role for glycosylation inhibitors in AIDS therapy.[70-76] Glucosidase inhibitors exhibit particularly strong anti-viral activity; this, coupled with a relative lack of toxicity towards lymphocytes, suggests their possible use as anti-AIDS agents.[76]

Lectins also show promise as agents for immunotherapy in addition to their use in studies of HIV infectivity. Mannose binding proteins, Concanavalin A, lentil lectin, wheat germ agglutinin and other lectins inhibit HIV infectivity.[77-79] These results indicate that high mannose type N-glycans and other glycans are important in this process.

Highly anionic compounds can inhibit HIV-1 binding to T lymphocytes, and may inhibit HIV replication and syncytium formation.[80,81] Sulfated alkyl oligosaccharides were potent inhibitors of HIV-cell binding as well as other sulfated polysaccharides such as dextran sulfate and heparin.[81,82]

Gal-ceramide has been identified as an alternate receptor for HIV-1 in the small intestine and colon.[83-85] CD4 negative cells were identified as targets when Gal-ceramide was present.

Further studies are required to delineate the relationship of HIV glycan structures to the infectious process, the biosynthesis of HIV glycoproteins in infected cells, the effect of HIV infection on activities of glycosyltransferases and the consequence of changed glycoprotein biosynthesis for the infectious process.

17.D. PARASITIC INFECTIONS

Trypanosomes have cell surface glycolipids which play an important role in the infectious process of parasite-induced disease. In addition, they also have surface mucin-like and other glycoproteins including the variant surface glycoprotein VSG which may contain high mannose N-glycans.[86,87] N-glycans may stabilize the protein coat and render it resistant to protease digestion. Certain surface glycoproteins are anchored to the membrane by glycosylphosphatidyl inositol (GPI) anchors.[87,88] During the differentiation process from epimastigote to infective metacyclic trypomastigote forms, the cells acquire a high content of ceramide-GPI, much of which is shed at the time of host invasion.

Certain parasites express O-glycans and mucin-like molecules. For example, *Schistosoma mansoni* expresses a Thr/Ser-rich glycoprotein in the epithelial cells of the reproductive tract of female worms.[89] The O-glycans of cell surface glycoproteins include simple T and Tn antigens and more complex core 2 structures. O-GlcNAc linkages are also present. Malaria proteins, for example from *Plasmodium falciparum*, may carry O-GlcNAc residues. Erythrocytes infected with these parasites have an active O-GlcNAc-transferase which is not found in uninfected erythrocytes.[90,91]

Some trypanosomal N-glycans are antigenic. The sera of patients with parasitic infections may contain high titers of anti Galα1-3Gal antibodies[92] resulting from Galα1-3Gal determinants on the parasite surface glycoproteins which are absent from human glycoproteins. Kinetoplast protozoa, *Leishmania* or trypanosome infections introduce the Galα1-3Gal epitope into humans and the resulting antibodies produced may be useful as a diagnostic tool. Studies on the adhesin of *Bordetella pertussis* and of malaria sporozoites and circumsporozoite proteins have suggested that sulfated glycoconjugates on epithelial cell surfaces may function as specific receptors for these adhesins.[93,94] Sporozoite invasion of hepatocytes could be blocked by fucoidan, dextran sulfate and heparin.

Sialic acid is important for trypanosome-host cell interactions as a ligand either on the parasite or on the host cell surface since sialidase or sialic acid binding lectins inhibit this interaction.[95] Epimastigotes and trypomastigotes of *Trypanosoma cruzi*, a parasite causing Chaga's disease, contain a surface trans-sialidase that utilizes host cell surface α2-3-linked sialic acid of glycoproteins and glycolipids as a substrate to sialylate trypomastigote molecules, probably GPI-anchored mucins. This process seems to be independent of CMP-sialic acid.[96] Sialylation may contribute to cell invasion by inhibiting complement activation or macrophage binding.[97,98]

Intestinal sulfomucins are involved in the establishment of certain intestinal helminths.[99] Treatment of rats with reserpine to increase sulfomucins[100] results in selective actions against implanted parasites. While reserpine treatment is associated with expulsion of species *Strongyloides venezuelensis*, a second species, *Nippostrongylus brasiliensis*, is retained.[101]

References

1. Tikkanen K, Haataja S, François-Gerard C et al. Purification of a galactosyl-1-4-galactose-binding adhesin from the gram positive meningitis-associated bacterium streptococcus suis. J Biol Chem 1995; 270:28874-28878.
2. Lingwood C. Glycolipids as receptors. Adv Lipid Res 1991; 1:39-55.
3. Karlsson K, Angstrom J, Bergstruom J et al. Microbial interaction with animal cell surface carbohydrates. APMIS 1992; 100 Suppl 27:71-83.
4. Hoskins L. Degradation of mucus glycoproteins in the gastrointestinal tract. In: Horowitz M, Pigman W eds. The Glycoconjugates Vol 2. New York:Academic Press, 1978; 235-253.
5. Springer G. Blood group activity in certain bacteria and plants. In:Kuhns W, ed. Blood Groups and Blood Transfusion. New York: Better Bellevue Association, 1967; 137-168.
6. Liukkonen J, Haataja S, Tikaanan K et al. Identification of N-acetylneuraminyl α2-3 poly-N-acetyllactosamine glycans as the receptors of sialic acid binding Streptococcus suis strains. J Biol Chem 1992; 267:21105-21111.
7. Haataja S, Tikkanen K, Liukkonen J et al. Characterization of a novel bacterial adhesion specificity of Streptococcus suis recognizing blood group P receptor oligosaccharides. J Biol Chem 1993; 268:4311-4317.
8. Kuehn M, Heuser J, Normark S et al. P pili in uropathogenic *E. coli* are composite fibers with distinct fibrillar adhesive tips. Nature 1992; 356:252-255.
9. Sheth H, Lee K, Wong W et al. The pili of Pseudomonas aeruginosa strains PAK and PAO bind specifically to the carbohydrate sequence βGalNAc(1-4)Gal found in glycosphingolipids asialo-GM_1 and asialo-GM_2. Molec Microbiol 1994; 11:715-723.
10. Deal C, Krivan H. Lacto- and ganglio- series glycolipids are adhesion receptors for Neisseria gonorrhoeae. J Biol Chem 1990; 265:12774-12777.
11. Lingwood C. Bacterial adhesins/glycolipid receptors. Curr Opin Struct Biol 1992; 2:693-700.
12. Tyrrell G, Ramotar K, Toye B et al. Alteration of the carbohydrate binding specificity of verotoxins from Galα1-4Gal to GalNAcβ1-3Galα1-4Gal and vice versa by site-directed mutagenesis of the binding subunit. Proc Natl Acad Sci USA 1992; 89:524-528.
13. Mobassaleh M, Koul O, Mishra K et al. Developmentally regulated Gb_3 galactosyltransferase and α-galactosidase determine Shiga toxin receptors in intestine. Am J Physiol 1994; 267:G618-G624.
14. Kiarash A, Boyd B, Lingwood C. Glycosphingolipid receptor function is modified by fatty acid content. J Biol Chem 1994; 269:11138-11146.
15. Garber N, Guempel U, Belz A et al. On the specificity of the D-galactose-binding lectin (PA-1) of Pseudomonas aeruginosa and its strong binding to hydrophobic derivatives of D-galactose and thiogalactose. Biochem Biophys Acta 1992; 1116:331-333.
16. Gilboa-Garber N. Pseudomonas aeruginosa lectins. Meth Enzym 1982; 83:378-385.
17. Sauter S, Rutherfurd S, Wagener C et al. Identification of the specific oligosaccharide sites recognized by type 1 fimbriae from *E. coli* on non-specific cross-reacting antigen, a CD66 cluster granulocyte protein. J Biol Chem 1993; 268:15510-15516.
18. Ponniah S, Abraham S, Endres R. T-cell-independent stimulation of immunoglobulin secretion in resting human B lymphocytes by the mannose-specific adhesin of *E. coli* type 1 fimbriae. Infec and Immun 1992; 60:5197-5203.
19. Estabrook M, Christopher N, Griffiss J et al. Sialylation and neutrophil killing of group C Neisseria meningitidis. J Inf Dis 1992; 166:1079-1088.
20. Cho J-W, Troy FA. Polysialic acid engineering: synthesis of polysialylated neoglycosphingolipids by using the polysialyltransferase from neuroinvasive Escherichia coli K1. Proc Natl Acad Sci USA 1994; 91:11427-11431.
21. Schweinle J, Nishiyasu M, Qiang Ding T et al. Truncated forms of mannose-binding protein multimerize and bind to mannose-rich Salmonella montevideo but fail to ac-

tivate complement in vitro. J Biol Chem 1993; 268:364-370.
22. Spear P. Glycoproteins specified by herpes simplex virus. In:Roizman B. The Herpes Viruses vol 3. New York:Plenum Press, 1985:315-336.
23. Radding W, Pan ZQ, Hunter E et al. Expression of HIV-1 envelope glycoprotein alters cellular calmodulin. Biochem Biophys Res Comm 1996; 218:192-197.
24. Freed EO, Martin MA. The role of human immunodeficiency virus type 1 envelope glycoproteins in virus infection. J Biol Chem 1995; 270:23883-23886.
25. Bour S, Schubert U, Peden K et al. The envelope glycoprotein of human immunodeficiency virus type 2 enhances viral particle release:a Vpu-like factor? J Virol 1996; 70:820-829.
26. Müthing J, Unland F, Heitman D et al. Different binding capacities of influenza and Sendai viruses to gangliosides from human granulocytes. Glycoconj J 1993; 10:120-126.
27. Johnson D, Frame M, Ligas M et al. Herpes simplex virus immunoglobulin G Fc receptor activity depends on a complex of two viral glycoproteins gE and gI. J Virol 1988; 62:1347-1354.
28. Fries L, Friedman H, Cohen G et al. Glycoprotein C of herpes simplex virus 1 is an inhibitor of the complement cascade. J Immunol 1986; 137:1636-1641.
29. Olofsson S. Carbohydrates in herpesviruses infection. APMIS 1992; 100 Supp 27:84-95.
30. Glick G, Toogood P, Wiley D et al. Ligand recognition by influenza virus; the binding of bivalent sialosides. J Biol Chem 1991; 266:23660-23669.
31. Markwell M, Paulson J. Sendai virus utilizes specific sialyloligosaccharides as host cell receptor determinants. Biochemistry 1980; 77:5693-5697.
32. Herrler G, Rott R, Klenk H-D et al. The receptor-destroying enzyme of influenza C virus is neuraminate-O-acetylesterase. EMBO J 1985; 4:1503-1506.
33. Herrler G, Gross HJ, Imhoff A et al. A synthetic sialic acid analogue is recognized by influenza C virus as a receptor determinant but is resistant to the receptor-destroying enzyme. J Biol Chem 1992; 267:12501-12506.
34. Li S, Schulman J, Itamura S et al. Glycosylation of neuraminidase determines the neurovirulence of influenza A/WSN/33 virus. J Virol 1993; 67:6667-6673.
35. Corfield A, Wagner S, O'Donnell L et al. The roles of enteric bacterial sialidase, sialate O-acetyl esterase and glycosulfatase in the degradation of human colonic mucin. Glycoconj J 1993; 10:7-15.
36. Miyoshi E, Ihara Y, Hayashi N et al. Transfection of N-acetylglucosaminyltransferase III gene suppresses expression of hepatitis B virus in a human hepatoma cell line, HB611. J Biol Chem 1995; 270:28311-28315.
37. Barré-Sinoussi F, Chermann J-C, Rey F et al. Isolation of a T-lymphotropic retrovirus from a patient at high risk for acquired immune deficiency syndrome (AIDS). Science 1983; 220:868-871.
38. Dalgleish A, Beverley P, Clapham P et al. The CD4 (T4) antigen is an essential component of the receptor for the AIDS virus. Nature 1984; 312:763-767.
39. Chao B, Costopoulos D, Curiel T et al. A 113-amino acid fragment of CD4 produced in Escherichia coli blocks human immunodeficiency virus-induced cell fusion. J Biol Chem 1989; 264:5812-5817.
40. Willey R, Bonifacino J, Potts B et al. Biosynthesis, cleavage and degradation of the human immunodeficiency virus 1 envelope glycoprotein gp160. Proc Natl Acad Sci USA 1988; 85:9580-9584.
41. Capobianchi MR, Ankel H, Ameglio F et al. Recombinant glycoprotein 120 of human immuno-deficiency virus is a potent interferon inducer. AIDS Res Hum Retroviruses 1992; 8:575-579.
42. Leonard C, Spellman M, Riddle L et al. Assignment of intrachain disulfide bonds and characterization of potential glycosylation sites of the type 1 recombinant human immunodeficiency virus envelope glycoprotein (gp120) expressed in Chinese Hamster Ovary cells. J Biol Chem 1990; 265:10373-10382.
43. Hansen J, Nielson C, Arendrup M et al. Broadly neutralizing antibodies targeted to

mucin-type carbohydrate epitopes of human immunodeficiency virus. J Virol 1991; 65:6461-6467.
44. Hansen J. Carbohydrates of Human immunodeficiency virus. APMIS 1992; 100 Suppl 27:96-108.
45. Hansen J, Clausen H, Hu S et al. An O-linked carbohydrate neutralization epitope of HIV-1 gp120 is expressed by HIV-1 env gene recombinant vaccinia virus. Arch Virol 1992; 126:11-20.
46. Hansen J-ES, Nielsen CM, Nielsen C et al. Correlation between carbohydrate structures on the envelope glycoprotein gp 120 of HIV-1 and HIV-2 and syncytium inhibition with lectins. AIDS 1989; 3:635-641.
47. Bernstein H, Tucker S, Hunter E et al. Human immunodeficiency virus type 1 envelope glycoprotein is modified by O-linked oligosaccharides. J Virol 1994; 68:463-468.
48. Hansen J, Clausen H, Nielsen C et al. Inhibition of human immunodeficiency virus (HIV) infection in vitro by anticarbohydrate monoclonal antibodies:peripheral glycosylation of HIV envelope glycoprotein gp120 may be a target for virus neutralization. J Virol 1990; 64:2833-2840.
49. Lifson J, Coutré S, Huang E et al. Role of envelope glycoprotein carbohydrate in human immunodeficiency virus (HIV) infectivity and virus-induced cell fusion. J Exp Med 1986; 164:2101-2106.
50. Giordanengo V, Limouse M, Du Roure LD et al. Autoantibodies directed against CD43 molecules with an altered glycosylation status on Human Immunodeficiency Virus Type I (HIV-1)-infected CEM cells are found in all HIV-1+ individuals. Blood 1995; 86:2301-2311.
51. Narasimhan S, Lee J, Cheung R et al. β-1,4-mannosyl glycoprotein β-1,4-N-acetylglucosaminyl transferase III in human B and T lymphocyte lines and in tonsillar B and T lymphocytes. Biochem Cell Biol 1988; 66:889-900.
52. Geyer H, Holschback C, Hunsmann G et al. Carbohydrates of human immunodeficiency virus:structures linked to the envelope glycoprotein 120. J Biol Chem 1988; 263:11760-11767.
53. Mizuochi T, Spellman M, Larkin M et al. Carbohydrate structures of the human immunodeficiency virus (HIV) recombinant envelope glycoprotein gp120 produced in Chinese Hamster Ovary cells. Biochem J 1988; 254:599-603.
54. Mizuochi T, Matthews T, Kato M et al. Diversity of oligosaccharide structures on the envelope glycoprotein gp120 of human immunodeficiency virus from the lymphoblastoid cell line H9. J Biol Chem 1990; 265:8519-8524.
55. Doe B, Steimer K, Walker C. Induction of HIV-1 envelope (gp120)-specific cytotoxic T lymphocyte responses in mice by recombinant CHO cell-derived gp120 is enhanced by enzymatic removal of N-linked glycans. Eur J Immunol 1994; 24:2369-2376.
56. Shilatifard A, Merkle R, Helland D et al. Complex-type N-linked oligosaccharides of gp120 from human immunodeficiency virus type 1 contain sulfated N-acetyl glucosamine. J Virol 1993; 67:943-952.
57. Adachi M, Hayami M, Kashiwagi N et al. Expression of Le^y antigen in human immunodeficiency virus-infected human T-cell lines and in peripheral lymphocytes of patients with acquired immune deficiency syndrome (AIDS) and AIDS-related complex (ARC). J Exp Med 1988; 167:323-331.
58. Liedtke S, Adamski M, Geyer R et al. Oligosaccharide profiles of HIV-2 external envelope protein:dependence on host cells and virus isolates. Glycobiology 1994; 4:477-484.
59. Lefebvre J, Giordanengo V, Limouse M et al. Altered glycosylation of leukosialin, CD43, in HIV-1 infected cells of the CEM line. J Exp Med 1994; 180:1609-1617.
60. Fenouillet E, Gluckman J, Bahraoui E. Role of N-linked glycans of envelope glycoproteins in infectivity of human immunodeficiency virus type 1. J Virol 1990; 64:2841-2848.
61. Bahraoui E, Benjouad A, Guetard D et al. Study of the interaction of HIV-1 and HIV-2 envelope glycoproteins with the CD4 receptor and role of N-glycans. AIDS Res Hum Retrovir 1992; 8:565-573.
62. König R, Ashwell G, Hanover J. Glycosylation of CD4:tunicamycin inhibits surface

expression. J Biol Chem 1988; 263:9502-9507.

63. Fennie C, Lasky L. Model for intracellular folding of the human immunodeficiency virus type 1 gp120. J Virol 1989; 63:636-646.
64. Benjouad A, Gluckman J-C, Rochat H et al. Influence of carbohydrate moieties on the immunogenicity of human immunodeficiency virus type l recombinant gp 160. J Virol 1992; 66:2473-2483.
65. Gram GJ, Hemming A, Bolmstedt A et al. Identification of an N-linked glycan in the V1-loop of HIV-1 gp120 influencing neutralization by anti-V3 antibodies and soluble CD4. Arch Virol 1994; 139:253-261.
66. Lee W, Syu W, Du B et al. Nonrandom distribution of gp120 N-linked glycosylation sites important for infectivity of human immunodeficiency virus type 1. Proc Natl Acad Sci USA 1992; 89:2213-2217.
67. Willey R, Smith D, Lasky L et al. In vitro mutagenesis identifies a region within the envelope gene of the human immunodeficiency virus that is critical for infectivity. J Virol 1988; 62:139-147.
68. Dash B, McIntosh A, Barrett W et al. Deletion of a single N-linked glycosylation site from the transmembrane envelope protein of human immunodeficiency virus type 1 stops cleavage and transport of gp160 preventing env-mediated fusion. J Gen Virol 1994; 75:1389-1397.
69. Li Y, Luo L, Rasool N et al. Glycosylation is necessary for the correct folding of human immunodeficiency virus gp 120 in CD4 binding. J Virol 1993; 67:584-588.
70. Dedera D, Vander Heyden N, Ratner L. Attenuation of HIV-1 infectivity by an inhibitor of oligosaccharide processing. AIDS Res Hum Retrovir 1990; 6:785-793.
71. Shimizu H, Tsuchie H, Yoshida K et al. Inhibitory effect of novel 1-deoxynojirimycin derivatives on HIV-1 replication. AIDS 1990; 4:975-979.
72. Pal R, Hoke G, Sarngadharan M. Role of oligosaccharides in the processing and maturation of envelope glycoproteins of human immunodeficiency virus type 1. Proc Natl Acad Sci USA 1989; 86:3384-3388.
73. Pal R, Mumbauer S, Hoke G et al. Brefeldin A inhibits the processing and secretion of envelope glycoproteins of human immunodeficiency virus type 1. AIDS Res Hum Retrovir 1991; 7:707-712.
74. Montefiori D, Robinson W, Mitchell W. Role of protein N-glycosylation in pathogenesis of human immunodeficiency virus 1. Proc Natl Acad Sci USA 1988; 85:9248-9252.
75. Dewar R,Vasudevachari M, Natarajan V. Biosynthesis and processing of human immunodeficiency virus type 1 envelope glycoproteins:effects of monensin on glycosylation and transport. J Virol 1989; 63:2452-2456.
76. Karlsson G, Butters T, Dwek R et al. Effects of the imino sugar N-butyldeoxynojirimycin on the N-glycosylation of recombinant gp120. J Biol Chem 1993; 268:570-576.
77. Favero J. Lectins in AIDS research. Glycobiology 1994; 4:387-396.
78. Balzarini J, Schols D, Neyts J et al. α-(1-3)- and α-(1-6)-D-mannose-specific plant lectins are markedly inhibitory to human immunodeficiency virus and cytomegalovirus infections in vitro. Antimicrob Agents Chemother 1991; 35:410-416.
79. Robinson Jr. E, Montefiori D, Mitchell W. Evidence that Mannosyl Residues are involved in Human Immunodeficiency Virus Type I (HIV-1) pathogenesis. AIDS Res Hum Retro 1987; 3:265-281.
80. Mitsuya H, Looney D, Kuno S et al. Dextran sulfate suppression of viruses in the HIV family:inhibition of virion binding to CD4+ cells. Science 1988; 240:646-649.
81. Uryu T, Ikushima N, Katsuraya K et al. Sulfated alkyl oligosaccharides with potent inhibitory effects on human immunodeficiency virus infection. Biochem Pharmacol 1992; 43:2385-2392.
82. Mbemba E, Gluckman J, Gattegno L. Interactions of HIV-1 and HIV-2 envelope glycoproteins with sulfated polysaccharides and mannose-6-phosphate. Glycobiology 1994; 4:13-21.
83. Bhat S, Mettus RV, Reddy EP et al. The galactosyl ceramide/sulfatide receptor binding region of HIV-1 gp120 maps to amino acids 206-275. AIDS Res Hum Retrovir 1993; 9:175-181.

84. Fantini J, Cook D, Nathanson R et al. Infection of colonic epithelial cell lines by type 1 human immunodeficiency virus is associated with cell surface expression of galactosylceramide, a potential alternative gp120 receptor. Proc Natl Acad Sci USA 1993; 90:2700-2704.
85. Yahi N, Baghdiguian S, Moreau H et al. Galactosyl ceramide (or a closely related molecule) is the receptor for human immunodeficiency virus type 1 on human colon epithelial HT29 cells. J Virol 1992; 66:4848-4854.
86. Zamze S. Glycosylation in parasitic protozoa of the Trypanosomatidae family. Glycoconj J 1991; 8:443-447.
87. Seranno AA, Schenkman S, Yoshida N et al. The lipid structure of the glycosylphosphatidylinositol-anchored mucin-like sialic acid acceptors of trypanosoma cruzi changes during parasite differentiation from epimastigotes to infective metacyclic trypomastigote forms. J Biol Chem 1995; 270:27244-27253.
88. Gerold P, Schuppert-Dieckmann A, Schwarz RT. Glycosyl-phosphatidyl-inositols synthesized by asexual erythrocytic stages of the malaria parasite Plasmodium falciparum. Candidates for plasmodial glycosyl-phosphatidylinositol membrane anchor precursors and pathogenic factors. J Biol Chem 1994; 269:2597-2606.
89. Cummings RD, Nyame AK. Glycobiology of schistosomiasis. FASEB J 1996; 10:838-848.
90. Hoessli DC, Davidson EA, Schwarz RT et al. Glycobiology of plasmodium falciparum:an emerging area of research. Glycoconj J 1996; 13:1-3.
91. Dieckmann-Schuppert A, Bause E, Schwarz RT. Studies on O-glycans of Plasmodium-falciparum-infected human erythrocytes. Evidence for O-GlcNAc and O-GlcNAc-transferase in malaria parasites. Eur J Biochem 1993; 210:779-788.
92. Avila J, Rojas M, Towbin H. Serological activity against Galactosyl-α(1-3) Galactose in sera from patients with several Kinetoplastida infections. J Clin Microbiol 1988; 26:126-132.
93. Menozzi F, Mutombo R, Renauld G et al. Heparin-inhibitable lectin activity of the filamentous hemagglutinin adhesin of Bordetella pertussis. Infec Immun 1994; 62:769-778.
94. Pancake S, Holt G, Mellouk S et al. Malaria sporozoites and circumsporozoite proteins bind specifically to sulfated glycoconjugates. J Cell Biol 1992; 117:1351-1357.
95. Vermelho AB, Meirelles MN. Sialoglycoconjugates in Trypanosoma cruzi-host cell interaction:possible biological models—a review. Memorias do Instituto Oswaldo Cruz. 1994; 89:69-79.
96. Takahashi N, Lee KB, Nakagawa H et al. Enzymatic sialylation of N-linked oligosaccharides using an α-(2,3)-specific trans-sialidase from trypanosoma cruzi:structural identification using a three-dimensional elution mapping technique. Anal Biochem 1995; 230:333-342.
97. Schenkman S, Ferguson M, Heise N et al. Mucin-like proteins linked to the membrane by glycosylphosphatidylinositol anchor are the major acceptors of sialic acid in a reaction catalyzed by trans-sialidase in metacyclic forms of Trypanosoma cruzi. Molec Biochem Pharmacol 1993; 59:293-304.
98. Schenkman S, Eichinger D, Pereira M et al. Structural and functional properties of Trypanosoma trans-sialidase. Ann Rev Microbiol 1994; 48:499-523.
99. Shi B-B, Ishikawa N, Itoh H et al. Goblet cell mucins of four genera of the subfamily cricetinae with reference to the protective activity against stronglyoides venezuelensis. Parasite Immunol 1994; 16:553-559.
100. Forstner J, Maxwell B, Roomi N et al. Intestinal secretion of mucus in chronically reserpine treated rats. Am J Physiol 1981; 241:G443.
101. Ishikawa N, Shi B-B, Khan AI et al. Reserpine-induced sulphomucin production by goblet cells in the jejunum of rats and its significance in the establishment of intestinal helminths. Parasite Immunol 1995; 17:581-586.

CHAPTER 18

Glycoproteins in Degenerative Disease and Xenograft Rejection

A considerable proportion of degenerative diseases is age-related. Advanced protein glycosylation increases with age and involves nonenzymatic covalent addition of carbohydrate to tissue proteins, and is a prominent feature of disorders affecting the normal biological functions of connective tissue, lens, blood vessels and nerves. In diabetes, tissue damage occurs over much shorter time periods than in normal individuals; thus, experimental studies have focused on diabetic models since the lesions are similar to those of old age.[1] Advanced glycation end products modify structures and functions of the extracellular matrix, cell surface molecules including vitronectin, laminin, lens crystallin and collagens[2] and may promote a generalized form of vasculopathy.[3] Gangliosides have also been found to accumulate in senile cataracts; among them, sialyl-Le^x neolacto-series gangliosides have been identified.[4]

In contrast to carbohydrate deficient glycoprotein syndrome where transferrin is undersialylated, Parkinson's disease is characterized by an increase in transferrin-bound sialic acid. However, this shift is not present in cerebrospinal fluid. The increase of free iron associated with decreased binding capabilities of over sialylated transferrin is thought to be significant in the pathogenesis of Parkinson's disease.[5]

18.A. BATTEN'S DISEASE

Neuronal ceroid-lipofuscinosis (Batten's disease) is a progressive neurological disorder in humans and animals characterized by the accumulation of lipopigments in the lysosomes of the brain and other tissues. Dolichol-pyrophosphoryl-oligosaccharides containing Man also accumulate and a small increase in glucosamine incorporation into glycoproteins is observed.[6] This condition is apparently not due to increased oligosaccharyltransferase activity[7] nor to abnormal dolichol metabolism[8] and requires further investigation.

18.B. OSTEOARTHRITIS

Osteoarthritis (OA) is a destructive disease of the articular surfaces of joints. It has uncertain etiology, is progressive in nature, and is accompanied by numerous biochemical alterations of glycoconjugates, including proteoglycans and gangliosides.[9] Early studies showed that the glycosaminoglycan content of OA cartilage is only slightly altered compared to that from normal individuals. In the former, there is a significant increase in chondroitin-4-sulfate and a decrease in keratan sulfate, probably due to altered synthesis.[10] In contrast to normal cartilage chondrocyte homogenates, those obtained from OA demonstrated significant increases in OA in several glycosyltransferases acting on glycoproteins and presumably also the glycoprotein-like portions of proteoglycans.[11] These include Gal-transferase acting on ovomucoid substrate, sialyltransferase acting on asiaofetuin substrate, Fuc-transferase acting on asialoglycoprotein substrates and Man-, Glc- and GlcNAc-transferases incorporating sugars into dolichol-phosphate derivatives (Table 4).

18.C. ALZHEIMER'S DISEASE

Alzheimer's disease is the most common type of progressive dementia affecting older persons.[12] Membrane β-amyloid precursor protein, βAPP, a ubiquitous protein, is processed to yield β/A4 protein which is deposited in various extracellular regions of the brain as amyloid deposits and produces characteristics of the disease. βAPP is both N- and O-glycosylated, is recognized by NCAM and has growth stimulating and trypsin inhibitory properties. The oligosaccharides of two soluble βAPP species from C6 glioma cells include bi- or tri-antennary N-glycans, bisecting and core α1-6 fucosylated structures. The two species differ in substitutions of Galβ1-4 residues by Fuc α1-2 and sialic acid α2-6 residues. It has therefore been postulated that processing of βAPP is regulated via the heterogeneity of its oligosaccharides.[13]

Human brain samples from patients with Alzheimer's disease were found to possess elevated levels of O-GlcNAc in comparison with controls. This upregulation was noted on βAPP, as well as microtubule-associated tau protein.[14] Glycosylated amyloid peptides may possess anti-protease effects.

Treatment of βAPP with α3-specific and nonspecific neuraminidases as well as lectin binding studies revealed that its O-glycans consist of 85% α6-sialylated and 15% α3-sialylated core 1 structures. In addition, O-glycans of β-amyloid precursor protein are regulated by thyroid stimulating hormone in thyrocytes and the degree of sialylation is increased upon stimulation with the hormone.[15]

Amyloid peptides expressed in COS cells lose protease inhibitory activity upon enzymatic N- or O-deglycosylation. When treated with neuraminidase, one of the recombinant proteins appears to exhibit lower inhibition.[16] This suggests that improper glycosylation promotes protein catabolism and the chronicity of Alzheimer's disease. The single α6-sialylated bi-antennary N-glycan in serum amyloid precursor and in tissue amyloid is marked by the low degree of microheterogeneity characteristic of glycoproteins.[17] In vivo, there appears to be little degradation of amyloid protein by proteases and glycosidases, thus contributing to its accumulation. Sulfate moieties of glycosaminoglycans may also affect the aggregation and deposition of amyloid fibrils.[18]

18.D. XENOGRAFT REJECTION

The acute shortage of organ donors has stimulated research on xenotransplantation, which is the transplantation of organs from discordant species. Xenografts from alternate species provide a potentially limitless supply of organs for humans but cause hyperacute and delayed xenograft rejection. Immunological intervention has resulted in successes in many forms of organ transplantation. However, non-human vertebrates, including pigs, possess a cell membrane Galα(1-3) Gal epitope which is absent in humans[19-21] and is capable of reacting with human naturally occurring anti-αGal IgM and IgG.[22-24] These preformed antibodies can combine with porcine endothelial carbohydrate structures to induce complement mediated destruction, lysis and necrosis, and result in organ dysfunction after xenotransplantation.

Porcine endothelial xenoantigens are seen on a number of glycoproteins and may be removed by cleavage of the N-glycan chains.[25] Similar antigens are also present on porcine platelets.[26] Endothelial activation and destruction including thromboses result from antigen-antibody reactions and complement activation. This process may be aggravated by upregulation of porcine E-selectin, binding of human neutrophils to E-selectin or migration of inflammatory cells into the graft tissue.[27] The binding of purified human IgM to pig aorta EC is inhibited by several monosaccharide derivatives and mucins containing GalNAc and other sugars.[28] These results suggest that a number of xenogeneic carbohydrate ligands may play an important role in xenograft rejection.

The importance of the Galα1-3Gal epitope in xenoreactivity was reiterated in recent studies of knockout mice which lack this antigen, following the inactivation of the α3-Gal-transferase gene. These mice are viable and have normal organs, except for the eyes which developed cataracts. Considerably less anti-Galα1-3 antibodies from human serum bind to their tissues in comparison with normal controls. In addition, human complement is activated less on cells from mice lacking the Galα1-3Gal epitope compared to normal mice.[29]

Recently, a new mechanism of EC activation by human blood components has been discovered which involves apoptosis.[30] Consequently, it seems likely that several mechanisms induce xenograft rejection based on the binding to xenoreactive carbohydrate ligands.

Circumvention of xenograft rejection may be accomplished by depletion of host naturally occurring antibodies by plasmapheresis, use of anti-IgM or anti-idiotypic antibodies,[31,32] genetic manipulation of donor organ carbohydrate synthesis, removing the α3-Gal-transferase;[33] expression of α-galactosidase on the endothelium of normal blood vessels;[34] expression of human complement inhibiting factors in pigs,[35] immunosuppression by chemotherapy,[36] inhibition of antigen-antibody binding by carbohydrate analogues, or inhibition of xenoantigen synthesis by glycosylation inhibitors. Pigs possess human-like major blood groups and donor pigs have to be compatible with the recipient's blood group status. This difficulty may be overcome by using blood group O (non-A) pigs. Recently, transgenic pigs have been developed by introducing α2-Fuc-transferase gene into pig embryos to reduce the antigenicity of Galα1-3Gal epitopes.[37] These transgenic pigs overexpress the H antigen and show increased resistance to human serum; their usefulness in xenotransplantation will have to be demonstrated.

References

1. Brownlee M. Advanced protein glycosylation in diabetes and aging. Ann Rev Med 1995; 46:223-234.
2. Charonis AS, Reger LA, Dege JE et al. Laminin alterations after in vitro nonenzymatic glycosylation. Diabetes 1990; 39:807-814.
3. Cohen MP, Clements RS, Cohen JA et al. Glycated albumin promotes a generalized vasculopathy in the db/db mouse. Biochem Biophys Res Comm 1996; 218:72-75.
4. Ogiso M, Okinaga T, Ohta M et al. Identification and synthetic pathway of sialyl-Lewis^x-containing neolacto-series gangliosides in lens tissues. 1. Characterization of gangliosides in human senile cataractous lens. Biochim Biophys Acta 1995; 1256:166-174.
5. van Kamp GJ, Mulder K, Kuiper M et al. Changed transferrin sialylation in Parkinson's disease. Clin Chim Acta 1995; 235:159-167.
6. Hall NA, Patrick AD. Accumulation of dolichol-linked oligo-saccharides in ceroid-lipofuscinosis (Batten disease). Am J Med Genetics Supplement 1988; 5:221-223.

7. van Dessel G, Lagrou A, Hilderson H et al. Dolichyl-pyrophosphoryloligosaccharide protein oligosaccharide transferase in neuronal ceroid-lipofuscinosis. Biochem Cell Biol 1991; 70:515-518.
8. Paton BC, Poulos A. Dolichol metabolism in cultured skin fibroblasts from patients with "neuronal" ceroid lipofuscinosis (Batten's disease). J Inher Metab Dis 1984; 7:112-116.
9. David MJ, Portoukalian J, Rebbaa A et al. Characterization of gangliosides from normal and osteoarthritic human articular cartilage. Arth Rheum 1993; 36:938-942.
10. Mankin HJ, Lippiello L. The glycosaminoglycans of normal and arthritic cartilage. J Clin Invest 1971; 50:1712-1719.
11. Richard M, Vignon E, Peschard M et al. Glycosyltransferase activities in chondrocytes from osteoarthritic and normal human articular cartilage. Biochem Int 1990; 22:535-542.
12. Schorderet M. Alzheimers disease: fundamental and therapeutic aspects. Experientia 1995; 51:99-105.
13. Saito F, Tani A, Miyatake T et al. N-linked oligosaccharide of β-amyloid precursor protein (βAPP) of C6 glioma cells:putative regulatory role in βAPP processing. Biochem Biophys Res Comm 1995; 210:703-710.
14. Griffith LS, Schmitz B. O-linked N-acetylglucosamine is upregulated in Alzheimer brains. Biochem Biophys Res Comm 1995; 213:424-431.
15. Graebert KS, Popp GM, Kehle T et al. Regulated O-glycosylation of the Alzheimer β-A4 amyloid precursor protein in thyrocytes. Eur J Cell Biol 1995; 66:39-46.
16. Godfroid E, Octave J-N. Glycosylation of the amyloid peptide precursor containing the Kunitz protease inhibitor domain improves the inhibition of trypsin. Biochem Biophys Res Comm 1990; 171:1015-1021.
17. Pepys MB, Rademacher TW, Amatayakul-Chantler S et al. Human serum amyloid P component is an invariant constituent of amyloid deposits and has a uniquely homogeneous glycostructure. Proc Natl Acad Sci USA 1994; 91:5202-5206.
18. Fraser PE, Nguyen JT, Chin DT et al. Effects of sulfate ions on Alzheimer β/A4 peptide assemblies:implications for amyloid fibril-proteoglycan interactions. J Neurochem 1992; 59:1531-1540.
19. Good AH, Cooper DKC, Malcolm AJ et al. Identification of carbohydrate structures that bind human antiporcine antibodies: Implications for discordant xenografting in humans. Transplant Proc 1992; 24:559-562.
20. Oriol R, Ye Y, Koren E et al. Carbohydrate antigens of pig tissues reacting with human natural antibodies as potential targets for hyperacute vascular rejection in pig-to-man transplantation. Transplant 1993; 56:1433-1442.
21. Oriol R, Ye Y, Koren E et al. Carbohydrate antigens of vascular endothelium and other pig tissues reacting with human natural antibodies. Transplant Proc 1994; 26:1398.
22. Galili U, Macher B, Buehler J et al. Human natural anti-α-galactosyl IgG, II The specific recognition of α(1-3) linked galactose residues. J Exp Med 1985; 162:573.
23. Galili U, Shohet SB, Kobrin E et al. Man, apes and old world monkeys differ from other mammals in the expression of α-galactosyl epitopes on nucleated cells. J Biol Chem 1988; 263:17755-17762.
24. Thibaudeau K, Anegon I, Lemauff B et al. Human antibodies to porcine platelets. Transplantation 1994; 57:1110-1115.
25. Platt JL, Lindman BJ, Chen H et al. Endothelial cell antigens recognized by xenoreactive human natural antibodies. Transplantation 1990; 50:817-822.
26. Platt JL, Holzknecht ZE. Porcine platelet antigens recognized by human xenoreactive natural antibodies. Transplantation 1994; 57:327-335.
27. Rollins S, Evans M, Johnson K et al. Molecular and functional analysis of porcine E-selectin reveals a potential role in xenograft rejection. Biochem Biophys Res Comm 1994; 204:763-771.
28. Young DSF, Asano H, Takahashi M et al. Unpublished.
29. Tearle RG, Tange MJ, Zannettino ZL et al. The α-1,3-galactosyltransferase knockout mouse. Implications for xenotransplantation.Transplantation 1996; 61:13-19.

30. Young DSF, Kadokura M, Brockhausen I et al. Human lectins induce apoptosis- another pathway to xenograft rejection. Transplantation Proceed 1996; 28:611-612.
31. Soares M, Latinne D, Elsen M et al. In vivo depletion of xenoreactive natural antibodies with an anti-μ monoclonal antibody. Transplant 1993; 56:1427-1433.
32. Young DSF, Asano H, Brockhausen I et al. Isolation and use of specific human preformed antibodies recognizing pig xenoantigens. Proceed XVth World Congress of Transplantation 1994; Kyoto.
33. Gustafsson K, Strahan K, Preece A. α1-3 Galactosyltransferase:a target for in vivo genetic manipulation in xenotransplantation. Immun Rev 1994; 141:59-67.
34. Lemarchand P, Jones M, Yamada I et al. In vivo gene transfer and expression in normal uninjured blood vessels using replication-deficient recombinant adenovirus vectors. Circulation Res 1993; 72:1132-1138.
35. Fodor W, Williams B, Matis L et al. Expression of a functional human complement inhibitor in a transgenic pig as a model for the prevention of xenogeneic hyperacute organ rejection. Proc Natl Acad Sci USA 1994; 91:11153-11157.
36. Golub E, Green D. In: Sunderland MA. eds Immunology:A Synthesis. Sinauer 2nd Ed. 1991:664-666
37. Koike C. Kannagi R, Takuma Y et al. Introduction of α(1,2)-fucosyltransferase and its effect on α-Gal epitopes in transgenic pig. Xenotransplantation 1996; 3:81-86.

CHAPTER 19

Glycosylation in Cancer and Oncogenic Transformation

Oncogenic transformation in many tissues is associated with altered glycosylation or the appearance of oncofetal glycoforms (Table 6).[1,2] The development of colon cancer occurs in several steps including accumulating activation of oncogenes and inactivation of suppressor genes.[3] Glycosylation changes arising during carcinogenesis most likely originate as a consequence of malignant transformation and changing growth and differentiation of the cancer cell. However, these glycosylation changes are critical for the biology of the cancer cell and influence their immunogenicity, cell adhesion and other properties. A mutated gene found in colon cancer appears to code for a member of the adhesion molecule family; this supports the idea that extracellular interactions are important in the control of cell growth.[4] The regulation of glycosyltransferase gene expression may be linked to the activation of oncogenes. A major challenge is to understand the control of glycosylation during malignant transformation. As discussed in chapter 8, many factors contribute to the biosynthesis of glycans.

Glycosyltransferases may be induced or suppressed due to increased transcriptional activity or changed mRNA stability induced by tumor-associated growth factors.[5] Differentiating agents may alter cell surface phenotypes. For example, retinoic acid induces α3-Gal-transferase at the transcriptional level in embryonal carcinoma cells; this results in the suppression of Le^x antigens by masking this epitope with Galα1-3 residues.[6]

Generally, glycosyltransferase activities in cancer have been measured in patients' blood or tumors, or in tumor-derived or transformed cell lines. Glycosyltransferase activities in body fluids are relatively low and their origin is uncertain. The changes observed may be due to switches in the type of related enzymes, changes in the expression of terminally acting or competing enzymes, or those that act early in the pathway and

Table 6. Glycosylation changes in human cancer

Disease	Carbohydrate Abnormality	Glycosyltransferase Abnormality	Chapter/ References
Cancer, general	↑ SA		19/80,81
	↑ O-acetyl SA		19/86
			20/26
	Neu5Gc		19/87,90
	truncated O-glycans		21/11
	↑ N-glycan branching		19/33
		↑ serum sialyl-T	19/82,83,85,123,129
			21/24
		↑ serum α3Fuc-T	19/114,115
		↑ serum Gal-T	19/82,85,119,120
		Gal-T isoenzymes	19/126,127-129
			22/3,5
	blood groups	↓ α2Fuc-T	19/113, 149, 151
	↑ T, Tn, SA-Tn		19/151
	antigens		20/60
			21/11
			24/5
	↑ SA-Tn		19/72,73
			20/22
	↓ sulfation		19/63
			20/10,11
	p^{185} kinase activation		19/68,69
	by sialomucin		
Bladder cancer		↑ Gal-T	19/118
	loss of ABO blood groups		22/32
		↓ A and B transferases	
		↓ α3Fuc-T	
	↑ T and Tn		22/37
Breast cancer	↑ branching	↑ GlcNAc-T V	24/31-33
		↑ serum Gal-T,	19/123
	↑ T and Tn	↓ core 1 β3Gal-T	21/11
			24/9
	↑ Fuc		19/108
	↓ Le^b		24/10
		↑ α3Fuc-T	24/11
	↑ Galβ1-3GlcNAc disaccharide		24/35
	↓ sulfation		24/44
	Neu5Gc		19/90
Choriocarcinoma, endometrial cancer	↑ core 2,		22/19,20
	unusual O-glycans,		22/22
	↑ branching		
	↑ α6 Fuc		
	unusual N-glycans		22/22
		↑ Fuc-T	22/16
	Type 1 chains		22/17

Table 6 continues on next page.

Disease	Carbohydrate Abnormality	Glycosyltransferase Abnormality	Chapter/ References
Colon cancer	↑ branched N-glycans		20/8
	truncated O-glycans		19/37
			21/11
	↓ sulfation of mucins		19/63,64
	blood groups		19/151
		↓ H-α2Fuc-T	19/113
	blood groups	↓ A-α3GalNAc-T	20/68
		↓ B-α3Gal-T	
	Le^x +/-SA		20/39
	extended Le^x +/-SA		20/41
	extended Le^y		
	tissue and serum SA-Le^a		19/36
			21/22
	↑ T		24/5
		↓core 3 β3GlcNAc-T	19/98
		↑ α3SA-T (O-glycan)	
		↑ α2Fuc-T	
		↓polypeptide αGalNAc-T	
		↓core 1 3-sulfo-T	
		↓ α6SA-T I	19/98
			20/17
		↓ core 1 β3Gal-T	20/17
		↓ β4Gal-T	
		↓ polypeptide GalNAc-T	
		↓ B-α3Gal-T	
		↓ α2/3Fuc-T	
	↑ Tn, sialyl-T		19/73
			20/22
			24/5
	↑ SAα6	↑ α6SA-T (N)	20/36,73
	↑ Neu5Gc		20/33-35
	↓ O-acetylation of SA		20/28,29
Gastric cancer	unusual P_1 pentasaccharide		20/138
	↑ T, Tn, sialyl-T and Tn		20/129,143
	incompatible A or A-like antigen		
	↑ mucin core 6		20/137
	↑ O-acetylation of SA		20/136
Hepatoma	↑ linear i chains		20/63
	↑ branching of N-glycans		21/17
	↑ Le^y		21/21
		↑ serum α2/3 Fuc-T	21/16,23
		↑ α6Fuc-T,	
		↓ tissue α2/3Fuc-T	
		↑ GlcNAc-T III	21/39
		↑ serum Fuc-T	21/24
		↓ SA-T	
		↓ Gal-T	
	serum SA-Le^a		21/22
Kidney tumors	Le^x, multimeric Le^x strong in differentiated absent in poorly differentiated cancer		22/34

Table continues on next page

Disease	Carbohydrate Abnormality	Glycosyltransferase Abnormality	Chapter/ References
Liver metastases	↑ Tn		21/11
	↑ Le^x		19/106,107,148
	↑ serum SA	Gal-T isoenzyme	21/27
		↑ serum α6Fuc-T	21/24
	↑ sialyl-Le^x, sialyl dimeric-Le^x		19/106
			26/33
Lung cancer	↓ A,B antigens, ↑ I antigen		23/2
	incompatible A antigen		23/3
	↑ T antigen		23/11
	changed Lewis antigens	↑ serum α3Fuc-T	23/1,9
		↑ β4Gal-T	19/122
Ovarian cancer	↑serum Fuc		24/11
Cervical cancer		↑ β4Gal-T	22/4,9
		↑ SA-T	
		↑ Fuc-T	
		Gal-T isoforms	22/3
	↑ sialyl-Tn		22/11,13,14
Pancreatic cancer	↑ Tn		21/11
		Gal-T isoenzyme	19/130
	↑ Le^x		21/7
	↑ sialyl-Tn		21/12,13
Parotid gland cancer		↑ β4Gal-T	19/121
Prostate cancer	↑ Le^y, sialyl-Le^x		22/28,29
	changed blood groups		
Polyps	extended Le^x and Le^y, blood groups altered, sialyl-Le^x		20/44,118
Rectal cancer	↑ H antigen	↑ α2Fuc-T	20/64
	core 5		20/77
		core 5 α3GalNAc-T	20/78
Salivary gland tumors	Tn, sialyl-Tn		19/72
Squamous cell cancer	↓ ABO antigens		25/3,4

Abbreviations: ↓ decreased; Fuc, fucose; ↑ increased; Gal, galactose; GalNAc, N-acetylgalactosamine; GlcNAc, N-acetylglucosamine; Neu5Gc, N-glycolyl-neuraminic acid; Man, mannose; SA, sialic acid; T, transferase.

therefore have a profound impact on the overall structures found. In some cases, the changes occur at the level of mRNA encoding glycosyltransferases.[7,8]

Neoplastic transformation alters a number of cell surface characteristics. Among the most significant of these are the loss of density-dependent growth inhibition and the occurrence of anchorage-independence.[9,10] Binding of plant lectins such as wheat germ agglutinin and Concanavalin A to transformed cells has revealed that these cells behave differently from normal and exhibit significant surface carbohydrate changes.[11,12]

Recent studies have indicated that overexpression of galectin genes may occur in neoplastic cells and may be correlated with tumor progression.[13-15] The expression of galectins on vascular endothelial cells suggests that they are involved in the adhesion and invasion of tumor cells. In vitro, antibodies against galectin-1 inhibit the adhesion of lymphoma cells to human microvascular endothelial cells.[16] The Gal-counter-receptors for galectins are ligands located on carcinoembryonic antigen and other glycoconjugates in human colon cancer cells.[17] Changes in glycosaminoglycans of the extracellular matrix have a shaping influence on tumor cell behavior[18] and quantitative increases in glycosaminoglycans occur following transformation.[19]

Alterations in sialic acid are found frequently as a distinguishing feature of cell transformation, and involve glycolipids and glycoproteins.[20-25]

The development of monoclonal antibodies made it possible to determine that certain oligosaccharides were tumor-associated oncofetal antigens.[26,27] Tumor-related changes may mirror reversions to structures present in early development.

Analysis of the detailed structures of glycopeptides and oligosaccharide chains has been useful in evaluating discrete alterations in tumors.[28-30] In many cases, cancer-associated carbohydrate alterations include increased expression of truncated forms of O-glycans, increased branching of complex N-glycans associated with increased poly-N- acetyl-lactosamines and expression of GlcNAc-transferase V, and altered sialylation, sulfation and fucosylation and other types of glycosylation. These changes may be a result of altered expression of glycosyltransferases and sulfotransferases. Alternatively, peptide moieties of glycoproteins may change in cancer and lead to increased peptide epitope expression as well as changes in peptide-specific glycosylation.

Cell surface tumor antigens vary between different types of cancers, the tissue, and the secretor-gene status. Antigens may be expressed in cancer that are not normally present in a particular organ, but may occur elsewhere in the body, or occur at a different period of growth and differentiation. Changes in cell surface oligosaccharides in cancer are largely quantitative rather than qualitative.

Glycopeptides from virus or chemically transformed cells from many species are generally larger than normal and carry more sialic acid.[31,32] The higher degree of sialylation may not only be due to increased sialyltransferase activities but also to an increase in N-glycan branching which provides substrate for sialylation.[33] Both increases and decreases in size have been observed in transformed cells for O-linked oligosaccharides.[34-38] A high degree of sialylation is often associated with the presence of shorter, simpler O-glycan chains.

NMR and lectin binding studies on glycopeptides from NIH 3T3 mouse fibroblasts show that cells acquire more sialylα2-3 and lose Galα1-3 terminal residues upon H-ras oncogene transfection.[39] There is, therefore, a reciprocal relationship between neutral and sialylated N-glycan chains in transformed and normal cells. The switch from Galα1-3 to sialylα2-3 is apparently not due to decreased expression of the α3-Gal-transferase, but rather due to a significant increase in the α6-sialyltransferase activity acting on N-glycans, whereas the α3-sialyltransferases acting on N- and O-glycans are not increased. Lectin binding studies demonstrate an increase in sialylα2-6/sialylα2-3 ratio in cellular glycoproteins in ras transformed NIH3T3 cells.[40] Infection of rat fibroblasts with c-ras increase mRNA levels for the α6-sialyltransferase acting on N-glycans.[7] In contrast, the α3-sialyltranferase acting on O-glycans is decreased.[8,41] Upon polyoma virus and Rous sarcoma virus transformation of BHK cells, an increase in N-glycan branching and a reduction in high mannose structures are observed. The tetra-antennary chains released from cells carry sialic acid in a α2-3 linkage.[42,43] These structural changes could be explained in part by the increase in GlcNAc-transferase V synthesizing the GlcNAcβ1-6 branch of tetra-antennary N-glycans.[44] A GlcNAc-transferase V increase in parallel with an increase in O-glycan branching by core 2 β6-GlcNAc-transferase has also been observed in T24H ras-transfected rat 2 fibroblasts compared to nontransfected cells. A similar difference was found in metastatic sublines of SP 1 mouse mammary carcinoma cells compared to benign carcinoma cells.[45]

Neoplasia of human breast and colon tissue is associated with increased GlcNAcβ1-6Manα1-6 Man- branches of N-glycans shown by L-PHA lectin staining.[46] Yousefi et al[45] suggested that β6-branching regulates the levels of polylactosamine chains in N- and O-glycans. Increased branching in transformed or cancer cells may in turn cause excessive coating of cell surfaces with a network of extended oligosaccharides.

19.A. ROLE OF MUCINS IN CANCER

Increases in the expression of mucins, primarily sialomucins, are a hallmark of many forms of cancer and leukemia cells,[47] and can enable cancer cells to circumvent natural forms of immune surveillance and factors determining their half life. Circulating mucins often appear in the serum of tumor patients[48] and may elicit an immune response which can be used clinically for diagnosis and prognosis. The concentrations of circulating mucins depend on many factors, including clearance by the liver.

The expression of MUC1,2,3, 4 and 5 varies in different cell types and in cancer cells.[49] The finding that the peptide moiety of glycosyltransferase substrates has a major impact on O-glycosylation[50,51] offers the possibility that mucins with different peptide backbones in the tandem repeat region are differently glycosylated. MUC1 is the cell surface mucin most commonly found on epithelial cells and increases in MUC1 and changes in its glycosylation have been reported in cancer. Many of these studies have been carried out using anti-peptide antibodies to cirumvent crossreactivity due to glycan heterogeneity, although the reactivity of these antibodies may be influenced by the presence of oligosaccharides. Changed glycosylation in cancer can alter the apparent expression of mucin peptide.[52] The differential expression of mucins in cancer may mobilize an immune response.[53] The appearance of highly immunogenic peptides forms the basis for immunotherapy using antigenic peptide, in particular those derived from the ubiquitous MUC1.[54-57]

Tumors often synthesize large amounts of mucins which results in increased secretion and shedding from the cell surface. Studies in athymic nude mice have shown a definite correlation between growth phases of HT29 colon cancer cells and mucin synthesis. Increased in vivo tumorigenicity occurs in cells in log phase growth inoculated in nude mice compared to injections of post confluent cells; mucin production is significantly increased in exponentially growing cells.[58]

Although some tumors secrete sulfomucins, sialomucins are the predominant species secreted from various premalignant or malignant tissues.[47] Undersulfation of proteoglycans, glycoproteins and mucins occurs frequently in oncogenic transformation and cancers in human and other species as shown by [^{35}S]-sulfate labeling, histochemical studies and anti-sulfomucin antibodies.[59-64]

The following studies suggest that surface glycoprotein in cancer may possess signal transduction properties as well as anti-recognition effects.[65-69] p185, a phosphoprotein receptor kinase encoded by the proto-oncogene c-erb B-2/HER2, is complexed with an N-glycosylated transmembrane subunit and a sialomucin subunit of rat sialomucin on rat ascites mammary adenocarcinoma cells. The rat mammary cell surface sialomucin has a potential role in inhibiting the final stage of attack by natural killer cells.[67] The transmembrane subunit has been found in some human breast cancers.[68] Activation of tumor-associated p185 kinase when complexed with the mucin was suggested by excessive kinase product formation in a number of tumors or tumor extracts tested by Western blot and immunocytochemistry.[68,69] The gene coding for the complex is regulated in a tissue-specific manner with an overexpression in some tumors. This association at the cell surface could provide an autonomous growth device and remove normal growth constraints. These events may lead to cell invasion and metastasis, and may be important for tumor progression.

Mucin-type oligosaccharides are often truncated in cancer, thereby expressing blood group-associated antigens T, sialyl T, Tn and sialyl-Tn (Table 1). The appearance of T antigen is observed in all epithelial cancers but not in benign disease.[38] T- and Tn-antigens are normally masked by the presence of other sugars or sulfate and are rarely detectable except for immunologically privileged sites such as human milk.[70]

Sialyl-Tn antigen has been shown to be present in well differentiated gastric and colonic tumors but not in less differentiated tumors.[71] Both sialyl-Tn and Tn antigens are expressed in salivary gland tumors.[72] Sialyl-Tn and Tn expression has been suggested by Springer et al to be a marker for poorly differentiated adenocarcinomas and mucinous carcinomas.[38] These antigens may be of prognostic value since they are often associated with advanced tumor stage and poor clinical outcome in several types of cancer.[73-75]

A possible reason for the appearance of Tn antigen may be that at a later stage of cancer, transferases extending O-glycans are turned off; there may also be extensive degradation in cancer tissue. The resulting truncated O-glycans may influence Muc peptide conformation which could in turn be related to T-cell-mediated tumor growth suppression.[76] Human colon cancer cell lines are useful for controlled studies on biosynthetic pathways for Tn and sialyl-Tn antigens and definition of their roles in cell function.[77]

The cancer-associated antigens may elicit an immune response. In general, human sera contain anti-T and anti-Tn antibodies; these may occur in response to antigens from the intestinal microflora. Boosted immune responses may destroy cancer cells in the presence of complement, and thus T- and Tn-antigens may be involved in the immunogenicity and biological behavior of tumor cells. This has made it possible to devise tests for the immunodiagnosis of cancer[78] and to prepare an effective and safe T/Tn antigen vaccine for the treatment of breast cancer.[79]

19.B. SIALYLATION

Analysis of glycopeptides isolated from the cell surfaces of cancer or transformed cells suggests that cancer cells are hypersialylated.[80,81] Increased sialyltransferase activities have been observed in the sera of a high proportion of cancer patients and animals bearing tumors.[82-85] Sialic acid occurs in various derivatives and linkages on cancer glycoproteins. Not only the amount but also the type of sialic acid changes in cancer. For example, the amount of O-acetyl derivatives[86] and N-glycolyl-neuraminic acid (Neu5Gc), the Hanganutziu-Deicher antigen, appears in cancer cells.[87-90] The expression of the monooxygenase synthesizing CMP-Neu5Gc from CMP-Neu5Ac controls the appearance of the Hanganutziu antigen which may elicit an antibody response. Human embryonal carcinoma cells contain long poly-n-acetyl-lactosamine chains with sialylα2-9sialylα2-3 termini that may be involved in development and differentiation.[91] Other postulated functions of cell surface sialic acid and sialyltransferase include the activation of platelets, the development of thrombosis, the masking of carbohydrate determinants and the regulation of cellular half life.[92,93]

Lectin binding and cell surface sialylation have been studied on related mouse T-lymphoma tumor cell lines with high and low metastatic potential. All low metastatic lines express receptor sites for soybean agglutinin (which binds Gal and GalNAc) and *Vicia villosa* agglutinin (which detects GalNAc). In highly metastatic lines these lectin binding sites are blocked by sialic acid, probably due to increased sialyltransferase activity measured with asialo-fetuin as a substrate.[94]

Ehrlich ascites tumor cells synthesize O- and N-linked sialoglycoproteins; the cells can be selected to grow as adherent or non-adherent forms. α3-Sialyltransferase (O) acting on Galβ1-3GalNAc- substrate is present at similar levels in both cell types. Using Galβ1-4GlcNAc substrate, non-adherent cells contain mostly α3-sialyltransferase activity whereas α6-sialyltransferase is prominent in adherent cells.[95]

The various enzyme products using low molecular weight non-glycoprotein acceptor substrates can be separated by high pH anion exchange chromatography.

Studies of human colorectal carcinoma cell-derived sialyltransferases show an increase in N-glycan specific α6-sialyltransferase in colon cancer[96] and metastatic tumors. α3-Sialyltransferase (O) activity is increased in early carcinomas,[97] colon cancer tissue,[98] breast cancer cells[99] and leukemia cells.[100]

The sialyl-Tn antigen (Table 1) can be detected using monoclonal antibody and has been useful as a marker in a number of adenocarcinomas, i.e. of the pancreas, ovary, uterus, stomach, colo-rectum and gallbladder, but not hepatocellular carcinomas, renal cell carcinomas or thyroid papillary carcinomas[101] (Table 6). Serum or plasma sialic acid increases have been described up to two times normal levels in certain forms of brain and thyroid tumors[102] and serum sialic acid levels have been proposed to be a marker for uterine cervical cancer.[103] However, the levels are not tumor-specific.

Polysialyl-NCAM, which is critical for the development of the nervous system, has been identified as a neurodeterminant in malignant lymphoma and AML with nervous system involvement.[104,105]

19.C. FUCOSYLATION

The relative proportions of various fucosylated oligosaccharide structures are often significantly altered in cancer. For example, the Le^x determinant is associated with several types of tumors.[106,107] The fucose content of haptoglobin is high in progressive ovarian and breast cancer but low when the disease is benign or in remission.[108] Increased oligosaccharide branching has been suggested as a cause for increased fucosylation since the GlcNAc content is high in patients with early but not advanced stages of breast cancer. Genetic and hormone-related influences require consideration when comparing tumor and normal epitopes in secretory tissue such as endometrium; for example, carcinomas from non-secretors showed lower expression of Le^y and H antigens than carcinomas from secretors.[109] This is liable to reflect differential expression of a membrane-type versus secretor-type α2-Fuc-transferase.[110,111] Estrogen stimulation may alter cell cycle patterns which may in turn create shifts in cell surface antigen expression.

α2-Fuc-transferase activities are changed in the sera or tissues from patients with several types of cancers, including colorectal cancers, with resultant alterations of H, Le^b and Le^y antigens.[98,112-113]

Monoclonal antibodies prepared against sialyl Le epitopes have defined a number of organ-specific antigenic structures which are altered in cancer. α3-Fuc-transferase activities that synthesize Le^x and Le^y antigens have been measured in sera and are significantly higher in patients with lung cancer and other neoplastic diseases compared to normal.[114,115]

Sialyl Le^x can be synthesized by several α3-Fuc-transferases; this can possess functional differences and result in different adhesion efficiencies to E-selectin.[116,117] Such distinctions may be made using specific cDNA probes as diagnostic reagents.

19.D. GALACTOSYLATION

β4-Gal-transferase has been extensively studied in cancer patients. The enzyme may be altered in cancer with respect to its activity, size, glycosylation and secretion. The serum or tissue enzyme is activated in patients with breast, lung, colon, ovarian cancer with and without liver metastases, parotid gland and other cancers.[118-122] Ip and Dao[123] found significantly increased serum Gal-transferase in breast cancer patients but only when metastatic disease was present. The increase becomes more significant as the clinical stage of breast cancer progresses from stage I to stage IV and returns to normal after mastectomy. The increase appears to be non-specific since it is also found in a variety of other diseases, including chronic renal failure, but may be of use in monitoring previously diagnosed patients.[124]

β4-Gal-transferase activity in tumor tissues and sera of rats with prostate and mammary adenocarcinomas is also significantly elevated compared to normal controls.[125]

A structurally and kinetically distinct isoenzyme found in sera of cancer patients, and especially in those patients with metastatic disease, may originate from the tumor[126-129] although the issue is controversial. The differences in properties between various species of Gal-transferase could be due to peptide or glycosylation differences. It has been suggested that the isoenzyme occurs in pancreatic cancer[130] although this is not specific for the disease; the presence of the enzyme correlates with tumor burden and disappears after surgical removal of the tumor.

19.E. BLOOD GROUPS

Blood groups are distributed on a number of tissues other than erythrocytes and for this reason are often referred to as histo-blood groups. Erythrocytic blood groups are related to cancer susceptibility, however, the mechanism is not well understood.[131] As summarized by Hakomori and Kannagi,[132] a variety of substrates can function as substrates for the addition of blood group carbohydrates. They include N- and O-glycans of glycoproteins and glycosphingolipids. The blood groups found on cancer glycoproteins include ABO (H), Lewis type, blood group P system, Cad, i, I, Tn and T antigens (Table 1). These antigens are detected by polyclonal and monoclonal antibodies[133] and are often altered in cancer. Lewis antigens are expressed in a tissue-specific fashion and change during development; they may be inappropriately expressed in cancer. Lewis antigens play an important role in cell adhesion (see chapter 9).

Tests of large population samples indicate that gastric and ovarian carcinomas are more likely to develop in blood group A than in group O individuals (ratio of 1.2 to 1). Blood group A and also B is strongly associated with salivary gland cancer. Cancer of the lip, gastric cancer and, less so, cancer of the large intestines is associated with blood group A but not B.[131] A retrospective study of 141 bladder cancer patients showed that blood group O patients have higher grade tumors and higher mortality rates than group A patients.[134]

Certain blood group antigens are expressed in the embryo (for example Le^x),[135] but not in the normal adult and may be re-expressed in the course of tumorigenesis. The effects of transformation are often the production of relatively simplified glycolipid or glycoprotein structures in comparison to the normal, and the synthesis of new or unusual types of glycoconjugates.[22] Deleted blood group expression in gastrointestinal and other cancers has been matched by decrements in the respective blood group A and B glycosyltransferases which convert glycolipids or glycoproteins with the H antigen to blood group A or B antigen.[136,137] Lack of blood group A or B expression may also be due to a decrease in H-dependent α2-Fuc-transferase.[113]

The blood group P system (Table 1) is highly expressed on Burkitt lymphomas[138] whereas sialyl Le^a is present in high amounts in gastrointestinal cancer.[139] Di- and tri-fucosylated type 2 chains with Le^x determinants as defined by monoclonal antibodies, also show accumulations in colon and liver adenocarcinomas and are present in small or undetectable amounts in normal mucosa and liver.[140] The rare blood group antigen Cad exists either on O-glycans or as a sialylparagloboside, and is frequently found in cancer tissues, particularly from patients with hepatocellular carcinoma.[141]

Tumors may on occasion express a histo-blood group which is not compatible with the erythrocytic blood group. For example, A or A-like antigen may be found in cancers of O or B individuals.[142,143] The cause for an abnormal expression of blood groups may be an aberrant synthesis by blood group-dependent glycosyltransferase. Glycosyltransferases have been shown in vitro to be capable of synthesizing incompatible blood groups, for example, the B-transferase can synthesize blood group A if presented with UDP-GalNAc instead of UDP-Gal.[144] Underlying or precursor structures may be exposed in cancers such as backbone structures for ABH blood groups[145] or precursor Lewis antigens.[146-148] The loss of A and B blood groups in malignant bladder and oral epithelia is correlated with loss of A and B transferases[149] as determined by anti A and B blood group and anti A and B α3-GalNAc- and Gal-transferase antibodies. This is in contrast to the reappearance of A and B antigens in colonic cancers where the transferases are not correlated, suggesting that the expression of blood group antigens is regulated by other factors than A and B transferase activities.[150,151]

Increased circulating blood group antigens reactive with a number of anti blood group antibodies are found in many cancer types and may correlate with the disease stage.[152] These antigens may be shed from cancer cell surfaces or may be derived from secreted glycoproteins or mucins. If there is high specificity, these antigens can be used as markers for the disease. This may be the case for the sialyl-Tn antigen in colorectal, gastric and ovarian cancer or sialyl-Le^a in gastric, non-small cell carcinoma of the lung and in pancreatic cancer.

References

1. De Vouge M, Mukherjee B. Transformation of normal rat kidney cells by v-K-ras enhances expression of transin 2 and an S-100 related calcium-binding protein. Oncogene 1992; 7:109-119.
2. Mas E, Abouakil N, Roudani S et al. Human fetoacinar pancreatic protein:an oncofetal glycoform of the normally secreted pancreatic bile-salt-dependent lipase. Biochem J 1993; 289:609-615.
3. Vogelstein B, Fearon ER, Hamilton SR et al. Genetic alterations during colorectal tumor development. New Eng J Med 1988; 319:525-532.
4. Fearon ER, Cho KR, Nigro JM et al. Identification of a chromosome 18q gene that is altered in colorectal cancers. Science 1990; 247:49-54.
5. Miyoshi E, Nishikawa A, Ihara Y et al. Transforming growth factor β up-regulates expression of the N-acetylglucosaminyltransferase V gene in mouse melanoma cells. J Biol Chem 1995; 270:6216-6220.
6. Cho SK, Yeh J, Cho M et al. Transcriptional regulation of α1,3-galactosyltransferase in enbryonal carcinoma cells by retinoic acid. J Biol Chem 1996; 271:3238-3246.
7. Le Marer N, Stéhelin D. High alpha-2, 6-sialylation of N-acetyllactosamine sequences in ras-transformed rat fibroblasts correlates with high invasive potential. Glycobiology 1995; 5:219-226.
8. Dall'Olio F, Malagolini N, Guerrini S et al. Differentiation-dependent expression of human β-galactoside α2,6-sialyltransferase mRNA in colon carcinoma CaCo-2 cells. Glycoconj J 1996; 13:115-121.
9. Holland JF, Frei E, Bast RC Jr et al. In:Lea and Febiger eds. Cancer Medicine 3rd Ed Vol 1. Philadelphia PA:1993:132-134.
10. Holland JF, Frei E, Bast RC Jr et al. In:Lea and Febiger eds. Cancer Medicine 3rd Ed Vol 1. Philadelphia PA:1993:324.
11. Aub JC, Sanford BH, Cote MN. Studies on reactivity of tumor and normal cells to a wheat germ agglutinin. Proc Natl Acad Sci USA 1965; 54:396-399.
12. Burger MM. A difference in the architecture of the surface membrane of normal and virally transformed cells. Proc Natl Acad Sci USA 1969; 62:994-1001.
13. Xu XC, el-Naggar AK, Lotan R. Differential expression of galectin-1 and galectin-3 in thyroid tumors. Potential diagnostic implications. Am J Pathol 1995; 147:815-822.
14. Chiariotti L, Berlingieri MT, Battaglia C et al. Expression of galectin-1 in normal human thyroid gland and in differentiated and poorly differentiated thyroid tumors. Internat J Cancer 1995; 64:171-175.
15. Schoeppner HL, Raz A, Ho SB et al. Expression of an endogenous galactose-binding lectin correlates with neoplastic progression in the colon. Cancer 1995; 75:2818-2826.
16. Lotan R, Belloni PN, Tressler RJ et al. Expression of galectins on microvessel endothelial cells and their involvement in tumour cell adhesion. Glycoconj J 1994; 11:462-468.
17. Ohannesian DW, Lotan D, Lotan R. Concomitant increases in galectin-1 and its glycoconjugate ligands (carcinoembryonic antigen, lamp-1, and lamp-2) in cultured

human colon carcinoma cells by sodium butyrate. Cancer Res 1994; 54:5992-6000.
18. Liotta LA, Rao CN, Barsky SH. Tumor invasion and the extracellular matrix. Lab Invest 1983; 49:636-649.
19. Baker SR, Blithe DL, Buck CA et al. Glycosaminoglycans and other carbohydrate groups bound to proteins of control and transformed cells. J Biol Chem 1980; 255:8719-8728.
20. Warren L, Buck CA, Tuszynski GP. Glycopeptide changes and malignant transformation:a possible role for carbohydrate in malignant behavior. Biochim Biophys Acta 1978; 516:97-127.
21. Warren L, Buck CA. The membrane glycoproteins of the malignant cell. Clin Biochem 1980; 13:191-197.
22. Hakomori S. Structures and organization of cell surface glycolipids dependency on cell growth and malignant transformation. Biochim Biophys Acta 1975; 417:55-89.
23. Hakomori S. Aberrant glycosylation in cancer cell membranes as focused on glycolipids: overview and perspectives. Cancer Res 1985; 45:2405-2414.
24. Hakomori S. Aberrant glycosylation in tumors and tumor-associated carbohydrate antigens. Adv Cancer Res 1989; 52:257-331.
25. Hakomori S, Kannagi R. Glycosphingolipids as tumor-associated and differentiation markers. J Natl Cancer Inst 1983; 71:231-251.
26. Miyaki M, Zenita K, Tanaka O et al. Stage-specific expression of SSEA-1-related antigens in the developing lung of human embryos and its relation to the distribution of these antigens in lung cancers. Cancer Res 1988; 48:7150-7158.
27. Fukui S, Numata Y, Kurosaka A et al. Production of monoclonal antibodies directed against carbohydrate moieties of cell surface glycoproteins. Jpn J Cancer Res 1988; 79:1119-1129.
28. Hounsell EF, Lawson AM, Feizi T. Structure and antigenic diversity in mucin carbohydrate chains. In:Chantler EN, Elder JB, Elstein M eds. Mucus in Health and Disease–II. New York:Plenum Press, 1982:39-43.
29. Hounsell EF, Lawson AM, Feeney J et al. Structural analysis of the O-glycosidically linked core-region oligosaccharides of human meconium glycoproteins which express oncofoetal antigens. Eur J Biochem 1985; 148:367-377.
30. Hounsell E. Glycoprotein Analysis in Biomedicine. Humana Press, Totowa, NJ. 1993; 1-17.
31. Buck C. A comparative study of glycoproteins from the surface of control and Rous sarcoma virus transformed hamster cells. Biochemistry 1970; 9:4567-4576.
32. Buck C, Glick M, Warren L. Glycopeptides from the surface of control and virus-transformed cells. Science 1971; 172:169-171.
33. Kobata A. Structural changes induced in the sugar chains of glycoproteins by malignant transformation of producing cells and their clinical application. Biochimie 1988; 266:1575-1585.
34. Hull S, Laine R, Kaizu T et al. Structures of the O-linked oligosaccharides of the major cell surface sialoglycoprotein of MAT-B1 and MAT-C1 ascites sublines of the 13762 rat mammary adenocarcinoma. J Biol Chem 1984; 259:4866-4877.
35. Hull SR, Bright A, Carraway KL et al. Oligosaccharide differences in the DF3 sialomucin antigen from normal human milk and the BT-20 human breast carcinoma cell line. Cancer Commun 1989; 1:261-267.
36a. Feizi T, Kapadia A, Gooi HC et al. Human monoclonal auto-antibodies detect changes in expression and polarization of the Ii antigens during cell differentiation in early mouse embryos and teratocarcinomas. In:Muramatsu T, Gachelin G, Moscona AA, Ikawa Y, eds. Teratocarcinoma and Embryonic Cell Interactions. NY:Acad Press 1982:201-215.
36b. Feizi T. Antigenicity of mucins–their relevance to tumour associated and stage specific embryonic antigens. In:Chantler EN, Elder JB, Elstein M eds. Mucus in Health and Disease–II. New York:Plenum Press 1982:29-39.
37. Springer GF. Blood group T and Tn antigens are universal, clonal, epithelial, cell-adhesive, autoimmunogenic carcinoma markers. Progress in Clin & Biol Res 1983; 133:157-166.
38. Springer G. T and Tn, general carcinoma antigens. Science 1984; 224:1198-1206.

39. Santer U, DeSantis R, Hard K et al. N-linked oligosaccharide changes with oncogenic transformation require sialylation and multiantennae. Eur J Biochem 1989; 181:249-260.
40. Vandamme V, Cazlaris H, Le Marer N et al. Comparison of sialyl- and α1-3 galactosyltransferase activity in NIH 3T3 cells transformed with ras oncogen:increased β-galactoside α2-6 sialyltransferase. Biochimie 1992; 74:89-100.
41. Delannoy P, Pelczar H, Vandamme V et al. Sialyltransferase activity in FR3T3 cells transformed with ras oncogen:decreased CMP-Neu5Ac:Galβ1-3GalNAc α-2,3-sialyltransferase. Glycoconj J 1993; 10:91-98.
42. Yamashita K, Ohkura T, Tachibana Y et al. Comparative study of the oligosaccharides released from baby hamster kidney cells and their polyoma transformant by hydrazinolysis. J Biol Chem 1984; 259:10834-10840.
43. Pierce M, Arango J. Rous sarcoma virus-transformed baby hamster kidney cells express higher levels of asparagine linked tri- and tetra-antennary glycopeptides containing GlcNAc-β(1-6)Man-α(1-6)Man- and poly-N-acetyllactosamine sequences than baby hamster kidney cells. J Biol Chem 1986; 261:10772-10777.
44. Yamashita K, Tachibana Y, Ohkura T et al. Enzymatic basis for the structural changes of asparagine-linked sugar chains of membrane glycoproteins of baby hamster kidney cells induced by polyoma transformation. J Biol Chem 1985; 260:3963-3969.
45. Yousefi S, Higgins E, Daoling Z et al. Increased UDP-GlcNAc:Galβ1-3GalNAc-R(GlcNAc to GalNAc) β1-6 N-acetylglucosaminyltransferase activity in metastatic murine tumor cell lines. Control of polylactosaminoglycan synthesis. J Biol Chem 1991; 266:1772-1782.
46. Fernandes B, Sagman U, Auger M et al. β1-6 branched oligosaccharides as a marker of tumor progression in human breast and colon neoplasia. Cancer Res 1991; 51:718-723.
47. Xerri L, Payan M-J, Choux R et al. Predominance of sialomucin secretion in malignant and premalignant pancreatic lesions. Hum Pathol 1990; 21:927-931.
48. Lan MS, Batra SK, Qi W-N et al. Cloning and sequencing of a human pancreatic tumor mucin cDNA. J Biol Chem 1990; 265:15294-15299.
49. Roussel P, Lamblin G. Human mucosal mucins in diseases. 1996, In: Glycoproteins and Disease. Montreiul J, Vliegenthart JFG, Schaditer H eds. Elsevier, Amsterdam 1996; 351-393.
50. Granovsky M, Bielfeldt T, Peters S, et al. O-glycan core 1 UDP-Gal:GalNAc β3-galactosyltransferase is controlled by the amino acid sequence and glycosylation of glycopeptide substrates. Eur J Biochem 1994; 221:1039-1046.
51. Brockhausen I, Möller G, Merz G et al. Control of glycoprotein synthesis:The peptide portion of synthetic O-glycopeptide substrates influences the activity of O-glycan core 1 Uridine 5'-diphospho-galactose: N-acetylgalactosamineα-R β3-galactosyl-transferase. Biochemistry 1990; 29:10206-10212.
52. Taylor-Papadimitriou J, D'Souza B, Burchell J et al. The role of tumor-associated antigens in the biology and immunotherapy of breast cancer. Ann NY Acad Sci 1993; 698:31-47.
53. Kotera Y, Fontenot J, Pecher G et al. Humoral immunity against a tandem repeat epitope of human mucin MUC1 in sera from breast, pancreatic and colon cancer patients. Cancer Res 1994; 54:2856-2860.
54. Denton G, Sekowski M, Price MR. Induction of antibody responses to breast carcinoma associated mucins using synthetic peptide constructs as immunogens. Cancer Lett 1993; 70:143-150.
55. Ding L, Lalani EN, Reddish M et al. Immunogenicity of synthetic peptides related to the core peptide sequence encoded by the human MUC1 mucin gene:effect of immunization on the growth of murine mammary adenocarcinoma cells transfected with the human MUC1 gene. Cancer Immunol Immunother 1993; 36:9-17.
56. McKenzie IFC and Xing PX. Mucins in breast cancer:recent immunological advances. Cancer Cells 1990; 2:75-78.
57. Gendler SJ, Spicer AP, Lalani EN et al. Structure and biology of a carcinoma-associated mucin, MUC1. Am Rev Respir Dis 1991; 144:S42-S47.

58. Niv Y, Schwartz B, Amsalem Y et al. Human HT-29 colon carcinoma cells:mucin production and tumorigenicity in relation to growth phases. Anticancer Res 1995; 15:2023-2028.
59. Winterbourne D, Kahn Z. Heparan sulfate alterations in tumor cells. Biochem Soc Trans 1984; 12:540-542.
60. Iozzo R. Proteoglycans:structure, function, and role in neoplasia. Lab Invest 1985; 53:373-396.
61. Yamori T, Kimura H, Stewart K et al. Differential production of high molecular weight sulfated glycoproteins in normal colonic mucosa, primary colon carcinoma, and metastases. Cancer Res 1987; 47:2741-2747.
62. Shamsuddin A, Trump B. Colon Epithelium III. In vitro studies of colon carcinogenesis in Fischer 344 rats. N-methyl-N-nitro-N-nitrosoguanidine-induced changes in colon epithelium in explant culture. J Natl Cancer Inst 1981; 66:403-411.
63. Irimura T, Wynn D, Hager L et al. Human colonic sulfomucin identified by a specific monoclonal antibody. Cancer Res 1991; 51:5728-5735.
64. Filipe M, Potet F, Bogomoletz W et al. Incomplete sulfomucin-secreting intestinal metaplasia for gastric cancer. Preliminary data from a prospective study from three centres. Gut 1985; 26:1319-1326.
65. Padhy L, Shih C, Cowing D et al. Identification of a phosphoprotein specifically induced by the transforming DNA of rat neuroblastomas. Cell 1982; 28:865-871.
66. Coussens L, Yang-Feng T, Liao Y et al. Tyrosine kinase receptor with extensive homology to EGF receptor shares chromosomal location with neu oncogene. Science 1985; 230:1132-1139.
67. Moriarty J, Skelly CM, Bharathan S et al. Sialomucin and lytic susceptibility of rat mammary tumor ascites cells. Cancer Res 1990; 50:6800-6805.
68. Wu K, Salas P, Yee L et al. Tissue and tumor expression of a cell surface glycoprotein complex containing an integral membrane glycoprotein activator of $p185^{neu}$. Oncogene 1994; 9:3139-3147.
69. Carraway K, Wu K, Yee L et al. Tissue and tumor expression of a cell surface glycoprotein complex containing an integral membrane glycoprotein activator of $p185^{neu}$ phosphorylation. Proc Amer Assoc Cancer Res 1995; 36:560.
70. Pierce-Cretel A, Pamblanco M, Strecker G et al. Heterogeneity of the glycans O-glycosidically linked to the hinge region of secretory immunoglobulins from human milk. Eur J Biochem 1981; 114:169-178.
71. Nakasaki H, Mitomi T, Noto T et al. Mosaicism in the expression of tumor-associated carbohydrate antigens in human colonic and gastric cancers. Cancer Res 1989; 49:3662-3669.
72. Therkildsen MH, Mandel U, Christensen M et al. Simple Mucin-type Tn and Sialosyl-Tn carbohydrate antigens in salivary gland carcinomas. Cancer 1993; 72:1147-1154.
73. Itzkowitz S, Yuan M, Montgomery C et al. Expression of Tn. sialosyl-Tn and T antigens in human colon cancer. Cancer Res 1989; 49:197-204.
74. Inoue H. Clinical value of sialyl Tn antigen in patients with gynecologic tumors. Obstet Gynecol 1990; 75:1032-1036.
75. Takahashi I, Maehara I, Kusumoto T et al. Predictive value of preoperative serum sialyl Tn antigen levels in prognosis of patients with gastric cancer. Cancer 1993; 72:1836-1840.
76. Singhal A, Fohn M, Hakomori S. Induction of α-N-acetylgalactosamine-O-serine/threonine Tn) antigen-mediated cellular immune response for active immunotherapy in mice. Cancer Res 1991; 51:1406-1411.
77. Ogata S, Chen A, Itzkowitz S. Use of model cell lines to study the biosynthesis and biological role of cancer-associated sialosyl-Tn antigen. Cancer Res 1994; 54:4036-4044.
78. Springer G, Desai P, Wise W et al. Pancarcinoma T and Tn epitopes: autoimmunogens and diagnostic markers that reveal incipient carcimomas and help establish prognosis. In:Herberman R, Mercer D eds. Immunodiagnosis of Cancer 2nd ed. New York:Marcel Dekker, 1990:587-612.
79. Springer G, Desai P, Tegtmeyer H et al. T/Tn antigen vaccine is effective and safe in preventing recurrence of advanced human breast carcinoma. Cancer Biotherapy 1994; 9:7-15.
80. Warren L, Fuhrer JP, Buck CA. Surface glycoproteins of normal and transformed

cells:a difference determined by sialic acid and a growth-dependent sialyl transferase. Proc Natl Acad Sci USA 1972; 69:1838-1842.

81. Warren L. The malignant cell and its membranes. Am J Pathol 1974; 77:69-76.
82. Ip C, Dao TL. Increase in serum and tissue glycosyltransferases and glycosidases in tumor-bearing rats. Cancer Res 1977; 37:3442-3447.
83. Kessel D, Allen J. Elevated plasma sialyltransferase in the cancer patient. Cancer Res 1975; 35:670-672.
84. Kessel D, Sykes E, Henderson M. Glycosyltransferase levels in tumors metastatic to liver and in uninvolved liver tissue. J Natl Cancer Inst 1977; 59:29-32.
85. Silver HKB, Karim KA, Archibald EL et al. Serum sialic acid and sialyltransferase as monitors of tumor burden in malignant melanoma patients. Cancer Res 1979; 39:5036-5042.
86. Holzhauser R, Faillard H. Sialic acids in human lymphocytes. Qualitative and quantitative alterations in cancer cases. Carbohydr Res 1988; 183:89-95.
87. Ikuta K, Nishi Y, Simizu Y et al. Hanganutziu-Deicher type heterophile antigen-positive cells in human cancer tissue demonstrated by membrane immunofluorescence. Biken J 1982; 25:47-50.
88. Ohashi Y, Sasabe T, Nishida T et al. Hanganutziu-Deicher heterophile antigen in human retinoblastoma cells. Am J Ophthalmol 1983; 96:321-325.
89. Kawai T, Kato A, Higashi H et al. Quantitative determination of N-glycolylneuraminic acid expression in human cancerous tissues and avian lymphoma cell lines as a tumor-associated sialic acid by gas chromatography-mass spectrometry. Cancer Res 1991; 51:1242-1246.
90. Devine P, Clark B, Birrell G et al. The breast tumor-associated epitope defined by monoclonal antibody 3E1.2 is an O-linked mucin carbohydrate containing N-glycolylneuraminic acid. Cancer Res 1991; 51:5826-5836.
91. Fukuda MN, Dell A, Oates JE et al. Embryonal lactosaminoglycan. The structure of branched lactosaminoglycans with novel disialosyl (sialyl α2-9 sialyl) terminal isolated from PA1 human embryonal carcinoma cells. J Biol Chem 1985; 260:6623-6631.
92. Bastida E, Almirall L, Jamieson GA et al. Cell surface sialylation of two human tumor cell lines and its correlation with their platelet-activating activity. Cancer Res 1987; 47:1767-1770.
93. Scialla SJ, Speckart SF, Hant MJ et al. Alterations in platelet surface sialyltransferase activity and platelet aggregation in a group of cancer patients with high incidence of thrombosis. Cancer Res 1979; 39:2031-2035.
94. Altevogt P, Fogel M, Cheingsong-Popov R et al. Different patterns of lectin binding and cell surface sialylation detected on related high- and low-metastatic tumor lines. Cancer Res 1983; 43:5138-5144.
95. Shigeta S, Winter H, Goldstein I. α(2-3) and α(2-6) sialyltransferase activities present in three variants of Ehrlich tumor cells: identification of the products derived from N-acetyllactosamine and β-D-Gal(1-3)-α-D-GalNAc(1-0)-Bn. Carbohydr Res 1994; 264:111-121.
96. Dall'Olio F, Malagolini N, Di Stefano G et al. Increased CMP-NeuAc:Galβ1-4GlcNAc-R α2-6 sialyltransferase activity in human colorectal cancer tissues. Int J Cancer 1989; 44:434-439.
97. Kemmner W, Krück D, Schlag P. Different sialyltransferase activities in human colorectal carcinoma cells from surgical specimens detected by specific glycoprotein and glycolipid acceptors. Clin Exp Metastasis 1994; 12:245-254.
98. Yang J, Byrd JC, Siddiki B et al. Alterations of O-glycan biosynthesis in human colon-cancer tissues, Glycobiology 1994; 4:873-884.
99. Brockhausen I, Yang J-M, Burchell J et al. Mechanisms underlying aberrant glycosylation of MUC1 mucin in breast cancer cells. Eur J Biochem 1995; 233:607-617.
100. Baker M, Kanani A, Brockhausen I et al. Presence of cytidine 5' monophospho-N-acetylneuraminic acid:Galβ1-3GalNAc-R α(2-3)-sialyltransferase in normal human leukocytes and increased activity of this enzyme in granulocytes from chronic myelogenous leukemia patients. Cancer Res 1987; 47:2763-2766

101. Yonezawa S, Tachikawa T, Shin S et al. Sialosyl-Tn antigen:its distribution in normal human tissues and expression in adenocarcinomas. Am J Clin Pathol 1992; 98:167-174.
102. Kökoglu E, Sönmez H, Uslu E et al. Sialic acid levels in various types of cancer. Cancer Biochem Biophys 1992; 13:57-64.
103. Lagana A, Martinez BP, Marino A et al. Correlation of serum sialic acid fractions as markers for carcinoma of the uterine cervix. Anticancer Res 1995; 15:2341-2346.
104. Grogan T, Guptil J, Mullin J et al. Polysialylated NCAM as a neurodeterminant in malignant lymphoma (ML). Lab Invest 1994; 70:110A.
105. Scott AA, Kopecky KJ, Grogan TM et al. CD56:a determinant of extramedullary and central nervous system (CNS) involvement in acute myeloid leukemia. Lab Invest 1994; 70:120A.
106. Hoff S, Matsushita Y, Qta D et al. Increased expression of sialyl-dimeric Le^x antigen in liver metastases of human colorectal carcinoma. Cancer Res 1989; 49:6883-6888.
107. Shirahama T, Ikoma M, Muramatsu H et al. Reactivity to fucose-binding proteins of lotus tetragonolobus correlates with metastatic phenotype of transitional cell carcinoma of the bladder. J Urol 1992; 147:1659-1664.
108. Dargan E, Thompson S, Cantwell BMJ. Changes in the fucose content of haptoglobin in breast and ovarian cancer:association with disease progression. Glycosyl Dis 1994; 1:37-43.
109. Ravn V, Mandel U, Svenstrup B et al. Expression of type-2 histo-blood group carbohydrate antigens (Le^x, Le^y, and H) in normal and malignant human endometrium. Virchows Archiv 1994; 424:411-419.
110. Kelly RJ, Ernst LK, Larsen RD et al. Molecular basis for H blood group deficiency in Bombay (O_h) and para-Bombay individuals. Proc Natl Acad Sci USA 1994; 91:5843-5847.
111. Hitoshi S, Kusunoki S, Kanazawa I et al. Molecular cloning and expression of two types of rabbit β-galactoside α1,2-fucosyltransferase. J Biol Chem 1995; 270:8844-8850.
112. Yazawa S, Nakamura J, Asao T et al. Aberrant α1→2 fucosyltransferases found in human colorectal carcinoma involved in the accumulation of Le^b and Y antigens in colorectal tumors. Jpn J Cancer Res 1993; 84:989-995.
113. Kuhns WJ, Schoentag R. Carcinoma-related alterations of glycosyltransferases in human tissues. Cancer Res 1981; 41:2767-2772.
114. Asao T, Yazawa S, Nagamachi Y et al. Serum alpha (1-3)-L-fucosyltransferase, carcinoembryonic antigen, and sialyl Lewis X-i antigen levels in lung cancer. Cancer 1989; 64:2541-2545.
115. Yazawa S, Madiyalakan R, Izawa H et al. Cancer-associated elevation of α(1-3)-L-fucosyltransferase activity in human serum. Cancer 1988; 62:516-520.
116. Sueyoshi S, Tsuboi S, Sawada-Hirai R et al. Expression of distinct fucosylated oligosaccharides and carbohydrate-mediated adhesion efficiency directed by two different α1,3-fucosyltransferases. J Biol Chem 1994; 269:32342-32350.
117. Majuri M-L, Niemelä R, Tiisala S et al. Expression and function of α2,3-sialyl- and α1,3/1,4-fucosyltransferases in colon adenocarcinoma cell lines:role in synthesis of E-selectin counter-receptors. Int J Cancer 1995; 63:551-559.
118. Plotkin GM, Wides RJ, Gilbert SL et al. Galactosyl transferase activity in human transitional cell carcinoma lines and in benign and neoplastic human bladder epithelium. Cancer Res 1979; 39:3856-3860.
119. Madiyalakan R, Piscorz CF, Piver MS et al. Serum β-(1-4)-galactosyltransferase activity with synthetic low molecular weight acceptor in human ovarian cancer. Eur J Cancer Clin Oncol 1987; 23:901-906.
120. Nozawa S, Yajima M, Sakuma T et al. Cancer-associated galactosyltransferase as a tumor marker for ovarian clear cell carcinoma. Cancer Res 1990; 50:754-759.
121. Humphries-Beyer MG, Maeda N, Purushotham KR et al. Increased expression of the enzyme β1-4-galactosyltransferase is associated with human parotid neoplasms. Proc Soc Exp Biol Med 1990; 193:293-300.
122. Kijimoto-Ochiai S, Makita A, Kameya T et al. Elevation of glycoprotein galactosyltransferase activity in human lung cancer related to histological types. Cancer Res

1981; 41:2931-2935.
123. Ip C, Dao T. Alterations in serum glycosyltransferases and 5'-nucleotidase in breast cancer patients. Cancer Res 1978; 38:723-728.
124. Kirschbaum BB. Glycoprotein metabolism in human renal disease:serum glycoproteins and glycoprotein:glycosyl transferase levels in chronic renal failure. J Lab Clin Med 1975; 86:764-771.
125. Jenis DM, Basu S, Pollard M. Increased activity of a beta-galactosyltransferase in tissues of rats bearing prostate and mammary adenocarcinomas. Cancer Biochem Biophys 1982; 6:37-45.
126. Podolsky DK, Weiser MM. Galactosyltransferase activities in human sera: detection of a cancer-associated isoenzyme. Biochem Biophys Res Comm 1975; 65:545-551.
127. Podolsky DK, Weiser MM. Purification of galactosyltransferase "Isoenzymes" I and II. Comparison of cancer-associated and normal galactosyltransferase activities. J Biol Chem 1979; 254:3983-3990.
128. Boyle FA, Cook ND, Peters TJ. Separation and partial characterization of two galactosyltransferase isoforms from malignant ascitic fluid. Clin Chim Acta 1988; 171:187-196.
129. Chatterjee SK, Battacharya M, Barlow JJ. Biochemical and immunologic characterization of galactosyltransferase purified from the ascites of ovarian cancer patients. J Natl Cancer Inst 1985; 75:237-248.
130. Podolsky DK, McPhee MS, Alpert E et al. Galactosyltransferase isoenzyme II in the detection of pancreatic cancer:comparison with radiologic, endoscopic, and serologic tests. New Engl J Med 1981; 304:1313-1318.
131. Mourant A, Kopec A, Domaniewska-Sobczak K. Blood Groups and Diseases. Oxford University Press, Oxford, 1978; 13-15.
132. Hakomori S, Kannagi R. Glycosphingolipids as tumor-associated and differentiation markers. J Natl Cancer Inst 1983; 71:231-251.
133. Bremer EG, Levery SB, Sonnino S et al. Characterization of a glycosphingolipd antigen defined by the monoclonal antibody MBr1 expressed in normal and neoplastic cells of human mammary gland. J Biol Chem 1984; 259:14773-14777.
134. Srinivas V, Khan A, Hoisington S et al. Relationship of blood groups and bladder cancer. J Urol 1986; 135:50-52.
135. Gooi H, Feizi T, Kapadia A et al. Stage-specific embryonic antigen involves α1-3 fucosylated type 2 blood group chains. Nature 1981; 292:156-158.
136. Stellner K, Hakomori S. Enzymatic conversion of "H_1 -glycolipid" to A or B-glycolipid and deficiency of these enzyme activities in adenocarcinoma. Biochem Biophys Res Comm 1973; 55:439-445.
137. Ørntoft T, Wolf H, Watkins W. Activity of the human blood group ABO, Se, H, Le, and X gene encoded glycosyltransferases in normal and malignant bladder urothelium. Cancer Res 1988; 48:4427-4433.
138. Nudelman E, Kannagi R, Hakomori S et al. A glycolipid antigen associated with Burkitt lymphoma defined by a monoclonal antibody. Science 1983; 220:509-511.
139. Magnani J, Nilsson B, Brockhaus M et al. A monoclonal antibody-defined antigen associated with gastrointestinal cancer is a ganglioside containing sialylated lacto-N-fucopentaose II. J Biol Chem 1982; 257:14365-14369.
140. Hakomori S, Nudelman E, Levery S et al. Novel fucolipids accumulating in human adenocarcinoma. J Biol Chem 1984; 259:4672-4680.
141. Hiraiwa N, Tsuyuoka K, Li Y et al. Gangliosides and sialoglycoproteins carrying a rare blood group. Cancer Res 1990; 50:5497-5503.
142. Clausen H, Hakomori S, Graem N et al. Incompatible A antigen expressed in tumors of blood group O individuals: immunochemical, immunohistologic, and enzymatic characterization. J Immunol 1986; 136:326-330.
143. Metoki R, Kakudo K, Tsuji Y et al. Deletion of histo-blood group A and B antigens and expression of incompatible A antigen in ovarian cancer. J Natl Cancer Inst 1989; 81:1151-1157.
144. Yates AD, Feeney J, Donald AR et al. Characterization of a blood-group A-active tetrasaccharide synthesized by a blood-group B gene-specified glycosyltransferase. Carboh

Res 1984; 130:251-260.

145. Simmons D, Perlmann P. Carcinoembryonic antigen and blood group substances. Cancer Res 1973; 33:313-322.

146. Hakomori S, Andrews H. Sphingoglycolipids with Le^b activity, and the copresence of Le^a, Le^b glycolipids in human tumor tissue. Biochim Biophys Acta 1970; 202:225-228.

147. Blaszczyk M, Hansson GC, Karlsson K-A et al. Lewis blood group antigens defined by monoclonal anti-colon carcinoma antibodies. Arch Biochem Biophys 1984; 233:161-168.

148. Fukushi Y, Hakomori S, Nudelman E et al. Novel fucolipids accumulating in human adenocarcinoma II selective isolation of hybridoma antibodies that differentially recognize mono-, di, and trifucosylated type 2 chain. J Biol Chem 1984; 259:4681-4685.

149. Mandel U, Langkilde NC, Ørntoft TF et al. Expression of histo-blood-group-A/B-gene-defined glycosyltransferases in normal and malignant epithelia:Correlation with A/B carbohydrate expression. Int J Cancer 1992; 52:7-12.

150. Kim Y, Issacs R, Perdomo JM. Alterations of membrane glycopeptides in human colonic adenocarcinoma. Proc Natl Acad Sci USA 1974; 71:4869-4873.

151. Dabelsteen E, Graem N, Clausen H et al. Structural variations of blood group A antigens in human normal colon and carcinomas. Cancer Res 1988; 48:181-187.

152. Ørntoft TF, Bech E. Circulating blood group related carbohydrate antigens as tumour markers. Glycoconj J 1995; 12:200-205.

CHAPTER 20

INTESTINAL CANCER

The biochemical and immunological properties of colon cancer mucins differ considerably from the normal. Colon cancer tissue produces glycoproteins and mucins with abnormal glycosylation patterns that may resemble fetal patterns.[1-7] In addition, the genes encoding the peptide moieties of glycoproteins and mucins may undergo alterations of expression in colon cancer. These events may be related to malignant potential and development of the tumor; the mechanisms inducing these changes are still not understood.

Certain features characterize the glycans which appear in colon cancer tissue. They include higher branched N-linked oligosaccharides which are often associated with the metastatic phenotype, premature sialylation of O-linked glycans which give rise to truncated structures, decrease in the sulfation of mucins, changes in the antigenic properties of glycoproteins and expression of cancer-associated carbohydrate and peptide epitopes (Table 6).[8-11]

Monoclonal antibodies have been of great value in correlating the differences in antigen expression with tumor progression in normal and cancerous colonic tissue as well as in other tissues.[12,13] Glycosylation may show regional differences along sections of the intestines and differences along the colonic crypt.[14,15] Moreover, glycoconjugate structures may be affected by microbial degradation.[16]

In general, colonic mucins from cancer tissues have lower molecular weights and less total carbohydrate[17] than their normal counterparts. The oligosaccharides show a reduction in number and mean chain length as well as changes in charge properties.[18,19] Thus O-glycans are reduced in size and complexity. Normal adult mucosa does not express T, Tn or sialyl-Tn antigens (Table 1) which are highly expressed in fetal colon and colon cancer with intermediate expression in the transitional mucosa (Fig. 9).[20] T antigen has been found especially in well-differentiated carcinomas.[21] Sialyl-Tn antigen, expressed in colon cancer, is associated with a poor prognosis.[20-22] Its synthesis precludes the formation of O-glycan core structures or further elongation.

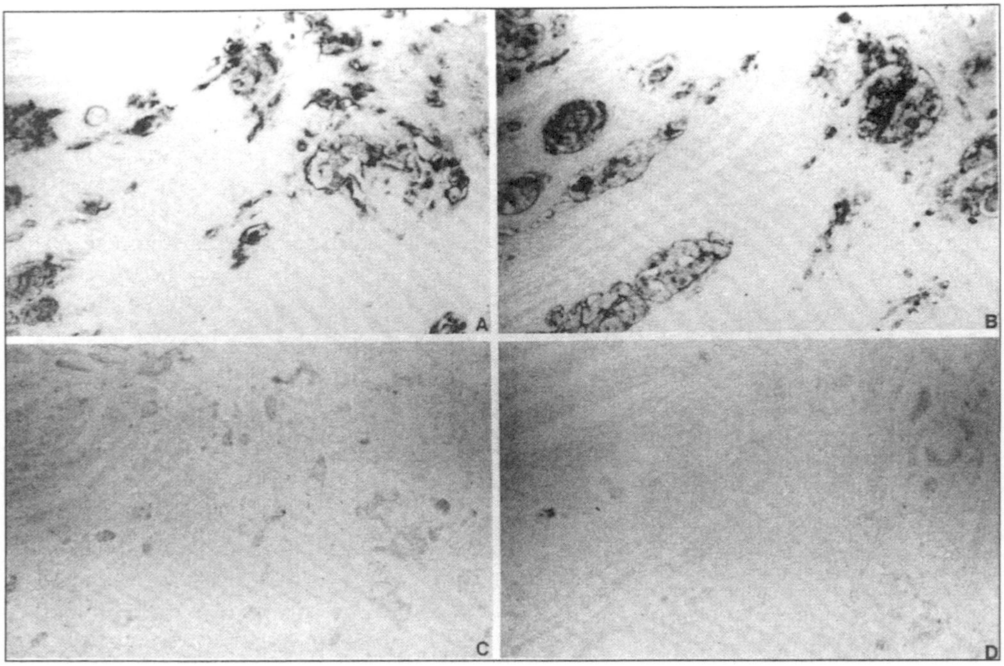

Fig. 9.I.

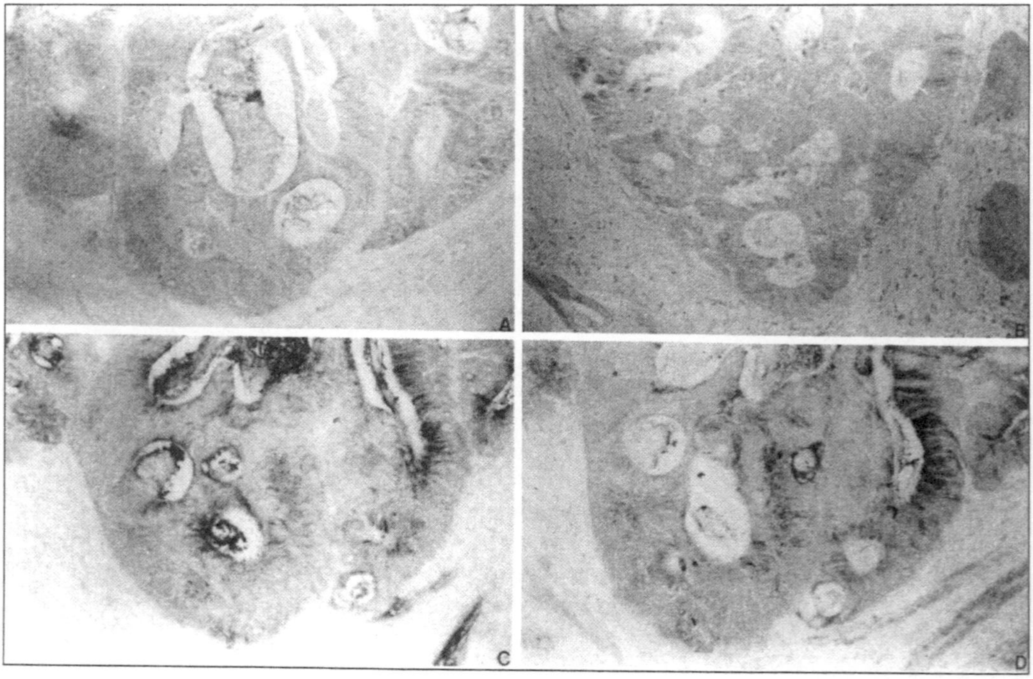

Fig. 9.II.

20.A. SIALYLATION IN COLON CANCER

Sialic acid residues may be modified by tissue-specific O-acetylation by O-acetyltransferase at carbon 4, 7, 8 or 9. A large proportion of the sialyl residues in the normal colon may be O-acetylated.[23] While mono O-acetylation is found ubiquitously, poly O-acetylation is characteristic of colonic tissue in man.[24] Histochemical and chromatographic methods demonstrate that the degree and type of O-acetylation of sialic acid is altered in cancer of the colon.[25-27] O-acetylation is frequently decreased in cancer cells as determined by HPLC, gas chromatography/mass spectrometry or by histochemistry using monoclonal antibodies recognizing O-acetylated sialic acid.[28,29] The loss of the O-acetyltransferase may occur early in the adenoma-carcinoma sequence since premalignant cells often lack O-acetylation.[27,30]

The presence of O-acetyl groups may protect colonic mucins from degradation by intestinal bacteria.[27] O-acetylation has the effect of masking the sialyl-Tn antigen since anti sialyl-Tn antibodies may not react with the O-acetylated form of sialic acid. Thus an increase in sialyl-Tn antigen may result from decreased O-acetylation rather than increased amounts of sialylα2-6GalNAc structures.[23,31,32] Other modifications of sialic acid include the switch from Neu5Ac to Neu5Gc that may be associated with colon cancer.[33-35]

The expression of the sialyl-Tn antigen (Fig. 9) is associated with a poor prognosis in colon cancer and is discussed in chapter 20.F. Sialylation of N-glycans is also changed in colon cancer, in particular there is an increase of sialylα2-6 Gal residues.[36] This is discussed in chapter 20.D.

20.B. LEWIS ANTIGENS

Colon cancer and human colonic cancer cell lines express a number of mucin type glycoproteins containing Lewis and sialyl-Lewis epitopes (Tables 1, 6). Monoclonal antibodies made against these epitopes can discriminate between MUC-1 mucins and other glycans which carry sialyl-Le^a.[37] The expression of Le^a is decreased throughout the colon in cancer, probably due to masking by sialic acid.[38] Le^x antigens are expressed in early development and not normally in the adult colon but are re-expressed as oncofetal antigens in colonic cancer with intermediate levels of expression in transitional tissue, similar to T and Tn antigens.[39] Other cancer-associated Lewis antigens are sialyl-Le^x and Le^y.[40,41]

Unusual Lewis structures can occur in colon cancer, such as extended Le^y,[41,42] internal Le^x or several Le^x structures per chain,[43] as well as their sialylated derivatives.[13,44] The occurrence of the sialyl-Le^x epitope appears to be increased in mucins from colonic tumor tissue[45] and has been associated with poor survival in colon cancer patients due to the high incidence of tumor recurrence.[46] While the sialyl-Le^x epitope is located on MUC1 in cancer tissue, serum contains MUC2 mucins with sialyl-Le^x.[13] Sialyl-dimeric Le^x antigen expression is increased in liver metastases of colon carcinomas.[47]

The appearance of extended Le^x determinants on O-glycans (sialyl-dimeric Le^x) has been found to correlate with colorectal tumor progression.[48] It has been suggested that the oncofetal expression of Le^x observed in human adenocarcinoma is regulated by the synthesis of the type 2 chain which is the substrate for the α3-Fuc-transferase, rather

Fig. 9. (see opposite page) Antigen expression in colon cancer tissue. Histologic sections from individuals with colon cancer. Staining was carried out with monoclonal antibodies or lectins to demonstrate the occurrence of tumor-associated carbohydrate antigens. In the first group of four photographs sections derived from a patient (I) with mucinous colon carcinoma are seen, strongly expressing sialyl-Tn (A) and Tn antigens (B), whereas in the same area of the tumor, blood group H (C) and sialyl-Le^a (D) are absent. In a second patient (II) with well differentiated colon carcinoma, core region antigens sialyl-Tn and Tn antigens are not expressed, but peripheral region antigens blood group H and Le^y are positive. Often there is a reciprocal relationship between core and peripheral antigens. Reprinted from Ogata et al[84] with permission.

than by fucosylation itself.[49] Colon cancer tissue probably expresses several enzymes of the α3-Fuc-transferase family (Table 2).[50]

20.C. MUCIN EXPRESSION

Glycoprotein peptide sequence composition, conformation and gene overexpression may greatly influence the nature and extent of the carbohydrate structures which they bear. It appears that the major mucin secreted in the colon is MUC2.[51] MUC2 expression in cancer differs according to the type of cancer, with an overexpression in adenomatous tissue and often a decreased expression in de novo carcinomas.[52] In rectosigmoid villous adenoma MUC2 mRNA is overexpressed, but in addition, high levels of MUC5AC mRNA are found. In contrast, this marker is not detected in adenocarcinomas or normal colon, and is believed to be of use in the early detection of recurrences of adenomas.[53]

Normal epithelial cells also have cell surface mucin MUC1 which is thought to be a marker for the progression and metastatic behavior of colon cancer cells.[54] Normal colonic MUC1 mucin does not express Le^x determinants whereas colon cancer MUC1 does,[13] indicating that changes in MUC1 glycosylation contribute to its cancer-associated expression. Mucin gene expression and mucin epitope variability is seen in normal intestinal tissue and in cancer,[55] and there may be an ectopic expression of MUC genes found normally in another tissue. For example, MUC5AC normally found in respiratory and reproductive mucosa is aberrantly expressed in rectosigmoid villous adenoma.[53] Many novel mucin epitopes defined by monoclonal antibodies have been reported in cancer.

20.D. BIOSYNTHESIS OF OLIGOSACCHARIDES IN COLON CANCER

The origin of glycosylation changes in colon cancer has been investigated by measuring the activities of several glycosyltransferases of the O-glycan pathway in normal and cancerous colonic tissue (Table 6).[17] α6-Sialyltransferase (O) synthesizing the sialyl-Tn antigen, core 1 β3-Gal-transferase, β4-Gal-transferase, blood group B α3-Gal-transferase, α2- and α3-Fuc-transferases and polypeptide αGalNAc-transferase are decreased in colon cancer tissue in comparison with normal tissue. Recent studies on tissue samples from 20 colon cancer patients[56] confirmed the decrease in polypeptide GalNAc-transferase and α6-sialyltransferase (O) but not of the other enzymes, possibly due to different assay conditions since in the latter studies Vmax values were compared. The decrease in polypeptide GalNAc-transferase may lead to incomplete O-glycosylation thereby leaving potential Ser or Thr sites unsubstituted. These results are consistent with the observed loss of mucin glycosylation in cancer.[18,19]

Although there is increased expression of the T-antigen (O-glycan core 1) in cancer tissue,[57] the enzymes synthesizing core 1, i.e. polypeptide GalNAc-transferase and core 1 β3-Gal-transferase, are not increased. The T antigen expression may therefore be due to a rerouted synthesis, for example, lack of extension leading to increased core 1 exposure in cancer. However, α3-sialyltransferase (O) acting on core 1 and α2-Fuc-transferase synthesizing the H-antigen, enzymes which are expected to mask the T-antigen by substituting core 1, show considerable activity increases in cancer tissue.[56] Moreover, these masking reactions do not favor increased exposure of the T antigen. It is possible that intracellular rerouting of enzymes and substrates prevents these reactions and thus can lead to the exposure of precursor structures.

The α3-sialyltransferase (O) activity is also increased in leukemia[58] and in breast cancer cells[59] and may prematurely terminate O-glycan synthesis since the sialylated core 1 structure is no longer a substrate for elongation and branching enzymes. However, the competition between α3-sialylation and other core 1 processing enzymes can only occur if the enzymes reside in the same Golgi compartment.

A possible explanation for the increase in T-antigen[60,61] comes from the finding that core 3 β3-GlcNAc-transferase, the enzyme synthesizing core 3, is decreased in most samples of colon cancer tissue.[56,62] Core 3 synthesis is probably competing with core 1 synthesis although the intracellular localization of these enzymes remains to be determined. Thus a decrease of core 3 synthesis may promote the synthesis of core 1 and expression of the T antigen. Core 2 synthesis is decreased in human colon cancer tissue which reduces the further processing of core 1 (Fig. 1). In addition, there is less core 4 synthesis and GlcNAcβ1-6 branching to form the I antigen (Table 1). It appears therefore that it is β6-GlcNAc-transferase M (see chapter 4.C.2) that is primarily affected in colon cancer. The poly-n-acetyl-lactosamine structures are expected to be primarily unbranched in colon cancer; this was confirmed by the use of monoclonal antibodies specific for these structures.[63]

In contrast to the truncation of O-glycans, the N-glycans in colon cancer cells appear to be more highly branched[8] as seen in the increased staining with L-PHA which could be attributed to an increase in GlcNAc-transferase V activity.

Several studies suggest that the α2-Fuc-transferase may be expressed in a cell type-specific manner along the intestines. The appearance of antigens, especially in the intestines, may also be theoretically subject to modification by intestinal bacteria which may degrade oligosaccharides through their glycosidases. However, within cancer tissue, glycosidase activities appeared to be at normal levels.[16]

The H- and Se-gene-dependent α2-Fuc-transferase activities are increased in rectal cancer compared to normal tissue.[64] This is consistent with the finding that *Ulex europeus* lectin ligands (Fucα1-2Galβ1-4GlcNAc-) are increasingly expressed in transformed distal colon and rectum.[65] Higher activities of α2-Fuc-transferase in colon cancer tissue compared to normal tissue may result in increased H type 1 and 2 chains and Lewis^b and Lewis^y structures.[56,66] The malignant progression to colon adenocarcinoma in human polyps and cancer tissue is associated with an increase in the expression of the α2-Fuc-transferase. Much higher levels of mRNA for α2-Fuc-transferase were found in adenocarcinomas and colon cancer cell lines compared to normal colon mucosa.[67]

The latter findings are consistent with immunohistochemical studies of ABH expression in 68 samples of colon carcinoma in which normal adjacent tissue was available for comparison. There is a net loss of ABH expression in the proximal cancerous colon and a net gain in the distal cancerous colon when compared with antigen patterns in normal tissue.[68] Blood group A and B transferases are reduced in activity in colonic cancer tissue extracts.[17,69]

The α3/4-Fuc-transferase activity is increased in carcinoma tissue compared with normal tissue from the same patient,[70] but this could not be confirmed by another study.[56] This variation may be due to different assay conditions, or due to a different spectrum of patients and sites of tissue sections.

The observed decrease in α6-sialyltransferase (O) in colon cancer tissue[56] is unexpected, considering the occurrence of cancer-associated sialyl-Tn antigen.[21,22,71] A lack of α6-sialyltransferase (O) activity was observed in breast cancer cells[59] and tumor cells derived from familial polyposis coli cells which produce sialomucins.[30,72] It is possible that assay conditions and substrate structures for this activity may be different in cancer due to the expression of other members of this gene family. Several different cDNA species of the α6-sialyltransferase family have been cloned and more members of this gene family may occur in various tissues and in cancer (see chapter 4.E.1). Other possibilities for sialyl-Tn expression may be decreased O-acetylation of sialic acid, decreased core 1 β3-Gal-transferase, increased α6-sialyltransferase (O) or other factors.

Neoplastic colonic mucosa but not normal mucosa expresses sialylα2-6 Gal structures as visualized by *Sambucus nigra* lectin[36] while sialylα2-3 residues as detected by *Maackia amurensis* lectin are more generally distributed in normal and cancer cells. Consistent with these structural data, Dall'Olio et al found that α6-sialyltransferase (N) acting on N-glycans is specifically increased in colorectal cancer.[73] The higher activity may be due to a lower K_m for CMP-sialic acid of the α6-sialyltransferase (N) in cancer cells. Differences in enzyme kinetics are commonly observed between cell types and demonstrate that not only mRNA levels but also enzyme kinetics and specificity may be regulated by tissue or cell-specific factors.

Another important alteration in colonic cell glycosylation is the switch from sulfated to sialylated glycoproteins during tumor progression,[74] shown by the decrease in the amount of sulfate found in mucins as visualized by anti sulfomucin antibody staining.[10,11] In tissue samples from colon cancer patients this decrease is accompanied by decreased core 1 3-sulfotransferase activity but not GlcNAc sulfation.[56] A similar decrease of core 1 3-sulfotransferase is seen in tumor cells derived from cultured cells of a patient with familial polyposis. The decrease correlates with the sulfation index (incorporation of radioactive sulfate per radioactive glucosamine) of mucins produced in these cells.[30] This suggests that core 1 3-sulfotransferase[75] is primarily responsible for colonic mucin sulfation and is downregulated during cancer.

Sialylated O-glycan core 5, sialylα2-6(GalNAcα1-3)GalNAc-, has been described in meconium[76] and therefore is produced in the intestines before birth. Sialylated core 5 is also found in human rectal adenocarcinoma mucins[77] and probably occurs as on oncofetal structure. The α3-GalNAc-transferase responsible for the synthesis of core 5 has been found in one out of six cancer tissue specimens[78] which suggests that the expression of this enzyme may reoccur only in some cancer types.

Carcinoembryonic antigen (CEA) is developmentally regulated since it is present during embryogenesis but virtually absent from the normal adult. It is strongly re-expressed in carcinomas, particularly colon carcinoma, and occurs as a number of variants. The higher glycosylated variants tend to be synthesized in cancer cells.[2] The biosynthesis of CEA in the presence of tunicamycin produced one molecular weight form both in normal and cancer cells suggesting that the abnormal forms of CEA found in cancer are glycosylation variants. It is of interest that in colorectal carcinomas cell surface H antigen and CEA often occur simultaneously.[79] CEA may thus provide a favorable substrate for ABH blood group glycosyltransferases.[80] CEA contains complex N-glycan chains with type 1 and 2 chains and Lewis antigens;[81] these appear to be modified early in cancer making this glycoprotein a sensitive model to study glycosylation changes in cancer.[2]

20.E. MODELS FOR STUDYING COLON CANCER-RELATED GLYCOSYLATION CHANGES

The discrepancies often observed in studies with cancer tissue may lie in the fact that tissue represents a heterogeneous cell population. Glycosyltransferase activities in tissue hemogenates reflect an average of all of these cell types. This problem may be partly circumvented by studying individual cell types in culture, however caution must be exercised in interpreting these data due to the possibility that cultured cells may lose their natural phenotype and may undergo various routes of differentiation. Another problem with cell lines is that the expression of glycosyltransferases fluctuates during cell cycling and this affects the phenotype. Nevertheless, cultured cells represent good models for the study of cancer since they do express glycoproteins and enzymes similar to those in cancer tissues; the examination of biosynthetic mechanisms in a number of cancer cell lines has been successful.

Mucin production in cultured human colon cancer cells HT29 depends on the growth phase and is a determinant of tumor growth when cells are injected into nude mice.[82] Human cancer cells and cell lines often express mucin genes such as those encoding MUC1, MUC2 and

MUC3.[83,84] Their expression levels as well as their antigenicity with respect to peptide and to carbohydrate differs between the various cancer cell lines.[85] mRNA for MUC1 is expressed at higher levels in human colon cancer LS174T cells but only at very low levels in HT29, Caco-2, T84 and LS180 cells or in the ileocaecal adenocarcinoma NCI-H498 cells.[86] MUC2 and MUC3 mRNA are expressed in all of these cell lines at higher levels while MUC4 mRNA is only detected in LS 174T and Caco-2 cells.

The mRNA levels for mucin MUC2 and MUC3 genes in HT29 human colon cancer cells change depending on the Gal and Glc content of the growth medium, indicating that the environment of cells influences mucin gene expression.[87] Glucose-free and galactose-substituted medium induces phenotypic differentiation and changes the ratio of MUC2 versus MUC3 expression. MUC2 gene expression is induced in HT29 cells by triggers of protein kinase C- and A-dependent signal transduction pathways[88] but not by sodium butyrate as a differentiating agent. Nuclear run off analyzes suggest that post transcriptional rather than transcriptional control mechanisms appear to be important in MUC2 induction. Mucin expression in HT29 human colon cancer cells is characteristic of subpopulations and stages of growth.[89] Different subpopulations express characteristic spectra of the various mucin genes. The levels of mRNA expression and maturity of mucins based on immunoreactivity fluctuate during growth and differ between populations. The levels of mucin production appear to be related to the metastatic potential of human colon cancer LS174T cell variants in nude mice.[90] Mucinous carcinomas which secrete large amounts of mucins often have a poor prognosis, perhaps because they are refractory to immune surveillance.[91]

Enhanced secretion of mucins and expression of MUC1, MUC2, MUC3 and possibly other mucins is found in methotrexate-resistant subpopulations of HT29 cells.[92] Dall'Olio et al[93] reported that methotrexate resistance was associated with a selective increase in α3-sialyltransferase (N) acting on N-glycans and a virtual disappearance of α6-sialyltransferase (N), both measured with N-acetyl-lactosamine substrates. Several other sialyltransferase activities toward asialoglycoprotein substrates are not correlated with methotrexate resistance.

HT29 cells produce sialomucin such as MUC1 and the expression of this mucin may be regulated by soluble factors. Conditioned medium from normal human colonic connective tissue cultures induces the production of sialomucins but leaves the O-glycan processing apparently unaffected.[94] The stably differentiated HT29 cell line CL.16E produces oligosaccharides mainly with O-glycan core 1, 2 and 4 structures.[95] These may be elongated and terminated with sialic acid in α2-3 linkage to Gal with the major structure sialylα2-3 Galβ1-4 GlcNAcβ1-6 (sialylα2-3) Galβ1-3 GalNAc-. Sulfate was also found on these O-glycans linked to the 6-position of the same Gal residues that were 3-substituted by sialic acid.

Human colonic cancer cells in culture may differ from normal colonic tissue in their glycoprotein production, glycosylation and glycosyltransferase expression. O-glycan core 3 β3-GlcNAc-transferase is present in normal human colon but is low in colon cancer tissue and is greatly reduced in colon cancer cell lines.[96,97] This could be due to an effect of cell culturing by which the expression of this enzyme is lost, but may also indicate that a selected cell population in colon cancer tissue is capable of synthesizing core 3. Core 3 β3-GlcNAc-transferase is undetectable in leukemia and certain breast, ovarian and intestinal cell lines. The absence of the enzyme in most cells including cancer and transformed cells may be explained by the tissue-specific downregulation of core 3 β3-GlcNAc-transferase.

Human colonic Caco-2 cells spontaneously differentiate to enterocytes and other cell types in culture. The amount of fucosylated poly-N-acetyl-lactosamine structures (i.e. those attached to lamp-1 glycoproteins) decreases during differentiation.[98,99] Concomitantly, a general increase in N-glycan branching enzymes (GlcNAc-transferases II, III, IV and V) occurs.[96] Increased branching would be expected to promote the addition of poly-n-acetyllactosamine structures. However, the action of GlcNAc-transferase III[100] which introduces the bisected GlcNAc residue β1-4-linked to Manβ- has the effect of turning off further branching and chain elongation, probably due to its effect on the three-dimensional structure of the N-glycan chain. Thus the increase of GlcNAc-transferase III activity in Caco-2 cells may block poly N-acetylactosamine synthesis while the increase in other branching enzymes may be a compensatory mechanism.

Caco-2 cells show some interesting differences when compared with normal colon. Caco-2 cells cannot synthesize core 3 and have changed the specificity of core 2 β6-GlcNAc-transferase from the M-type in normal colon to the L-type.[101] Thus the enzyme in Caco-2 cells cannot synthesize the I antigen or core 4 effectively. This change in core 2 β6-GlcNAc-transferase specificity is seen in several but not all colon cancer cell lines.[97] The expression of the L or M type enzyme may depend on whether cells are differentiating to mucin-secreting or other cell types.

During enterocytic differentiation of Caco-2 cells core 2 β6-GlcNAc-transferase increases while core 1 β3-Gal-transferase[96] decreases. α6-Sialyltransferase (N) expression which is high in colon cancer and in fetal and newborn rats as opposed to normal adult rats[73,102] is under developmental regulation. During differentiation of Caco-2 cells the activity[103] and mRNA expression[104] increases.

Caco-2 cells also secrete a soluble form of α6-sialyltransferase into the culture medium. In addition, differentiated Caco-2 cells secrete a soluble form of β4-GalNAc-transferase synthesizing the Sd^a determinant (Table 1).

The expression of sialylα2-6 residues on N-glycans has been shown by the binding of the sialylα2-6Gal/GalNAc-binding *Sambucus nigra* lectin and this was correlated with α6-sialyltransferase (N) activity in colorectal cell lines COLO 205, SW 948 and a subline SW 948 FL.[105] These sialic acid residues may play a role in adhesion since colon cancer cell variants with high activities of α6-sialyltransferase (N) and high expression of N-glycans with sialylα2-6 linkage adhere less well to cell culture flasks than their counterparts with less activity. In contrast, adhesion of colon cancer cells to collagen type IV appears to be correlated with sialylation since sialidase treatment decreases cell adhesion.[106]

Human colonic cancer HT29 cells possess glycoproteins that bind to L-PHA indicating that these are branched due to the action of GlcNAc-transferase V. Examination of cells and cell medium by pulse labeling, peptide mapping and lectin binding suggest that the expression of various glycoforms of secreted or intracellular glycoproteins is a function of the differentiation status and progression of metastases.[107]

Poly N-acetyl-lactosaminoglycans have been implicated in cell adhesion phenomena of normal and tumor cells. Their association with antigens such as sialyl-Le^x in colon cancer and other tumors can indicate invasive and metastatic phenotypes. Inhibition studies with the N-glycan processing inhibitor deoxymannojirimycin in human colon cancer HT29 cells suggest that N-glycan processing is essential for the expression and the adhesion function of integrins. Altered glycosylation of integrin adhesion molecules occurs in human colorectal cancer cells along with decreased adhesion to extracellular matrix.[108]

Adhesion of sialyl-Le^a and sialyl-Le^x ligands to endothelial E-selectins[109,110] is thought to be responsible for tumor cell adhesion to activated endothelium and therefore involved in metastases. Sawada et al[111] have found that adhesion mediated by the E-selectin binding is proportional to the amount of cell surface lysosomal membrane glycoprotein lamp-1. Both lamp-1 and lamp-2 occur on colon cancer L4 cell surfaces with more of these glycoproteins expressed in highly metastatic cells compared to low metastatic cells. Highly metastatic colon cancer cell lines contain more poly-n-acetyl-lactosamine sequences attached to lamp glycoproteins and those cells expressing sialyl-Le^x are bound more efficiently to E-selectin than poorly metastatic lines. The E-selectin-dependent cell adhesion mediated by sialyl-Le^x ligands may therefore be critical in the metastatic spread of cancer cells.[112a,b]

Two different α3-Fuc-transferase activities synthesizing Le^x structures are present in human colon cancer cell lines HT29 and COLO205, distinguished by their divalent cation requirement, N-ethylmaleimide inhibition and substrate specificity. The enzyme from HT29 cells does not act on fetuin but acts on neutral substrates (asialo-fetuin) whereas the enzyme from COLO205 cells acts equally on sialylated and desialylated substrates, exhibits higher α4- than α3-Fuc-transferase activity, and may thus be equivalent to α3/α4-Fuc-transferase III.[113]

20.F. MODEL CELLS TO STUDY THE MECHANISM OF SIALYL-Tn EXPRESSION

Although sialylα2-6 GalNAc structures occur in normal submaxillary mucins from several species, the occurrence of this structure in other tissues is often associated with tumors. The mechanism for the appearance of the sialyl-Tn antigens in human colon cancer cells has been investigated in model cells isolated from human colon cancer LS174T cell line.[114] LSB cells contain larger oligosaccharides and do not express sialyl-Tn. They are more metastatic in nude mice whereas the LSC cells are less metastatic and do express sialyl-Tn and Tn antigens.[115] Both cell lines are similar in their sensitivity to lysis by natural killer cells. Investigations of the O-glycan biosynthetic pathways revealed that both cell lines have the potential to make elongated, fucosylated, sialylated and sulfated O-glycans of core 1 and 2 but not 3 and 4 structures (Fig. 10). However, LSC cells show a defect in core 1 synthesis because of the absence of significant core 1 β3-Gal-transferase activity. The activity is not reactivated by Na-butyrate differentiation. The absence of core 1 and 3 synthesis explains why cells are committed to synthesizing GalNAc and sialyl-GalNAc chains.

An alternative mechanism for sialyl-Tn expression is seen in the rat colon cancer cell model LMCR. Two cell lines differ in sialyl-Tn expression; LMCR+ cells express sialyl-Tn and the enzymes synthesizing and elongating core 1 and 2 but not core 3. Thus the appearance of the sialyl-Tn antigen is possible in the presence of core 1 synthesis, due to kinetic factors or due to the intracellular organization of enzymes. The lack of sialyl-Tn expression in LMCR- cells is associated with a loss of the α6-sialyltransferase (O) which is reactivated by Na-butyrate differentiation.[116] The α6-sialyltransferase (O) is therefore another enzyme which is regulated during differentiation.

20.G. POLYPS

Human colorectal polyps, especially those from patients with familial polyposis coli, tend to develop into carcinomas. Carbohydrate alterations are less likely observed in hyperplastic polyps with minimal malignant potential;[44] but adenomatous polyps do possess tumorigenic tendency and may demonstrate atypical glycan synthesis and structure. Glycan abnormalities may result from discordance between mucin gene mRNA expression and their effects upon subsequent oligosaccharide biosynthesis.

During the development to cancer there is an increase in siayl-Le^x in sialomucins as detected by monoclonal antibodies,[117] together with the appearance of extended Le^y and Le^x antigens which correlates with the malignant potential of polyps.[41,118] Cancer-associated changes in mucins may be found in a proportion of apparently benign polyps[14] and may therefore indicate that some cells are becoming malignant.[44]

Differences in glycosyltransferase activities have been reported between normal, and cancerous mucosa and polyps from patients with familial polyposis.[119] However, in this study glycoprotein substrates with multiple potential acceptor sites were used. A model of clonogenic and tumorigenic cell lines has been developed from polyps of a patient with polyposis coli to

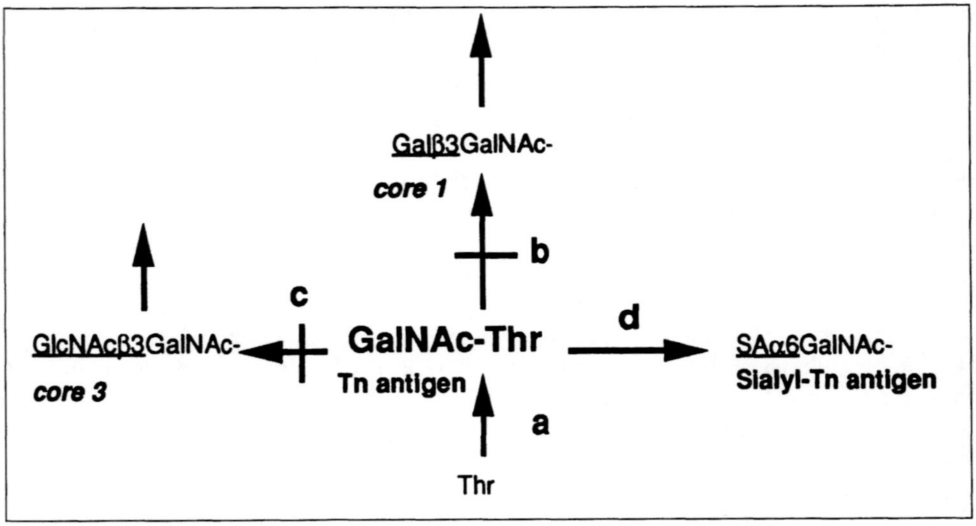

Fig. 10. Biosynthesis of O-glycans in the human colonic cancer cell line LSC.[116] The letters denote the transferases catalyzing the conversions: a, polypeptide α-GalNAc-transferase; b, core 1 β3-Gal-transferase; c, core 3 β3-GlcNAc-transferase; d, α6-sialyltransferase (O). Blocked pathways are indicated by a bar across the arrow. LSC cells do not express the enzymes synthesizing O-glycan core 1 and 3 which are the precursor structures for branched and elongated O-glycans found in normal colonic mucins. The lack of common core structures leads to the appearance of the cancer-associated Tn and sialyl-Tn antigens.

study the progression to cancer. In this adenoma-carcinoma sequence marked antigen expression and decreased sulfation of secreted mucins occurs.[120] These changes are accompanied by fluctuations in glycosyltransferase and sulfotransferase activities.[30] Core 3 synthesis, α6-sialyltransferase (O), core 4 and I β6-GlcNAc-transferase activities are lost in the tumorigenic cells which could explain the increased exposure of MUC1 peptide epitopes. Although the α3-Fuc-transferase is active in all cell lines, the expression of sialyl-Le^x antigen is lost in the clonogenic and tumorigenic cells. α4-Fuc-transferase activity is very low in all cases explaining the lack of Le^a expression. The sulfation index as well as staining with anti sulfomucin antibodies is greatly reduced in the tumorigenic cells. This correlates with a decrease in sulfotransferases acting on O-glycan core 1 and GlcNAcβ1-6GalNAc-R (core 6) substrates. Many of these findings correspond to those from studies on colon cancer tissue which suggests that these polyposis-derived cells are a good model to study glycosylation changes associated with carcinogenesis.

20.H. BIOLOGICAL BEHAVIOR OF COLON CANCER CELLS

Experiments to determine whether the onco-developmentally regulated H antigens directly affect tumor behavior were carried out in a rat model of colon cancer. In these cancer cells, the H antigen is carried by spliced variants of the adhesion molecule CD44. Transfection of a highly tumorigenic clone was carried out using an antisense fragment of the gene coding for rat α2-Fuc-transferase. The transfected subclones had decreased α2-Fuc-transferase activity and cell surface H antigen and were far less tumorigenic in syngeneic hosts relative to controls.[121]

In a number of rat and human colon cancer cells, increased resistance to NK-mediated cytotoxicity seems to be associated with higher than normal blood group A and H cell surface expression, whereas increased susceptibility is associated with lower levels of A and H as well as truncated structures such as T and Tn antigens.[122] The presence of larger glycans containing A, H and Lewis antigens appears to protect cell surfaces making them resistant to NK cell lysis. Thus cells expressing these more highly processed oligosaccharides carry cell adhesion ligands which increase their survival and also their invasive potential.

In addition to surface carbohydrates, human colon cancer tissues may show changes in the expression of endogenous lectins at the cell surface. These may function in cell adhesion and metastasis.[123] Galectins, Gal binding lectins, have been implicated in neoplastic progression of colon cancer.[124] Differentiating agents may modulate the expression of galectins. Differentiation of human colon cancer cells KM12 is accompanied by galectin-1 neoexpression and increased CEA expression. Endogenous glycoproteins containing ligands for galectin binding were affinity purified and identified as CEA, lamp-1 and lamp-2.[125]

Differences in tumor cell-selectin binding are thought to be significant factors as related to variability in the host response to cancer progression. This problem was analyzed by studying the interactions of human colon cancer cell lines (LS180, T84, COLO205, COLO320, Caco-2) with recombinant purified E-, P-, and L-selectins. Each cell line displayed a characteristic binding pattern. Thus, COLO320 binds to P- and L-selectin but not E-selectin, HT29 cells bound E-selectin but not P- or L-selectin, LS180 and T84 cells react to different degrees to all selectins while Caco-2 cells show little or no adhesion to any selectin. It was noted that E-selectin binding to this panel correlates well with the expression of sialyl-Le^x and sialyl-Le^a, in contrast to the binding patterns with P- and L- selectins. For example, COLO320 cells, which exhibit low levels of these sialylated antigens, do not bind E-selectin.[126]

Premature truncation of mucin oligosaccharides can be induced in culture with GalNAcα-benzyl which is a competitive substrate for O-glycan core 1 and 3 formation and reduces sugar incorporation into mucins by competition with the natural substates.[127] Truncation of O-glycans significantly reduces the binding of colon cancer cells Colo 205 to E-selectin.[128]

20.I. GASTRIC CANCER

Mucin-type O-glycan structures may be truncated in gastric cancers as they are in other cancers resulting in Tn, sialyl-Tn, T and sialyl-T antigen expression which is much less common in normal gastric mucosa.[129] Several different mucin core proteins are expressed in gastric mucosa and they carry T antigens recognized by peanut lectin binding.[130]

The expression of T and Tn antigens in gastric tissue determined by immunohistochemistry is dependent on the cell type and precursors of gastric carcinomas such as metaplasia, dysplasia and hyperplastic lesions and is also associated with gastric tumors (Table 6).[131] Changes in glycosylation occur early in the carcinogenic sequence. In many cancers, sialyl-Tn expression is correlated with a poor clinical outcome and shortened patient survival.[132] The serum levels of sialyl-Tn may be useful as a prognostic indicator and to predict tumor stage.[133] These high serum levels appear to originate from high sialyl-Tn expressing tumors.

The expression of sialyl-Tn in tumor tissue has been associated with lymph node metastasis and peritoneal dissemination[134] as well as the invasive properties of the cancer.[135] The proliferative activity of the tumor is higher in the group expressing sialyl-Tn antigen.[134] The O-acetyl derivatives of sialic acid are absent from normal gastric mucosa but are present in a high number of gastric tumors.[136]

A novel tumor-related carbohydrate epitope F1α has been defined as O-glycan core 6 GlcNAcβ1-6GalNAcα-R.[137] It is thought that the expression of this epitope begins almost at the same time as does carcinogenesis in gastric epithelial cells. Staining patterns for sialyl-Tn and T antigens and F1α are not always correlated in that sialyl-Tn and Tn antigen expression is related to the degree of malignancy while F1α expression does not differentiate between primary and metastatic lesions. The synthesis of core 6 and sialyl-Tn is mutually exclusive, corroborating the fact that the expression of these antigens seldom coincides.

A cancer-type glycoconjugate distinct from the normal could be demonstrated in a stomach lesion from a patient whose blood group and tissue tested positively for a rare genotype. The tumor carbohydrate was identified as GalNAcβ1–3Galβ1-3GlcNAcβ1-3Galβ1-4Glcβceramide.[138] This is a rare variant of the P blood group system (Table 1). Specific absorption of the patient's own antibodies by her own tumor cells was interpreted as being due to a genetic mutation resulting in the synthesis of tumor antigens which are foreign to the host. Antibodies directed against glycolipid antigens are believed to cross-react with glycoprotein determinants carrying the same carbohydrate residues in similar linkages and sequences.[139]

Lewis antigens are highly expressed in gastric cancers. The expression is lower in normal gastric tissue and depends on the cell type and the secretor status.[140]

MUC gene expression is altered in neoplastic gastric epithelium as seen by mRNA analyzes. While normal gastric mucosa expresses MUC1, MUC5 and MUC6, high levels of MUC3 and MUC4 are expressed with decreased levels of MUC5 and MUC6 in gastric cancers.[141] The MUC1 tandem repeat peptide epitope is exposed in gastric cancer, as seen in other cancers. It has been suggested that a single terminal Fuc residue linked to O-glycan core 1 attached to Thr near the MUC1 peptide epitope could mask this epitope. Removal of the Fuc residue by α-fucosidase results in epitope exposure.[142]

Incompatible blood group A expression has been observed in gastric tumor tissue of blood group O individuals.[143] Antibodies against blood group A transferase stained three of 31 frozen tumor tissues by immunohistology. This suggests that the incompatible blood group A expression is due to the inappropriate synthesis of the enzyme responsible for the A or Forssman-like antigens.

Aberrant sulfation may occur in gastric cancers. A monoclonal antibody against a sulfated carbohydrate epitope strongly reacts by histologic staining with tissue from gastric cancer and intestinal metaplasia compared to normal tissues which are negative.[144] Immunostaining and RNA analysis indicates that MUC1 is absent and MUC2 is expressed at very low levels, suggesting that the sulfated carbohydrates reside on different glycoproteins in gastric cancer. A sulfotransferase that sulfates Gal-cerebroside is significantly elevated in gastric tumors in relation to the normal counterpart.[145] It is possible that this enzyme also acts on glycoproteins.[75]

Sulfatides of human gastric mucosa are believed to serve as major receptors for cell adhesion by *Helicobacter pylori*. Colonization of the stomach with *Helicobacter pylori* is closely associated with differentiated-type gastric cancer and may therefore increase the risk for gastric cancer. Monoclonal anti-sulfatide antibodies reduce *Helicobacter pylori* adhesion to the gastric cancer cell line KATO III.[146]

The cell adhesion molecule E-cadherin is believed to function in the suppression of cancer invasion. Reduced E-cadherin in gastric carcinoma cell lines may cause dysfunction of cell-cell adhesion and thus trigger invasion and metastasis.[147]

References

1. Hanisch F-G, Heimbüchel G, Baldus SE et al. Monoclonal antibody FW6 defines an epitope on α3/4 monofucosylated poly-lactosaminoglycans expressed by fetal and colon carcinoma-associated mucins. Cancer Res 1993; 53:4367-4375.
2. Garcia M, Seigner C, Bastid C et al. Carcinoembryonic antigen has a different molecular weight in normal colon and in cancer cells due to N-glycosylation differences. Cancer Res 1991; 51:5679-5686.
3. Cooper H, Malecha M, Bass C et al. Expression of blood group antigens H-2, Le^y, and sialylated-Le^a in human colorectal carcinoma.

Am J Pathol 1991; 138:103-110.

4. Lance P, Lev R. Colonic oligosaccharide structures deduced from lectin-binding studies before and after desialylation. Hum Pathol 1991; 22:307-312.
5. Pilbrow SJ, Hertzog PJ, Linnane AW. The adenoma-carcinoma sequence in the colorectum–early appearance of a hierarchy of small intestinal mucin antigen (SIMA) epitopes and correlation with malignant potential. Br J Cancer 1992; 66:748-757.
6. Yuan M, Itzkowitz S, Boland C et al. Comparison of T-antigen expression in normal, pre-malignant and malignant human colonic tissue using lectin and antibody immunohistochemistry. Cancer Res 1986; 46:4841-4847.
7. Itzkowitz S. Caarbohydrate changes in colon carcinoma. In:Carbohydrate Pathology. Dabelsteen E, Clausen H eds. Copenhagen: Munksgaard, 1992; APMIS Suppl. 27, 100:173-180.
8. Fernandes B, Sagman U, Auger M et al. β1-6 branched oligosaccharides as a marker of tumor progression in human breast and colon neoplasia. Cancer Res 1991; 51:718-723.
9. Dennis J, Laferté S, Yagel S et al. Asparagine-linked oligosaccharides associated with metastatic cancer. Cancer Cells 1989; 1:87-92.
10. Yamori T, Kimura H, Stewart K et al. Differential production of high molecular weight sulfated glycoproteins in normal colonic mucosa, primary colon carcinoma, and metastases. Cancer Res 1987; 47:2741-2747.
11. Yamori T, Ota DM, Cleary KR et al. Monoclonal antibody against human colonic sulfomucin:immunochemical detection of its binding sites in colonic mucosa, colorectal primary carcinoma, and metastases. Cancer Res 1989; 49:887-894.
12. Nakasaki H, Mitomi T, Noto T et al. Mosaicism in the expression of tumor-associated carbohydrate antigens in human colonic and gastric cancers. Cancer Res 1989; 49:3662-3669.
13. Hanski C, Hanski M-L, Zimmer T et al. Characterization of the major sialyl-Lewis^x positive mucins present in colon, colon carcinoma, and sera of patients with colorectal cancer. Cancer Res 1995; 55:928-933.
14. Boland C, Montgomery C, Kim Y. Alterations in human colonic mucin occurring with cellular differentiation and malignant transformation. Proc Natl Acad Sci USA 1982; 79:2051-2055.
15. Bresalier RS, Boland CR, Kim YS. Regional differences in normal and cancer-associated glycoconjugates of the human colon. J Natl Cancer Inst 1985; 75:249-255.
16. Hoskins L. Degradation of mucus glycoproteins in the gastrointestinal tract. In: Horowitz M, Pigman E eds. The Glycoconjugates Vol 2. New York:Academic Press, 1978:235-253.
17. Kim YS, Isaacs R. Glycoprotein metabolism in inflammatory and neoplastic diseases of the human colon. Cancer Res 1975; 35:2092-2097.
18. Boland CR, Deshmukh GD. The carbohydrate composition of mucin in colonic cancer. Gastroenterology 1990; 98:1170-1177.
19. Shimamoto C, Deshmukh GD, Rigot WL et al. Analysis of cancer-associated colonic mucin by ion-exchange chromatography: evidence for a mucin species of lower molecular charge and weight in cancer. Biochim Biophys Acta 1989; 991:284-295.
20. Ørntoft T, Harving N, Langkilde N. O-linked mucin-type glycoproteins in normal and malignant colon mucosa:lack of T antigen expression and accumulation of Tn and sialosyl-Tn antigens in carcinomas. Int J Cancer 1990; 45:666-672.
21. Itzkowitz S, Yuan M, Montgomery C et al. Expression of Tn. sialosyl-Tn and T antigens in human colon cancer. Cancer Res 1989; 49:197-204.
22. Itzkowitz SH, Bloom EJ, Kokal WA et al. Sialosyl-Tn:a novel mucin antigen associated with prognosis in colorectal cancer patients. Cancer 1990; 66:1960-1966.
23. Jass JR, Allison LM, Edgar S. Monoclonal antibody TKH2 to the cancer-associated epitope sialosyl Tn shows cross-reactivity with variants of normal colorectal goblet cell mucin. Pathol 1994; 26:418-422.
24. Varki A. Diversity in the sialic acids. Glycobiology 1992; 2:25-40.
25. Allen D, Connolly N, Biggart J. Mucin profiles in ulcerative colitis with dysplasia and carcinoma. Histopathology 1988; 13:413-424.
26. Hutchins J, Reading C, Giavazzi R et al.

Distribution of mono-, di- and tri-O-acetylated sialic acids in normal and neoplastic colon. Cancer Res 1988; 48:483-489.

27. Corfield A, Wagner S, Clamp J et al. Mucin degradation in the human colon. Production of sialidase, sialate O-acetylesterase, N-acetylneuraminate lyase, arylesterase and glycosulfatase activities by strains of fecal bacteria. Infec Immun 1992; 60:3971-3978.
28. Jass J, Roberton A. Colorectal mucin histochemistry in health and disease:a critical review. Pathol Internat 1994; 44:487-504.
29. Muchmore EA, Varki NM, Fukuda M et al. Developmental regulation of sialic acid modifications in rat and human colon. FASEB J 1987; 1:229-235.
30. Vavasseur F, Dole K, Yang J et al. O-glycan biosynthesis in human colorectal adenoma cells during progression to cancer. Eur J Biochem 1994; 222:415-424.
31. Jass JR, Edgar S. Unicryptal loss of heterozygosity in hereditary non-polyposis colorectal cancer. Pathology 1994; 26:414-417.
32. Ogata S, Ho I, Chen A et al. Tumor associated sialylated antigens are constitutively expressed in normal human colonic mucosa. Cancer Res 1995; 55:1869-1874.
33. Higashi H, Hirabayashi Y, Fukui Y et al. Characterization of N-glycolylneuraminic acid-containing gangliosides as tumor-associated Hanganutziu-Deicher antigen in human colon cancer. Cancer Res 1985; 45:3796-3802.
34. Hirabayashi Y, Kasakura H, Matsumoto M et al. Specific expression of unusual GM2 ganglioside with Hanganutziu-Deicher antigen activity on human colon cancers. Jap J Cancer Res 1987; 78:251-260.
35. Miyoshi I, Higashi H, Hirabayashi Y et al. Detection of 4-0-acetyl-N-glycolylneuraminyl lactosylceramide as one of tumor-associated antigens in human colon cancer tissues by specific antibody. Molec Immunol 1986; 23:631-638.
36. Sata T, Roth J, Zuber C et al. Expression of α2,6-linked sialic acid residues in neoplastic but not in normal human colonic mucosa. Am J Pathol 1991; 139:1435-1448.
37. Baeckstrom D, Hansson GC, Nilsson O et al. Purification and characterization of a membrane-bound and a secreted mucin-type glycoprotein carrying the carcinoma-associated sialyl-Le a epitope on distinct core proteins. J Biol Chem 1991; 266:21537-21547.
38. Schoentag R, Primus FJ, Kuhns W. ABH and Lewis blood group expression in colorectal carcinoma. Cancer Res 1987; 47:1695-1700.
39. Shi ZR, McIntyre J, Knowles BB et al. Expression of a carbohydrate differentiation antigen, stage-specific embryonic antigen 1, in human colonic adenocarcinoma. Cancer Res 1984; 44:1142-1147.
40. Fukushima K, Hirota M, Terasaki PI et al. Characterization of sialosylated Lewis^x as a new tumor-associated antigen. Cancer Res 1984; 44:5279-5285.
41. Kim YS, Yuan M, Itzkowitz SH et al. Expression of Le^x and extended Le^y blood group-related antigens in human malignant, premalignant, and nonmalignant colonic tissues. Cancer Res 1986; 46:5985-5992.
42. Naitoh H, Nakajima T, Nagamachi Y et al. A clinicopathological evaluation of anti-fucosylated antigens antibody (YB-2) in colorectal carcinoma. Glycosyl Dis 1994; 1:31-36.
43. Hakomori S, Nudelman E, Levery S et al. Novel fucolipids accumulating in human adenocarcinoma. I, glycolipids with di- or trifucosylated type 2 chain. J Biol Chem 1984; 259:4672-4680.
44. Itzkowitz S, Yuan M, Fukushi Y et al. Lewis^x and sialylated Lewis^x -related antigen expression in human malignant and non-malignant colonic tissue. Cancer Res 1986; 46:2627-2632.
45. Hanski C, Drechsler K, Hanisch F-G et al. Altered glycosylation of the MUC-1 protein core contributes to the colon carcinoma-associated increase of mucin-bound sialyl-Lewis^x expression. Cancer Res 1993; 53:4082-4088.
46. Nakamori S, Kameyama M, Imaoka S et al. Increased expression of sialyl Lewis^x antigen correlates with poor survival in patients with colorectal carcinoma:clinicopathological and immunohistochemical study. Cancer Res 1993; 53:3632-3637.
47. Hoff S, Matsushita Y, Qta D et al. Increased expression of sialyl-dimeric Le^x

antigen in liver metastases of human colorectal carcinoma. Cancer Res 1989; 49:6883-6888.

48. Matsushita Y, Nakamori S, Seftor EA et al. Human colon carcinoma cells with increased invasive capacity obtained by selection for sialyl-dimeric Le^x antigen. Exp Cell Res 1991; 196:20-25.
49. Holmes EH, Ostrander GK, Clausen H et al. Oncofetal expression of Le^x carbohydrate antigens in human colonic adenocarcinomas. J Biol Chem 1987; 262:11331-11338.
50. Stroup GB, Anumula KR, Kline TF et al. Identification and characterization of two distinct α-(1-3)-L-fucosyltransferase activities in human colon carcinoma. Cancer Res 1990; 50:6787-6792.
51. Tytgat K, Büller H, Opdam F et al. Biosynthesis of human colonic mucin:MUC2 is the prominent secretory mucin. Gastroenterology 1994; 107:1352-1366.
52. Blank M, Klussmann E, Krüger-Krasagakes S et al. Expression of MUC2 mucin in colorectal adenomas of different histological types. Int J Cancer 1994; 59:301-306.
53. Buisine M-P, Janin A, Maunoury V et al. Aberrant expression of a human mucin gene (MUC5AC) in rectosigmoid villous adenoma. Gastroenterol 1996; 110:84-91.
54. Nakamori S, Ota DM, Cleary KR et al. MUC1 mucin expression as a marker of progression and metastasis of human colorectal carcinoma. Gastroenterology 1994; 106:353-361.
55. Ho SB, Niehans GA, Lyftogt C et al. Heterogeneity of mucin gene expression in normal and neoplastic tissues. Cancer Res 1993; 53:641-651.
56. Yang J, Byrd JC, Siddiki B et al. Alterations of O-glycan biosynthesis in human colon-cancer tissues. Glycobiology 1994; 4:873-884.
57. Yuan M, Itzkowitz S, Boland C et al. Comparison of T-antigen expression in normal, pre-malignant and malignant human colonic tissue using lectin and antibody immunohistochemistry. Cancer Res 1986; 46:4841-4847.
58. Baker MA, Kanani A, Brockhausen I et al. Presence of cytidine 5'-monophospho-N-acetylneuraminic acid:Galβ1-3GalNAc-R α(2-3)-sialyltransferase in normal human leukocytes and increased activity of this enzyme in granulocytes from chronic myelogenous leukemia patients. Cancer Res 1987; 47:2763-2766.
59. Brockhausen I, Yang J-M, Burchell J et al. Mechanisms underlying aberrant glycosylation of MUC1 mucin in breast cancer cells. Eur J Biochem 1995; 233:607-617.
60. Springer GF. Blood group T and Tn antigens are universal, clonal, epithelial cell adhesive, autoimmunogenic carcinoma markers. Progress in Clin & Biol Res 1983; 133:157-166.
61. Cooper HS. Peanut lectin-binding sites in large bowel carcinoma. Lab Invest 1982; 47:383-390.
62. King M, Chan A, Roe R et al. Two different glycosyltransferase defects that result in GalNAcα-O-peptide (Tn) expression. Glycobiology 1994; 4:267-279.
63. Miyake M, Kohno N, Nudelman ED et al. Human IgG_3 mono-clonal antibody directed to an unbranched repeating Type 2 chain (Galβ1-4GlcNAcβ1-3Galβ1-4GlcNAcβ1-3Galβ1-R) which is highly expressed in colonic and hepatocellular carcinoma. Cancer Res 1989; 49:5689-5695.
64. Ørntoft TF, Greenwell P, Clausen H et al. Regulation of the oncodevelopmental expression of type 1 chain ABH and $Lewis^b$ blood group antigens in human colon by α-2-L-fucosylation. Gut 1991; 32:287-293.
65. Irimura T, Ota D, Cleary K. Ulex europeus agglutinin I-reactive high molecular weight glycoproteins of adenocarcinoma of distal colon and rectum and their possible relationship with metastatic potential. Cancer Res 1987; 47:881-889.
66. Yazawa S, Nakamura J, Asao T et al. Aberrant α1→2 fucosyltransferases found in human colorectal carcinoma involved in the accumulation of Le^b and Y antigens in colorectal tumors. Jpn J Cancer Res 1993; 84:989-995.
67. Sun J, Thurin J, Cooper HS et al. Elevated expression of H type GDP-L-fucose:β-D-galactoside α-2-L-fucosyltransferase is associated with human colon adenocarcinoma progression. Proc Natl Acad Sci USA 1995; 92:5724-5728.
68. Yuan M, Itzkowitz SH, Palekar A et al.

Distribution of blood group antigens A, B, H, Le^a and Le^b in normal, fetal, and malignant colonic tissue. Cancer Res 1985; 45:4499-4511.

69. Kuhns WJ, Schoentag R. Carcinoma-related alterations of glycosyltransferases in human tissues. Cancer Res 1981; 41:2767-2772.
70. Yazawa S, Madiyalakan R, Izawa H et al. Cancer-associated elevation of α(1-3)-L-fucosyltransferase activity in human serum. Cancer 1988; 62:516-520.
71. Ørntoft T, Harving N, Langkilde N. O-linked mucin-type glycoproteins in normal and malignant colon mucosa:lack of T antigen expression and accumulation of Tn and sialosyl-Tn antigens in carcinomas. Int J Cancer 1990; 45:666-672.
72. Corfield AP, Clamp JR, Casey AD et al. Characterization of a sialic acid-rich mucus glycoprotein secreted by a premalignant human colorectal adenoma cell line. Int J Cancer 1990; 46:1059-1065.
73. Dall'Olio F, Malagolini N, Di Stefano G et al. Increased CMP-NeuAc:Galβ1-4GlcNAc-R α2-6 sialyltransferase activity in human colorectal cancer tissues. Int J Cancer 1989; 44:434-439.
74. Irimura T, Wynn D, Hager L et al. Human colonic sulfomucin identified by a specific monoclonal antibody. Cancer Res 1991; 51:5728-5735.
75. Kuhns W, Jain R, Matta K et al. Characterization of a novel mucin sulfotransferase activity synthesizing sulfated O-glycan core 1, 3-sulfate-Galβ1-3GalNAcα-R. Glycobiol 1995; 5:689-697.
76. Hounsell EF, Lawson AM, Feeney J et al. Structural analysis of the O-glycosidically linked core-region oligosaccharides of human meconium glycoproteins which express oncofoetal antigens. Eur J Biochem 1985; 148:367-377.
77. Kurosaka A, Nakajima H, Funakoshi I et al. Structures of the major oligosaccharides from human rectal adenocarcinoma glycoprotein. J Biol Chem 1983; 258:11594-11598.
78. Kurosaka A, Funakoshi I, Matsuyama M et al. UDP-GalNAc: GalNAc-mucin α-N-acetylgalactosamine transferase activity in human intestinal cancerous tissues. FEBS Lett 1985; 190:259-262.
79. Schoentag R, Williams V, Kuhns W. The distribution of blood group substance H and CEA in colorectal carcinoma. Cancer 1984; 53:503-509.
80. Bali JP, Magous R, Lecou C et al. Presence of blood group H antigen on a carcinoembryonic antigen, and its enzymatic modification into blood group A and B specificities. Cancer Res 1976; 36:2124-2129.
81. Kuhns W, Primus F. Alteration of blood groups and blood group precursors in cancer. Prog Clin Biochem Med 1985; 2:49-98.
82. Niv Y, Schwartz B, Amsalem Y et al. Human HT-29 colon carcinoma cells:mucin production and tumorigenicity in relation to growth phases. Anticancer Res 1995; 15:2023-2028.
83. Dahiya R, Kwak K-S, Byrd JC et al. Mucin synthesis and secretion in various human epithelial cancer cell lines that express the MUC-1 mucin gene. Cancer Res 1993; 53:1437-1443.
84. Ogata S, Uehara H, Chen A et al. Mucin gene expression in colonic tissues and cell lines. Cancer Res 1992; 52:5971-5978.
85. Devine PL, Birrell GW, Whitehead RH et al. Expression of MUC1 and MUC2 mucins by human tumor cell lines. Tumor Biol 1992; 13:268-277.
86. Hollingsworth MA, Strawhecker JM, Caffrey TC et al. Expression of MUC1, MUC2, MUC3 and MUC4 mRNAs in human pancreatic and intestinal tumor cell lines. Int J Cancer 1994; 57:198-203.
87. Mack DR, Hollingsworth MA. Alteration in expression of MUC2 and MUC3 mRNA levels in HT29 colonic carcinoma cells. Biochem Biophys Res Comm 1994; 199:1012-1018.
88. Velcich A, Augenlicht LH. Regulated expression of an intestinal mucin gene in HT29 colonic carcinoma cells. J Biol Chem 1993; 268:13956-13961.
89. Lesuffleur T, Porchet N, Aubert J-P et al. Differential expression of the human mucin genes MUC1 to MUC5 in relation to growth and differentiation of different mucus-secreting HT-29 cell subpopulations. J Cell Sci 1993; 106:771-783.
90. Bresalier RS, Niv Y, Byrd JC et al. Mucin

production by human colonic carcinoma cells correlates with their metastatic potential in animal models of colon cancer metastasis. J Clin Invest 1991; 87:1037-1045.

91. Symonds D, Vickery A. Mucinous carcinoma of the colon and rectum. Cancer 1976; 37:1891-1900.
92. Dahiya R, Lesuffleur T, Kwak K-S et al. Expression and characterization of mucins associated with the resistance to methotrexate of human colonic adenocarcinoma cell line HT29. Cancer Res 1992; 52:4655-4662.
93. Dall'Olio F, Malagolini N, Guerrini S et al. Resistance to methotrexate is associated with selective changes of α2,6- and α2,3-sialyltransferase activities toward N-acetyllactosaminic sequences in human colon cancer cell line HT-29. Biochem and Biophys Res Comm 1993; 196:714-720.
94. Dohi D, Sutton R, Frazier M et al. Regulation of sialomucin production in colon carcinoma cells. J Biol Chem 1993; 268:10133-10138.
95. Capon C, Laboisse CL, Wieruszeski J-M et al. Oligosaccharide structures of mucins secreted by the human colonic cancer cell line CL.16E. J Biol Chem 1992; 267:19248-19257.
96. Brockhausen I, Romero PA, Herscovics A. Glycosyltransferase changes upon differentiation of CaCo-2 human colonic adenocarcinoma cells. Cancer Res 1991; 51:3136-3142.
97. Vavasseur F, Yang J-M, Dole K et al. Synthesis of O-glycan core 3:Characterization of UDP-GlcNAc:GalNAc β3-N-acetylglucosaminyltransferase activity from colonic tissues and lack of the activity in human cancer cell lines. Glycobiology 1995; 5:351-357.
98. Youakim A, Romero P, Yee K et al. Decrease in polylactosaminoglycans associated with lysosomal membrane glycoproteins during differentiation of CaCo-2 human colonic adenocarcinoma cells. Cancer Res 1989; 49:6889-6895.
99. Youakim A, Herscovics A. Differentiation associated decrease in the proportion of polyfucosylated polylactosaminoglycans of CaCo-2 human colonic adenocarcinoma cells. Biochem J 1987; 247:299-306.
100. Narasimhan S. Control of glycoprotein synthesis VII. UDP-GlcNAc:glycopeptide β4-N-acetylglucosaminyltransferase III, an enzyme from hen oviduct which adds GlcNAc in β1-4 linkage to the β-linked mannose of the trimannosyl core of N-glycosyl oligosaccharides. J Biol Chem 1982; 257:10235-10242.
101. Kuhns W, Rutz V, Paulsen H et al. Processing O-glycan core 1, Galβ1-3GalNAcα-R. Specificities of core 2, UDP-GlcNAc: Galβ1-3GalNAc-R(GlcNAc to GalNAc) β6-N-acetylglucosaminyltransferase and CMP-sialic acid:Galβ1-3GalNAc-R α3-sialyltransferase. Glycoconj J 1993; 10:381-394.
102. Dall'Olio F, Malagolini N, Di Stefano G et al. Postnatal development of rat colon epithelial cells is associated with changes in the expression of the β1,4-N-acetylgalactosaminyltransferase involved in the synthesis of Sd[a] antigen and of α2,6-sialyltransferase activity towards N-acetyl-lactosamine. Biochem J 1990; 270:519-524.
103. Dall'Olio F, Malagolini N, Serafini-Cessi F. The expression of soluble and cell-bound alpha 2,6 sialyltransferase in human colonic carcinoma CaCo-2 cells correlates with the degree of enterocytic differentiation. Biochem Biophys Res Comm 1992; 184:1405-1410.
104. Dall'Olio F, Malagolini N, Guerrini S et al. Differentiation-dependent expression of human β-galactoside α2,6-sialyltransferase mRNA in colon carcinoma CaCo-2 cells. Glycoconj J 1996; 13:115-121.
105. Dall'Olio F, Malagolini N, Stefano G et al. α2,6 Sialylation of N-acetyllactosaminic sequences in human colorectal cancer cell lines:relationship with non-adherant growth. Int J Cancer 1991; 47:291-297.
106. Morgenthaler J, Kemner W, Brossmer R. Sialic acid dependent cell adhesion to collagen IV correlates with in vivo tumorigenicity of the human colon carcinoma sublines HCT116, HCT116a and HCT116b. Biochem Biophys Res Commun 1990; 171:860-866.
107. La Ferté S, Loh L. Characterization of a family of structurally related glycoproteins expressing β1-6 branched asparagine-linked oligosaccharides in human colon carcinoma cells. Biochem J 1992; 283:193-201.
108. Lampe B, Stallmach A, Riecken EO. Al-

tered glycosylation of integrin adhesion molecules in colorectal cancer cells and decreased adhesion to the extracellular matrix. Gut 1993; 34:829-836.

109. Takada A, Ohmori K, Takahashi N et al. Adhesion of human cancer cells to vascular endothelium mediated by a carbohydrate antigen, sialyl-Lewis a. Biochem Biophys Res Comm 1991; 179:713-719.

110. Majuri M-L, Mattila P, Renkonen R. Recombinant E-selectin-protein mediates tumor cell adhesion via sialyl-Lea and sialyl-Lex. Biochem Biophys Res Comm 1992; 182:1376-1382.

111. Sawada R, Lowe JB, Fukuda M. E-selectin-dependent adhesion efficiency of colonic carcinoma cells is increased by genetic manipulation of their cell surface lysosomal membrane glycoprotein-1 expression levels. J Biol Chem 1993; 268:12675-12681.

112a. Sawada R, Tsuboi S, Fukuda M. Differential E-selectin-dependent adhesion efficiency in sublines of a human colon cancer exhibiting distinct metastatic potentials. J Biol Chem 1994; 269:1425-1431.

112b. Sawada T, Ho JJL, Chung Y-S et al. E-selectin binding by pancreatic tumor cells is inhibited by cancer sera. Int J Cancer 1994; 57:901-907.

113. Holmes EH, Levery SB. Biosynthesis of fucose containing lactoseries glycolipids in human colonic adenocarcinoma COLO 205 cells. Arch Biochem Biophys 1989; 274:633-647.

114. Ogata S, Chen A, Itzkowitz S. Use of model cell lines to study the biosynthesis and biological role of cancer-associated sialosyl-Tn antigen. Cancer Res 1994; 54:4036-4044.

115. Bresalier RS, Byrd JC, Itzkowitz SH et al. Liver metastasis and adhesion to the sinusoidal endothelium by human colon cancer cells is related to mucin carbohydrate chain length. Proceed AACR 1995; abstract 421.

116. Brockhausen I, Dickinson N, Ogata S et al. Enzymatic basis for the high Sialyl-Tn expression in a colon cancer cell line. International Symp Glycoconjugates, Seattle. Glycoconj J 1995; 12:566.

117. Hanisch FG, Hanski C, Hasegawa A. Sialyl Lewis(x) antigen as defined by monoclonal antibody AM-3 is a marker of dysplasia in the colonic adenoma-carcinoma sequence. Cancer Res 1992; 52:3138-3144.

118. Yuan M, Itzkowitz SH, Ferrell LD et al. Expression of Lewis^x and sialylated Lewis^x antigens in human colorectal polyps. J Natl Cancer Inst 1987; 78:479-488.

119. Slomski CA, Durham JP, Watne AL. Glycosyltransferase levels in familial polyposis coli. J Surgical Res 1986; 40:406-410.

120. Paraskeva C, Hague A, Rooney N et al. A single human colonic adenoma cell line can be converted in vitro to both a colorectal adenocarcinoma and a mucinous carcinoma. Int J Cancer 1992; 51:661-664.

121. Labarrière N, Piau JP, Otry C et al. H blood group antigen carried by CD44V modulates tumorigenicity of rat colon carcinoma cells. Cancer Res 1994; 54:6275-6281.

122. Blottière HM, Burg C, Zennadi R et al. Involvement of histo-blood-group antigens in the susceptibility of colon carcinoma cells to natural killer-mediated cytotoxicity. Int J Cancer 1992; 52:609-618.

123. Lotan R, Raz A. Endogenous lectins as mediators of tumor cell adhesion. J Cell Biochem 1988; 37:107-117.

124. Schoeppner HL, Raz A, Ho SB et al. Expression of an endogenous galactose-binding lectin correlates with neoplastic progression in the colon. Cancer 1995; 75:2818-2826.

125. Ohannesian DW, Lotan D, Lotan R. Concomitant increases in galectin-1 and its glycoconjugate ligands (carcinoembryonic antigen, lamp-1, and lamp-2) in cultured human colon carcinoma cells by sodium butyrate. Cancer Res 1994; 54:5992-6000.

126. Mannori G, Crottet P, Cecconi O et al. Differential colon cancer cell adhesion to E-, P-, and L-selectin:role of mucin-type glycoproteins. Cancer Res 1995; 55:4425-4431.

127. Kuan S, Byrd J, Basbaum C. Inhibition of mucin glycosylation by Aryl-N-acetyl-α-galactosaminides in human colon cancer cells. J Biol Chem 1989; 264:19271-19277.

128. Kojima N, Handa K, Newman W et al. Inhibition of selectin-dependent tumor cell adhesion to endothelial cells and platelets by blocking O-glycosylation of these cells. Biochem Biophys Res Comm 1992;

182:1288-1295.

129. David L, Nesland JM, Clausen H et al. Simple mucin-type carbohydrate antigens (Tn, Sialosyl-Tn and T) in gastric mucosa, carcinomas and metastases. APMIS 1992 Suppl 27; 100:162-172.

130. Masuzawa Y, Miyauchi T, Hamanoue M et al. A novel core protein as well as polymorphic epithelial mucin carry peanut agglutinin binding sites in human gastric carcinoma cells. J Biochem 1992; 112:609-615.

131. Carneiro F, Santos L, David L et al. T (Thomsen-Friedenreich) antigen and other simple mucin-type carbohydrate antigens in precursor lesions of gastric carcinoma. Histopathology 1994; 24:105-113.

132. Werther JL, Rivera-MacMurray S, Bruckner H et al. Mucin-associated sialosyl-Tn antigen expression in gastric cancer correlates with an adverse outcome. Br J Cancer 1994; 69:613-616.

133. Maeda K, Chung Y-S, Onoda N et al. Serum sialyl-Tn antigen level as a prognostic indicator in patients with gastric cancer. Int J Oncol 1994; 4:129-132.

134. Kakeji Y, Maehara Y, Morita M et al. Correlation between sialyl Tn antigen and lymphatic metastasis in patients with Borrmann type IV gastric carcinoma. Brit J Cancer 1995; 71:191-195.

135. Kushima R, Jancic S, Hattori T. Association between expression of sialosyl-Tn antigen and intestinalization of gastric carcinomas. Int J Cancer 1993; 55:904-908.

136. Mullen PJ, Carr N, Milton JD et al. Immunohistochemical detection of O-acetylated sialomucins in intestinal metaplasia and carcinoma of the stomach. Histopath 1995; 27:161-167.

137. Yamashita Y, Chung YS, Horie R et al. Alterations in gastric mucin with malignant transformation:novel pathway for mucin synthesis. J Natl Cancer Inst 1995; 87:441-446.

138. Kannagi R, Levine P, Watanabe K et al. Recent studies of glycolipid and glycoprotein profiles and characterization of the major glycolipid antigen in gastric cancer of a patient of blood group genotype pp (Tj^{a-}) first studied in 1951. Cancer Res 1982; 42:5249-5254.

139. Levine P. Blood group and tissue genetic markers in familial adenocarcinoma:potential specific immunotherapy. Semin Oncol 1978; 5:25.

140. Sakamoto J, Watanabe T, Tokumaru T et al. Expression of $Lewis^{a}$, $Lewis^{b}$, $Lewis^{x}$, $Lewis^{y}$, Sialyl-$Lewis^{a}$, and Sialyl-$Lewis^{x}$ blood group antigens in human gastric carcinoma and in normal gastric tissue. Cancer Res 1989; 49:745-752.

141. Ho SB, Shekels LL, Toribara et al. Mucin gene expression in normal, preneoplastic, and neoplastic human gastric epithelium. Cancer Res 1995; 55:2681-2690.

142. Bara J, Imberty A, Pérez S et al. A fucose residue can mask the MUC-1 epitopes in normal and cancerous gastric mucosae. Int J Cancer 1993; 54:607-613.

143. David L, Leitao D, Sobrinho-Simoes M et al. Biosynthetic basis of incompatible histo-blood group A antigen expression: Anti-A transferase antibodies reactive with gastric cancer tissue of type O individuals. Cancer Res 1993; 53:5494-5500.

144. Ohe Y, Hinoda Y, Irimura T et al. Expression of sulfated carbohydrate chains detected by monoclonal antibody 91.9H in human gastric cancer tissues. Jpn J Cancer Res 1994; 85:400-408.

145. Mitsuyama T, Gasa S, Taniguchi N et al. Elevation of sulfatide synthesis in human gastric adenocarcinoma:biochemical characteristics common to adenocarcinomas. J Exp Clin Cancer Res 1983; 2:25-30.

146. Tatsuta M, Iishi H, Okuda S et al. Association of Helicobacter pylori with differentiated-type early gastric cancer. Cancer 1993; 72:1841-1845.

147. Oda T, Kanai Y, Oyama T et al. E-cadherin gene mutations in human gastric carcinoma cell lines. Proc Natl Acad Sci USA 1994; 91:1858-1862.

CHAPTER 21

Pancreatic and Liver Cancer

21.A. PANCREATIC CANCER

For the cloning of MUC1 mucin, deglycosylated human pancreatic tumor mucin was used to prepare antiserum; this was then used to select cDNA clones from a pancreatic tumor cell expression library.[1] The pancreatic MUC1 mucin contains 80% carbohydrate and has a larger molecular weight (greater than 1 million Da) compared to breast mucin which has about 50% carbohydrate.[2] MUC1 is a type I membrane glycoprotein with five N-glycosylation sites and numerous O-glycans. It is thought that the exposure of core protein epitopes on MUC1 produced by tumors is a consequence of glycosylation changes.

The pancreatic mucin was found in the serum and other body fluids of patients with pancreatic and other adenocarcinomas.[3] Humoral IgM type immunity directed against MUC1 tandem repeat peptide epitope is present in pancreatic and other cancer patients.[4]

Balagué et al studied the expression of mucin genes in normal and malignant pancreatic tissues.[5] Normal pancreatic tissues generally express MUC1, as well as MUC3 and MUC5B in interlobal ducts. MUC2, MUC4 and MUC5A are barely detectable in any pancreatic tissue. In contrast, in pancreatic adenocarcinomas, MUC1, MUC3, MUC4, MUC5A and MUC5B are variably expressed.

Mucin expression appears to depend on the cell type. A majority of pancreatic tumor cell lines studied, including the metastatic pancreatic cell line Colo357, express relatively high levels of MUC1 and MUC4, but low levels of MUC2 and no MUC3; in contrast, most intestinal tumor cell lines express MUC2 and MUC3 but less MUC1 and MUC4.[6]

Le^x antigens are expressed far more frequently in pancreatic hyperplasia and cancer than in chronic pancreatitis (Table 6). This suggests that hyperplasia may represent a premalignant state. Le^x is not expressed in normal pancreatic tissue which, however, often expresses Le^y determinants in a cell type-specific manner.[7] The lowest Le^x expression is found in poorly differentiated cancers. Le^x and Le^y antigens are not always expressed concordantly, suggesting that the α2- and α3-Fuc-transferases involved in the synthesis of these antigens are regulated selectively in pancreatic cancers. Pancreatic tumor cells SW1990, CAPAN-2 and PANC-1 bind recombinant E-selectin very avidly; SW1990 cells exhibit the highest binding and cell surface sialyl-Le^a and sialyl-Le^x expression. Sialyl-Le^x and sialyl-Le^a are believed to be bound to MUC1 peptide.[8]

The cell surface expression of these Lewis antigens is inhibited by the O-glycan extension inhibitor GalNAcα-benzyl but not with tunicamycin, suggesting that these E-selectin ligands are found primarily on O-glycans. GalNAcα-benzyl treatment also exposes MUC1 tandem repeat peptide epitopes. Binding of pancreatic tumor cells to E-selectin is abrogated by mucins isolated from SW1990 xenografts, anti Le^a antibody (and anti sialyl-Le^x antibody to a lesser degree) as well as by sera from pancreatic tumor patients containing sialyl-Le^x but not by normal sera.[9] This strongly suggests an inhibitory effect of mucin carbohydrate antigens on the cell adhesion of pancreatic cancer cells. Measurement of sialyl-Le^x antigen may be of use in the diagnosis of pancreatic cancer.[10]

Similar to other cancers,[11] Tn and sialyl-Tn antigens are both elevated in pancreatic cancer, although sialyl-Tn defined by antibody staining is considered to be more specific in distinguishing malignant from normal tissue.[12] Pancreatitis and normal pancreas acinar cells but not normal ductal tissue stains with *Vicia villosa* lectin binding to GalNAc residues (Tn antigen). Sialyl-Tn antigen is totally absent in normal pancreas but is present in 97% of pancreatic cancers and in most samples of pancreatitis. The T antigen is less abundant in normal and cancerous pancreas.[13]

Glycoproteins isolated from the pancreatic cell line FG-Met-2 also contain N-glycans as shown by peptide N-glycosidase release of glucosamine-labeled chains. Sulfate labeling studies indicates that 50% of the sulfate label is incorporated into glucosamine labeled chains, some of which may be proteoglycan in nature. These sulfated glycans are partially sialylated as shown by neuraminidase treatment.[14]

21.B. HEPATOMA

Since liver has the potential to synthesize many of the N-glycosylated glycoproteins found in the serum, most of the aberrant glycosylation in liver tumors pertains to N-glycans, unlike the prominent O-glycan alterations found in breast, stomach, colon and ovarian tumors.

α-Fetoprotein (AFP) is synthesized in the liver and its serum level is elevated in liver cirrhosis, chronic hepatitis and hepatocellular carcinoma.[15] This N-glycan containing glycoprotein is found in human cord serum and is considered an oncofetal glycoprotein. α6-Fucosylation of the N-glycan chain core of AFP appears to be increased in primary hepatoma, and to a lesser extent in liver cirrhosis and chronic and acute hepatitis.[16] Structural analysis of the α2-6-sialylated bi-antennary chains of human cord serum AFP indicates that α6-fucosylation of the Asn-linked GlcNAc residue of AFP increases inversely with the fetal gestation period.[17] Thus, α6-fucosylation may be an indicator of the de-differentiation of human hepatocytes in hepatocellular carcinoma.

In human hepatoma cell lines HepG2 and HuH-7 and hepatoblastoma cell line HuH-6, AFP also contains α6-fucosylated bi-antennary, bisected and tri-antennary chains (Fig. 3). The ratios of structures differ in these cell lines and they are correlated with GlcNAc-transferase III, IV, V and α6-Fuc-transferase activities.[16]

N-glycans in transferrin purified from the sera of patients with hepatocellular carcinoma and liver cirrhosis were compared with those from normal individuals.[18] N-glycanase treated and derivatized samples analyzed by HPLC in combination with exoglycosidase digestion reveal an increase in α6-fucosylated bi-antennary complex-type sugar chains in the majority of patients with hepatocellular hepatoma compared to normal individuals. In another study, Yamashita et al[19] found highly branched N-glycan chains in transferrin from heptocellular carcinoma. Compared to normal transferrin, these N-glycans contain four times more sialylα2-3 Gal- groups, more Le^x determinants, and Fuc in the N-glycan core. In addition, a small amount of bisected structures are detected in cancer, but not in normal transferrin.

Fucosylation changes have also been noted in plasma membrane fractions of chemically-induced Morris hepatoma 7777 in the rat.[20] In contrast to human hepatoma, the α6-fucosylation of the proximal GlcNAc of N-glycans is slightly reduced compared to other rat tissues and serum. However, only the glycans from hepatoma contain Fucα1-2 linked to Gal.

Le^y antigens are expressed on the cell surface and in the cytoplasm of hepatocellular tumors (Table 6). This expression is linked to the proliferative activity and dedifferentiation of the cancer cells.[21] Sialyl-Le^a antigens are present in the sera of liver cancer patients but may not originate from cancer cells since they are not found in hepatocellular carcinoma or in parenchymal liver cells but are expressed in the intrahepatic bile ducts.[22] α2- and α3-Fuc-transferase activities towards asialo-fetuin substrate and α6-Fuc-transferase activities towards asialo-agalacto-fetuin substrate in the serum of patients with hepatocellular carcinoma are increased compared to plasma from normal or cirrhotic liver patients. However, in the tumor tissue, α2- and α3-Fuc-transferase activities are much lower than in normal tissue from healthy individuals.[23] This suggests that liver and liver cancer tissue are not necessarily the source for serum enzymes but that these may be released from other affected tissues such as hepatic bile duct which expresses Lewis type antigens.

β4-Gal- and Dol-P-Man-synthase activities are higher in liver cancer tissue compared to normal liver tissue. Other reports show that hepatoma or metastatic liver tissue has lower sialyltransferase and Gal-transferase activities but higher Fuc-transferase, however this may be in contrast to activity levels in the serum.[23-25]

Serum Gal-transferase in primary or metastatic liver cancer appears to be abnormal either due to the existence of different glycoforms or due to proteolytic cleavage.[26] Upon isoelectric focusing two enzyme species have been found in the cancer patients compared to three species in normal serum.[27] Similar findings of the existence of Gal-transferase isoforms in hepatoma were reported for other species including chicken.[28]

The N-glycans from poorly differentiated chemically transformed rat Zajdela hepatoma cells have more tri- and tetra-antennary and bisected structures in contrast to normal rat hepatocytes which carry a high proportion of bi-antennary and no bisected N-glycans.[29] The GlcNAc1-6 Man branch synthesized by GlcNAc-transferase V is primarily found in these hepatoma cells. Poly-n-acetyl-lactosamine structures terminated with Galα3 residues and unusual sialic acid linkages are also found in the hepatoma cells.

Sialyltransferases appear to be differently regulated in rat hepatoma; the activity of α6-sialyltransferase acting on N-glycans is increased in liver cancer.[30] Similar patterns of activities are seen in regenerating rat liver tissue. The human hepatoma cell line HepG2 is particularly rich in α6-sialyltransferase (N) acting on N-glycans. The high level of mRNA expression can be reduced by about 90% by Na-butyrate treatment, indicating that the enzyme is regulated during differentiation, probably by post translational mechanisms, because the rate of transcription is unaltered.[31]

HepG2 cells make transferrins with unusual highly branched N-glycan structures.[32] GlcNAc-transferases I to V synthesizing N-glycan branches up to the tetra-antennary structures are active in HepG2 cells. However, GlcNAc-transferase VI,[33] responsible for the synthesis of penta-antennary chains reported to occur in HepG2 cells[32] could not be detected.[34]

In the rat, normal liver γ-glutamyl-transpeptidase contains mainly non-bisected N-glycans, but in rat hepatoma cells this glycoprotein carries bisected N-glycans.[35]

The appearance of bisected structures in rat hepatoma is explained by the high activity of GlcNAc-transferase III in tumor tissue.[36,37] GlcNAc-transferase III and V are greatly increased in rat hepatoma and cultured rat hepatoma cells compared to normal tissue while GlcNAc-transferase IV is not significantly changed.[38] In human hepatoma and liver cirrhosis, GlcNAc-transferase III activity may possibly also be elevated.[39] However, N-glycans of γ-glutamyltranspeptidase from human hepatocellular carcinoma appear to have no bisected residues but more tri- and tetra-antennary structures as well as high mannose oligosaccharides which are not found on the normal liver enzyme.[35]

LEC rats develop spontaneous hepatomas and jaundice with symptoms similar to Wilson's disease.[40] LEC rats also exhibit high activities of GlcNAc-transferase III and V in hepatoma tissue.[41] Hepatoma tumor tissue from LEC rats accumulate unusual glycolipids isogloboside and ganglio-N-tetraosyl ceramide with blood group B determinants in contrast to normal liver tissue which possesses negligible amounts of these structures. In hepatoma-disposed LEC rats, the β6-GlcNAc-transferase which synthesizes blood group I and acts on the internal Gal residue of Galβ1-4GlcNAcβ1-3Galβ1-4Glc-R is virtually undetectable in liver tissue compared with control rat liver, although present in other organs and in serum as well as in fetal liver. Results using an antibody which recognizes the I antigen indicate that its expression in hepatoma is less prominent than in healthy liver.[42] Activities of i β3-GlcNAc-transferase and I β6-GlcNAc-transferase acting on GlcNAcβ1-3Galβ1-4Glc-R are comparable in normal and cancerous liver. The activity of the β6-GlcNAc-transferase may therefore be related to the genetic defect in LEC rats. In human hepatocellular carcinoma, the synthesis of the branched I antigen appears to be defective since increased amounts of linear unbranched poly-n-acetyl-lactosamine chains are detected by monoclonal antibody.[43]

References

1. Lan MS, Batra SK, Qi W-N et al. Cloning and sequencing of a human pancreatic tumor mucin cDNA. J Biol Chem 1990; 265:15294-15299.
2. Gendler SJ, Lancaster CA, Taylor-Papadimitriou J et al. Molecular cloning and expression of human tumor-associated polymorphic epithelial mucin. J Biol Chem 1990; 265:15286-15293.
3. Metzgar RS, Rodriguez N, Finn OJ et al. Detection of a pancreatic cancer-associated antigen (DU-PAN-2 antigen) in serum and ascites of patients with adenocarcinoma. Proc Natl Acad Sci USA 1984; 81:5242-5246.
4. Kotera Y, Fontenot J, Pecher G et al. Humoral immunity against a tandem repeat epitope of human mucin MUC1 in sera from breast, pancreatic and colon cancer patients. Cancer Res 1994; 54:2856-2860.
5. Balagué C, Audié JP, Porchet N et al. In situ hybridization shows distinct patterns of mucin gene expression in normal, benign and malignant pancreas tissues. Gastroenterology 1995; 109:953-964.
6. Hollingsworth MA, Strawhecker JM, Caffrey TC et al. Expression of MUC1, MUC2, MUC3 and MUC4 mRNAs in human pancreatic and intestinal tumor cell lines. Int J Cancer 1994; 57:198-203.
7. Kim YS, Itzkowitz SH, Yuan M et al. Le^x and Le^y antigen expression in human pancreatic cancer. Cancer Res 1988; 48:475-482.
8. Ho JJL, Siddiki B, Kim YS. Association of Sialyl-Lewis^a and Sialyl-Lewis^x with MUC-1 apomucin in a pancreatic cancer cell line. Cancer Res 1995; 55:3659-3663.
9. Sawada T, Ho JJL, Chung YS et al. E-selectin binding by pancreatic tumor cells is inhibited by cancer sera. Int J Cancer 1994; 57:901-907.
10. Ohta H, Sawabu N, Takemori Y et al. Measurement of sialylated stage-specific embryonic antigen-1 in pure pancreatic juice for the diagnosis of pancreatic cancer. Int J Pancreatol 1994; 15:35-41.
11. Springer G. T and Tn, general carcinoma antigens. Science 1984; 224:1198-1206.
12. Ching CK, Holmes SW, Holmes GKT et al. Blood group sialyl-Tn antigen is more specific than Tn as a tumor marker in the

pancreas. Pancreas 1994; 9:698-702.

13. Itzkowitz S, Kjeldsen T, Friera A et al. Expression of Tn, sialosyl-Tn and T antigens in human pancreas. Gastroenterology 1991; 100:1691-1700.
14. Sundblad G, Kajiji S, Quaranta V et al. Sulfated N-linked oligosaccharides in mammalian cells. J Biol Chem 1988; 263:8897-8903.
15. Taketa K, Sekiya C, Namiki M et al. Lectin-reactive profiles of alpha-fetoprotein characterizing hepatocellular carcinoma and related conditions. Gastroeneterology 1990; 99:508-518.
16. Ohno M, Nishikawa A, Koketsu M et al. Enzymatic basis of sugar structures of α-fetoprotein in hepatoma and hepatoblastoma cell lines: correlation with activities of α1-6 fucosyltransferase and N-acetylglucosaminyltransferases III and V. Int J Cancer 1992; 51:315-317.
17. Yamashita K, Taketa K, Nishi S et al. Sugar chains of human cord serum α-fetoprotein: characteristics of N-linked sugar chains of glycoproteins produced in human liver and hepatocellular carcinomas. Cancer Res 1993; 53:2970-2975.
18. Matsumoto K, Maeda Y, Kato S. Alteration of asparagine-linked glycosylation in serum transferrin of patients with hepatocellular carcinoma. Clin Chim Acta 1994; 224:1-8.
19. Yamashita K, Koide N, Endo T. Altered glycosylation of serum transferrin of patients with hepatocellular carcinoma. J Biol Chem 1989; 264:2415-2423.
20. Nuck R, Orthen B, Reutter W. Occurrence of α1-2 fucosylation in membrane glycoproteins of Morris hepatoma 7777 but not in liver–Aberrant type of fucosylation in a malignant tissue. Eur J Biochem 1992; 208:669-676.
21. Wakabayashi M, Shiro T, Seki T et al. Lewis Y antigen expression in hepatocellular carcinoma: an immunohistochemical study. Cancer 1995; 75:2827-2835.
22. Itai S, Nishikata J, Yoneda T et al. Tissue distribution of 2-3 and 2-6 sialyl Lewis A antigens and significance of the ratio of two antigens for the differential diagnosis of malignant and benign disorders of the digestive tract. Cancer 1991; 67:1576-1587.
23. Hutchinson WL, Du M-Q, Johnson PJ et al. Fucosyltransferases: d ifferential plasma and tissue alterations in hepatocellular carcinoma and cirrhosis. Hepatology 1991; 13:683-688.
24. Kessel D, Sykes E, Henderson M. Glycosyltransferase levels in tumors metastatic to liver and in uninvolved liver tissue. J Natl Cancer Inst 1977; 59:29-32.
25. Ingraham HA, Aldaheff JA. Characterization of sialyltransferase in noncancerous and neoplastic human liver tissue. J Natl Cancer Inst 1978; 61:1371-1374.
26. Liu CK, Schmied R, Waxman S. Characterization of galactosyltransferase released from human hepatoma cells. Enzyme 1982; 28:258-267.
27. Qian GX, Liu CK, Waxman S. Abnormal isoelectric focusing patterns of serum galactosyltransferase activity in patients with liver neoplasia (41759). Proc Soc Exp Biol Med 1984; 175:21-24.
28. Ivanov S, Gavazova E, Chelibonova-Lorer H. Galactosyltransferase:multiple forms in serum of normal and hepatoma Mc-29 bearing chickens and from liver and hepatoma microsomal and plasma membrane preparations. Cancer Biochem Biophys 1986; 9:41-51.
29. Goulut-Chassaing C, Bourrillon R. Structural differences between complex-type Asn-linked glycan chains of glycoproteins in rat hepatocytes and Zajdela hepatoma cells. Biochim Biophys Acta 1995; 1244:30-40.
30. Miyagi T, Koseki M, Tsuiki S. Comparative study of the levels of sialyltransferases responsible for the formation of sugar chains in glycoproteins and gangliosides in rat liver and hepatomas. Jpn J Cancer Res 1988; 79:742-749.
31. Shah S, Lance P, Smith TJ et al. n-Butyrate reduces the expression of β-galactoside α2,6-sialyltransferase in HepG2 cells. J Biol Chem 1992; 267:10652-10658.
32. Campion B, Léger D, Wieruszeski J-M et al. Presence of fucosylated tri-antennary, tetra-antennary and penta-antennary glycans in transferrin synthesized by the human hepatocarcinoma cell line Hep G2. Eur J Biochem 1989; 184:405-413.
33. Brockhausen I, Hull E, Hindsgaul O et al. Control of glycoprotein synthesis. Detection

and characterization of a novel branching enzyme from hen oviduct, UDP-N-acetylglucosamine:GlcNAcβ1-6(GlcNAcβ1-2)Manα-R (GlcNAc to Man) β-4-N-acetylglucosaminyltransferase VI. J Biol Chem 1989; 264:11211-11221.
34. Brockhausen I, unpublished.
35. Yamashita K, Hitoi A, Taniguchi N et al. Comparative study of the sugar chains of glutamyltranspeptides purified from rat liver and rat AH-66 hepatoma cells. Cancer Res 1983; 43:5059-5063.
36. Nishikawa A, Fujii S, Sugiyama T et al. High expression of an N-acetylglucosaminyltransferase III in 3'-methyl DAB-induced hepatoma and ascites hepatoma. Biochem Biophys Res Comm 1988; 152:107-112.
37. Narasimhan S, Schachter H, Rajalakshmi S. Expression of N-acetylglucosaminyltransferase III in hepatic nodules during rat liver carcinogenesis promoted by orotic acid. J Biol Chem 1988; 263:1273-1281.
38. Nishikawa A, Gu J, Fujii S et al. Determination of N-acetylglucosaminyltransferases III, IV and V in normal and hepatoma tissues of rats. Biochim Biophys Acta 1990; 1035:313-318.
39. Ishibashi K, Nishikawa A, Hayashi N et al. N-acetylglucosaminyltransferase III in human serum, and liver and hepatoma tissues:increased activity in liver cirrhosis and hepatoma patients. Clin Chim Acta 1989; 185:325-332.
40. Ariga T, Kasai N, Miyoshi I et al. Accumulation of isogloboside ganglio-N-tetraosyl ceramide having blood group B determinant in the hepatomas of female LEC rats. Biochim Biophys Acta 1995; 1254:257-266.
41. Miyoshi E, Nishikawa A, Ihara Y et al. N-acetylglucosaminyltransferase III and V messenger RNA levels in LEC rats during hepatocarcinogenesis. Cancer Res 1993; 53:3899-3902.
42. Gu J, Nishikawa A, Matsuura N et al. Deficiency of β(1-6) N-acetylglucosaminyltransferase involved in the biosynthesis of blood group I antigen in the liver of LEC rats. Jpn J Cancer Res 1992; 83:878-884.
43. Miyake M, Kohno N, Nudelman ED et al. Human IgG_3 mono-clonal antibody directed to an unbranched repeating Type 2 chain (Galβ1-4GlcNAcβ1-3Galβ1-4GlcNAcβ1-3Galβ1-R) which is highly expressed in colonic and hepatocellular carcinoma. Cancer Res 1989; 49:5689-5695.

CHAPTER 22

Cancer of the Urinary and Reproductive Tracts

22.A. OVARIAN CANCER

An association of ovarian cancer with ABO blood groups has been described in large population samples.[1] This was confirmed in a recent study of 1261 women with ovarian cancer. Comparison of blood group A patients with a large normal cohort confirmed earlier findings which stated that the cancer risk in the former group was 1.2 with 1 considered as the risk factor in normals.[2] A possible explanation is that many tumors which may express Forssman A-like antigen could be eliminated by naturally occurring anti-A in non-A persons; this form of immune surveillance would be absent in A individuals.

An elevation of the activity of β4-Gal-transferase and appearance of isoforms of the enzyme in human body fluids have been associated with ovarian tumors.[3] The enzyme is present in ovarian tumor effusion fluid[4,5] but its usefulness has been restricted because of false positives in benign diseases. Recent attempts to improve its tumor specificity have centered on the preparation of mouse monoclonal antibodies to β4-Gal-transferase. In this study several monoclonal antibodies were produced which seem particularly useful in distinguishing ovarian cancer from benign diseases. Cloned cDNA fragments of β4-Gal-transferase from a human ovarian cancer cell line RMG-1 encode some of the epitopes recognized by these antibodies.[6] The results suggest that proteolytic fragments of β4-Gal-transferase are produced in tumor cells, leading to the exposure of new epitopes.

Haptoglobin and alpha-1-proteinase inhibitor differ in the composition of N glycan chains and undergo different glycosylation changes in ovarian cancer.[7] Sugar composition and lectin binding analyzes suggest that the glycan branching patterns of these two glycoproteins, both synthesized in the liver, are not the same; this may be related to differences in their biological properties in ovarian cancer. Glycosylation differences may be attributed to a specific removal of one glycoform from the circulation, possibly by the liver.

Haptoglobin isolated by extraction from human blood with *Lotus tetragonobolus* lectin shows a significantly elevated fucose content in patients with ovarian cancer as determined by HPAEC.[8] This corresponds to the elevation of α2-, α3- and α4-Fuc-transferase activities in ovarian cancer tissue compared to one sample of normal ovarian tissue (Table 6).[9] The α4-Fuc-transferase in particular is dramatically increased in the tumor. In the sera of ovarian cancer patients the α3-Fuc-transferase activity towards the specific substrate 3-sulfo-Galβ1-4GlcNAc is elevated compared to the activity in normal sera. A number of synthetic acceptors with sialic acid, Fuc, sulfate and methyl substitutions were used to study patient samples and revealed differences in the substrate specificities of Fuc-transferases between ovarian cancer tissue and serum.

Antibodies which recognize epitopes on MUC1 core protein are used as markers in patients with ovarian cancer and may be used as indicators of the malignant potential.[10,11] However, the recognition by these antibodies may be modified due to changed MUC1 glycosylation in cancer cells and thus the interpretation of these data is difficult. Cytotoxic T cells from ovarian malignant tumors have been shown to be targeted against MUC1 mucin core peptides.[12] Transfected EBV-transformed B-cell lines expressing either MUC1 or MUC1 peptides from the tandem repeat region are lysed by these cytotoxic lymphocytes compared to control cells transfected with a frame shift mutant MUC1 cDNA. Lysis is inhibited by antibody recognizing the MUC1 tandem repeat region peptide. These results raise the possibility of specific immunotherapy.

The expression of sialyl-Tn antigen is greatly increased in various types of ovarian tumors.[13,14] Although the intensity of anti sialyl-Tn antibody staining is higher in malignant than in borderline tumors, there is no correlation with differentiation or histological classification. In some tumors there is concomitant expression of the Tn antigen. Sialyl-Tn antigens in the serum of ovarian cancer patients, as in other cancers, is associated with a poor prognosis.[14]

22.B. ENDOMETRIAL CANCER AND CHORIOCARCINOMA

Cancer-related carbohydrate phenotypes and their precursor stages have been studied by immunodetection methods using lectins or monoclonal antibodies. Endometrial cells are modulated by hormones and, similar to squamous cells, may show altered phenotypic expression during secretory and proliferative phases.[15]

ABH and Lewis antigens are found on surface epithelium but are more restricted in underlying layers. The expression of these blood group antigens depends on the genetic background, i.e. blood group and secretor status, and is regulated during the menstrual cycle. Monosialyl-Le^a appears to be a marker for secretory differentiation of the endometrium.

In uterine endometrial cancer, α2-, α3- and α4-Fuc-transferase activities are all increased compared to the activities in normal endometrial tissue. Cultured cancer cell lines derived from human endometrium express Le^b epitopes. Prominent levels of α2- and especially α4-Fuc-transferase activities are found in these cells compared to ovarian and cervical cancer cell lines. This suggests that the Le^b expression is regulated by α2- and α4-Fuc-transferases which are involved in the synthesis of the Le^b antigen.[16]

Type 1 chains, exposed after neuraminidase treatment, and H, Le^a and Le^b antigens appear to be especially prevalent in endometrial cancer rather than in normal endometrium.[17] Strong binding of *Griffonia simplicifolia* -A4 lectin recognizing GalNAc-R is present in practically all cases of endometrial adenocarcinoma compared to 18% of normal tissues while only 24% of the endometrial adenocarcinomas express the A antigen. A high level of H antigen expression was suggested by *Ulex europeus* agglutinin-1 binding to cancer tissues (recognizing Fuc) and by anti H antibodies.[18]

Immunostaining using monoclonal antibodies demonstrates sialyl-Tn antigen on most samples of endometrial carcinoma but not on normal proliferative endometrium or in hyperplastic tissue. Variable antibody staining defines a mixed cell population as expected in cycling tissue. Menopausal and normal secretory endometrium shows a high level of staining with increasing levels in the late secretory phase.[13]

Human chorionic gonadotrophin (hCG) produced in the placenta of pregnant women is an N- and O-glycosylated glycoprotein hormone excreted in the urine. Patients with hydatiform moles and normal women produce hCG with mostly core 1 O-glycan structures. In choriocarcinoma, hCG carries more O-glycans with core 2 structures; intermediate levels are found in invasive moles (Table 6).[19,20] The choriocarcinoma cell line BeWo produces hCG containing O-glycan structures found on normal hCG as well as sialylated core 2 structures carrying the sialyl-Le^x determinant.[21]

The N-glycans from BeWo cells consist of sialylated, bisected bi-and triantennary structures. N-glycans of hCG from choriocarcinoma are unusual biantennary structures lacking the branch synthesized by GlcNAc-transferase II. Choriocarcinoma hCG also contains triantennary chains not found in normal hCG and shows increased fucosylation but reduced sialylation.[22] This suggests that trophoblasts secreting hCG undergo significant changes in the expression of glycosyltransferases upon cancerous transformation. The biological activity of cancer-associated hCG is strongly reduced although there appears to be no impairment of receptor binding.[23]

22.C. PROSTATE CANCER

Cancer of the prostate is the leading cause of cancer in males; most cancers of the prostate are adenocarcinomas.[24] The course of the disease can be quite variable; metastases tend to lodge in lymph nodes and in the bone. Clonal karyotypic abnormalities in tumor cells are associated with an unfavorable outcome in patients.[25] Prostate specific antigen, identified as an epithelial glycoprotein synthesized in acini and ducts of the prostate gland with a kallikrein-like serine protease activity, has proved valuable as a tumor marker for monitoring established disease, although somewhat less useful as a screening test.[26,27]

A recent study has indicated that sialyl-Le^x may be an important marker in patients with metastatic prostate cancer. Upregulation of sialyl-Le^x is associated with a poor prognosis; increased expression of this epitope in tumor tissue is correlated with hormone-resistant, aggressive disease. Blood group ABO and Lewis antigens have been reported to change in prostatic cancer. Le^x antigen is not well expressed in normal or cancerous prostate tissue but Le^y and sialyl-Le^x antigens are absent or in low amounts in benign tissue and more highly expressed in cancer.[28,29]

In the rat prostate and seminal vesicles, sulfoglycoprotein-2 is produced at high levels during carcinogenesis induced by N-nitroso-N-methylurea and promoted by testosterone propionate. The expression of this glycoprotein is associated with apoptosis, for example, in the prostate following castration, and can be observed in early stages of carcinogenesis but is barely detectable in the normal prostate or testes. Sulfoglycoprotein-2 may thus serve as a marker for carcinogenesis.[30]

22.D. URINARY TRACT CARCINOMA

The urothelium expresses blood group ABO and Lewis antigens which may change in cancer.[31] ABO blood groups are often lost in cancers with invasive recurrence. This is associated with loss of activities of the A and B transferases in bladder cancer tissue while α2- and α4-Fuc-transferase activities are at normal levels and α3-Fuc-transferase is reduced.[32] Patients with erythrocyte blood group A and bladder cancer tend to have lower grade tumors than patients with blood group O.[33] Le^x and Le^y antigens are expressed in dedifferentiated and invasive cells, consistent with high activities of the α3- and α4-Fuc-transferase activities. Differentiated renal adenocarcinomas strongly express Le^x and multimeric Le^x although these markers have not been noted in poorly differentiated cells.[34] These alterations, defined by monoclonal antibodies, occur in the course of early cancer development.

In human urothelial cancer cell lines differing in tumorigenicity and metastatic potential, the expression of blood groups also differs as well as the degree of fucosylation.[35]

Bladder tumor transitional epithelium demonstrates losses of ABO antigens as well as the appearance of cryptic T and Tn antigens. While T antigen expression is increased in bladder tumors, Tn antigen is partially lost in invasive tumors.[36] Transitional cell carcinomas in the bladder often develop into invasive cancers when these do not express ABO blood groups. The degree of tumor invasiveness has been coupled to strong expression of T and Tn epitopes.[37]

Gal-transferase activities were reported to change in human bladder cancer cells[38] and in rat bladder cells upon carcinogenic stimulation.[39] Gal-transferase isoenzymes are present in cells from a human bladder transitional cell carcinoma line and in the cell medium, possibly produced by protease action on the membrane bound species.[40] These isoenzymes differ in their K_m values and metal ion requirements.

A monoclonal antibody (19A211), which identifies a tumor-associated antigen of superficial bladder tumors, reacts with a glycoform of CEA; active immunization of bladder cancer patients with this glycoform is under consideration for immunotherapy in cancer.[41]

References

1. Mourant A, Kopec A, Domaniewska-Sobczak K. Blood Groups and Diseases. Oxford University Press, Oxford, 1978:13-15.
2. Henderson J, Seagroatt V, Goldacre M. Ovarian cancer and ABO blood groups. J Epidemiol Commun Health 1993; 47:287-289.
3. Podolsky DK, MM, Isselbacher KJ et al. A cancer-associated galactosyltransferase isoenzyme. New Eng J Med 1978; 299:703-705.
4. Chatterjee SK, Battacharya M, Barlow JJ. Biochemical and immunologic characterization of galactosyltransferase purified from the ascites of ovarian cancer patients. J Natl Cancer Inst 1985; 75:237-248.
5. Uemura M, Sakaguchi T, Uejima T et al. Mouse monoclonal antibodies which recognize a human (β1-4)galactosyltransferase associated with tumor in body fluids. Cancer Res 1992; 52:6153-6157.
6. Uejima T, Uemura M, Nozawa S et al. Complementary DNA cloning for galactosyltransferase associated with tumor and determination of antigenic epitopes recognized by specific monoclonal antibodies. Cancer Res 1992; 52:6158-6163.
7. Turner GA, Goodarzi MT, Thompson S. Glycosylation of alpha-1-proteinase inhibitor and haptoglobin in ovarian cancer: evidence for two different mechanisms. Glycoconj J 1995; 12:211-218.
8. Thompson S, Cantwell BMJ, Matta KL et al. Parallel changes in the blood levels of abnormally-fucosylated haptoglobin and alpha 1,3 fucosyltransferase in relationship to tumour burden: more evidence for a disturbance of fucose metabolism in cancer. Cancer Letters 1992; 65:115-121.
9. Chandreskaran EV, Jain RK, Matta KL. Ovarian cancer α1,3-L-fucosyltransferase. J Biol Chem 1992; 267:23806-23814.
10. Devine PL, McGuckin MA, Quin RJ et al. Serum markers CASA and CA 15-3 in ovarian cancer; all MUC1 assays are not the same. Tumor Biol 1994; 15:337-344.
11. Tashiro Y, Yonezawa S, Kim YS et al. Immunohistochemical study of mucin carbohydrates and core proteins in human ovarian tumors. Hum Pathol 1994; 25:364-372.
12. Ioannides C, Fisk B, Jerome K et al. Cytotoxic T cells from ovarian malignant tumors can recognize polymorphic epithelial mucin core peptides. J Immunol 1993; 151:3693-3703.
13. Inoue M, Ogawa H, Tanizawa O et al. Immunodetection of sialyl-Tn antigen in normal, hyperplastic and cancerous tissues of the uterine endometrium. Virchows Archiv 1991; 48:157-162.

14. Kobayashi H, Terao T, Kawashima Y. Serum sialyl Tn as an independent predictor of poor prognosis in patients with epithelial ovarian cancer. J Clin Oncol 1992; 10:95-101.
15. Ravn V, Teglbjaerg CS, Mandel U et al. The distribution of type 1 chain ABH and related histo-blood group antigens in normal cycling human endometrium. Int J Gynec Path 1993; 12:70-79.
16. Kubushiro K, Tsukazaki K, Sakuma Y et al. Enzymatic basis for the accumulation of Lewis(b) antigen in uterine endometrial cancer. Jap J Cancer Res 1995; 86:361-367.
17. Shiozawa T, Tsukuhara Y, Nakayama J et al. Immunohistochemical localization of blood group substances in normal and neoplastic endometrial tissues–with special reference to type 1 core chain expression. Gynec Obstet Invest 1991; 32:185-188.
18. Memon RS, Yonezawa S, Hasui K et al. Expression of blood group antigens and lectin binding in adenocarcinomas of the endometrium. Acta Pathol Jpn 1990; 40:509-516.
19. Amano J, Nishimura R, Mochizuki M et al. Comparative study of the mucin-type sugar chains of human chorionic gonadotropin present in the urine of patients with trophoblastic diseases and healthy pregnant women. J Biol Chem 1988; 263:1157-1165.
20. Cole LA. The O-linked oligosaccharide structures are striking different on pregnancy and choriocarcinoma hCG. J Clin Endocrin Metab 1987; 65:811-813.
21. Hard K, Damm JB, Spruijt MP et al. The carbohydrate chains of the beta subunit of human chorionic gonadotropin produced by the choriocarcinoma cell line BeWo. Novel O-linked and novel bisecting-GlcNAc-containing N-linked carbohydrates. Eur J Biochem 1992; 205:785-798.
22. Mizuochi T, Nishimura R, Derappe C et al. Structures of the Asparagine-linked sugar chains of human chorionic gonadotropin produced in choriocarcinoma. J Biol Chem 1983; 258:14126-14129.
23. Nishimura R, Endo Y, Tanabe K et al. The biochemical properties of urinary human chorionic gonadotropin from the patients with trophoblastic diseases. J Endocrin Invest 1981; 4:349-358.
24. Holland JF, Frei E, Bast RC Jr et al. In: Lea and Febiger, Cancer Medicine 3rd Ed Vol 1. Philadelphia: 1993;132-134.
25. Holland JF, Frei E, Bast RC Jr et al. In:Lea and Febiger. Cancer Medicine 3rd Ed Vol 1. Philadelphia: 1993; 324.
26. Lundgren R, Heim S, Mandahl M et al. Chromosome abnormalities are associated with unfavorable outcome in prostate cancer patients. J Urol 1992; 147:784-878.
27. Oesterling JE. Prostate specific antigen:a critical assessment of the most useful tumor marker for adenocarcinoma of the prostate. J Urol 1991; 145:907-923.
28. Jorgensen T, Berner A, Kaalhus A et al. Upregulation of the oligosaccharide sialyl Lewis^x: new prognostic parameter in metastatic prostate cancer. Cancer Res 1995; 55:1817-1819.
29. Mårtensson S, Bigler SA, Brown M et al. Sialys-Lewis^x and related carbohydrate antigens in the prostate. Hum Pathol 1995; 26:735-739.
30. Kadomatsu K, Anzano M, Slayter M et al. Expression of sulfated glycoprotein-2 is associated with carcinogenesis induced by N-nitroso-N-methylurea in rat prostate and seminal vesicle. Cancer Res 1993; 53:1480-1483.
31. Ørntoft TF. Carbohydrate changes in bladder carcinoma. APMIS 1992; 100 Suppl 27:181-187.
32. Ørntoft TF, Wolf H, Watkins W. Activity of the human blood group ABO, Se, H, Le, and X gene-encoded glycosyltransferases in normal and malignant bladder urothelium. Cancer Res 1988; 48:4427-4433.
33. Srinivas V, Khan A, Hoisington S et al. Relationship of blood groups and bladder cancer. J Urol 1986; 135:50-52.
34. Fukushi Y, Orikasa S, Shepard T et al. Changes of Le^x and dimeric Le^x haptens and their sialylated antigens during development of human kidney and kidney tumors. J Urol 1986; 126:1048-1056.
35. Meldgaard P, Holmes EH, Bennett EP et al. Blood group ABO-related glycosylation of urothelial cell lines: Immunocytological, enzymatic, and genetic characterization. Cancer Res 1994; 54:2440-2447.

36. Langkilde NC, Wolf H, Clausen H et al. Nuclear volume and expression of T-antigen, Sialosyl-Tn-antigen and Tn-antigen in carcinoma of the human bladder. Cancer 1992; 69:219-227.
37. Coon JS, Weinstein RS. Blood group antigens in tumor cell membranes. Biomembranes 1983; 11:173-205.
38. Plotkin GM, Wides RJ, Gilbert SL et al. Galactosyl transferase activity in human transitional cell carcinoma lines and in benign and neoplastic human bladder epithelium. Cancer Res 1979; 39:3856-3860.
39. Plotkin GM, Gilbert SL, Wides RJ et al. Galactosyltransferase activity in rat bladder transitional cell carcinoma lines and in exfoliated cells in urine during carcinogenesis and reversible hyperplasia. Cancer Biochemistry Biophysics 1980; 4:251-256.
40. Gmeiner B, Wolf B. Comparison of cell-associated and soluble galactosyltransferase isoenzymes from a human bladder transitional cell carcinoma line. Cancer Res 1987; 47:2311-2316.
41. Bergeron A, LaRue H, Fradet Y. Identification of a superficial bladder tumor-associated glycoform of the carcinoembryonic antigen by monoclonal antibody 19A211. Cancer Res 1996; 56:908-915.

CHAPTER 23

Lung Cancer

Tissues and sera from lung cancer patients contain sialylated and fucosylated poly-n-acetyl-lactosamine type 1 and type 2 chain epitopes regarded as general carcinoma antigens. These epitopes have been found on oncofetal types of glycoproteins and glycolipids.[1] A and B blood group determinants may be lost in lung cancer and incompatible blood groups may appear[2] while the expression of the I antigen (Tables 1, 6) may increase. Deletion of tumor cell blood group A antigen independent of the person's blood group status is characteristic of patients with progressive disease.[3,4]

Monoclonal antibodies against chains possessing H/Le^y/Le^b with the common structure Fucα1-2Galβ1-R give positive reactions on stained tumor cells from patients with poor survival; such cells also have higher degrees of motility and invasiveness than tumors not expressing these antigens.[5] Not only the expression of blood group A, but also binding sites for ABH antigens have been correlated with improved survival of patients with bronchial cancer.[6]

Normal human lung expresses Le^x and Le^y antigens in a stage-specific manner.[7] Lung cancer tissue and the serum of lung cancer patients exhibit various levels of Le^x, Le^y, and sialylated and poly-Le^x antigens.[1] Monoclonal antibodies directed against Le^x/Le^y/H antigens inhibit tumor cell motility.[8] The expression of Le^x/Le^y/H antigens demonstrated by immunohistochemistry was found to inversely correlate with the five-year survival of lung cancer patients. α3-Fuc-transferase activities synthesizing the Le^x/Le^y epitopes in the sera of lung cancer patients are significantly increased compared to normal levels.[9]

Cloned cells from squamous lung carcinomas develop sublines which are either Le^a-X (Le^a and Le^x determinant in the same chain; see Table 1) positive or Le^a-X negative; the former secrete mucin gels associated with increased invasiveness. These mucins may exclude antibodies from underlying cells and therefore exert blocking effects. Consistent with this hypothesis is the finding that in nude rats, there is more inflammatory cell infiltration into regressing Le^a-X negative tumors compared to progressing Le^a-X positive tumors.[10] In fact, Le^a-X negative cells do not develop to tumors in nude mice. In comparison, after transplantation of cells into the brain which is an immunoprivileged site, tumors develop and progress regardless of the expression of Le^a-X.

As shown with many other cancers, the T antigen is expressed on lung cancers. Anti T antibodies raised against the lung cancer cell line A549 have been useful in demonstrating T antigen in many cases of lung cancer and other neoplasms.[11]

Polysialic acid, comprised of repeating α2-8 sialic acid units, have been implicated in the invasive behavior of lung cancer cells. Antibodies directed against the neural cell adhesion molecule NCAM (containing PSA) and against PSA alone, showed that both PSA and NCAM or a closely related molecule occur on small cell tumor cells H69. The glycoprotein occurs in two forms, based upon the extent of sialylation, where the heavily sialylated form is expressed in lung tumor cases with a poor prognosis.[12]

A relationship between the expression of highly branched N-glycans and cell adhesion has been demonstrated in Mv1Lu mink lung epithelial cells which were transfected with GlcNAc-transferase V.[13] Compared to control cells, GlcNAc-transferase V transfected cells adhere less to fibronectin and collagen type IV although levels of α5β1 and $\alpha_v\beta3$ integrins remain similar. The loss of contact inhibition of cell growth induced by GlcNAc-transferase V is inhibited by swainsonine which inhibits the processing of N-glycans by mannosidase II, preventing the synthesis of complex N-glycans. Integrins and lamp-2 glycoproteins are substrates for GlcNAc-transferase V in transfected cells as seen in their larger molecular weights. With increasing levels of GlcNAc-transferase V activities in various transfected cells, L-PHA sensitivity increases as well as the incidence of benign tumors resulting in nude mice.

References

1. Nilsson O. Carbohydrate antigens in human lung carcinomas. APMIS 1992; Suppl 27 V 100:149-161.
2. Hirohashi S, Ino Y, Kodama T et al. Distribution of blood group antigens A, B, H, and I(Ma) in mucus-producing adenocarcinoma of human lung. J Natl Cancer Inst 1984; 72:1299-1305.
3. Kovarik S, Davidsohn I, Stejskal R. ABO antigens in cancer:detection with the mixed cell agglutination reaction. Arch Pathol 1968; 86:12-21.
4. Lee J, Ro J, Sahin A et al. Expression of blood group antigen A–a favorable prognostic factor in non-small cell lung cancer. N Engl J Med 1991; 324:1084-1090.
5. Miyake M, Taki T, Hitomi S et al. Correlation of expression of H/Le^y/Le^b antigens with survival in patients with carcinoma of the lung. N Engl J Med 1992; 327:14-18.
6. Kayser K, Bovin NV, Korchagina EY et al. Correlation of expression of binding sites for synthetic blood group A-, B- and H-trisaccharides and for sarcolectin with survival of patients with bronchial carcinoma. Eur J Cancer 1994; 30A:653-657.
7. Miyake M, Zenita K, Tanaka O et al. Stage-specific expression of SSEA-1-related antigens in the developing lung of human embryos and its relationship to the distribution of these antigens in lung cancers. Cancer Res 1988; 48:7150-7158.
8. Miyake M, Taki T, Hitomi S et al. Correlation of expression of H/Le^y/Le^b antigens with survival in patients with carcinoma of the lung. N Engl J Med 1992; 327:14-18.
9. Asao T, Yazawa S, Nagamachi Y et al. Serum alpha (1 → 3)-L-fucosyltransferase, carcinoembryonic antigen, and sialyl Lewis X-i antigen levels in lung cancer. Cancer 1989; 64:2541-2545.
10. Stranahan PL, Howard RB, Pfenninger O et al. Mucin gel formed by tumorigenic squamous lung carcinoma cells has Le^a-X oligosaccharides and excludes antibodies from underlying cells. Cancer Res 1992; 52:2923-2930.
11. Stein R, Chen S, Grossman W et al. Human lung carcinoma monoclonal antibody specific for the Thomsen-Friedenreich antigen. Cancer Res 1989; 49:32-37.
12. Moolenaar CECK, Muller EJ, Schol DJ et al. Expression of neural cell adhesion molecule-related sialoglycoprotein in small cell lung cancer and neuroblastoma cell lines H69 and CHP-212. Cancer Res 1990; 50:1102-1106.
13. Demetriou M, Nabi IR, Coppolino M et al. Reduced contact-inhibition and substratum adhesion in epithelial cells expressing GlcNAc-Transferase V. J Cell Biol 1995; 130:383-392.

CHAPTER 24

BREAST CANCER

Aberrant glycosylation, sulfation and antigen expression is a common characteristic of breast cancer cells.[1-3] In breast cancer, as in other cancers, there are changes in ABO blood group determinants and a general reduction of O-glycan processing with expression of T and Tn antigens.[4] Both decreases and increases, and association with metastasis[5,6] of the T antigen have been observed in breast cancer cells. The density of Tn epitope expression and a high ratio of Tn over T antigen in breast cancer was found to be an indicator of high aggressiveness (Table 6).[7] Brooks and Leathem[8] reported a strong association between the presence of lymph node metastases and *Helix pomatia* binding, suggesting the synthesis of incomplete O-glycan chains (Tn antigen) in metastatic cells. Zhuang et al[9] measured the *Helix pomatia* reactivity and the enzyme synthesizing the T antigen, core 1 β3-Gal-transferase, in breast carcinoma samples. The appearance of the Tn antigen in extracted glycoproteins was inversely related to the enzyme activity, suggesting that the synthesis of core 1 may be regulating Tn antigen expression by masking GalNAc residues.

Changed fucosylation is apparent in mammary carcinomas. Staining of normal and cancerous breast tissue with monoclonal antibodies showed that the Le^b antigen is lost in 80% of breast cancer cases while normal breast tissue and benign lesions express the Le^b antigen.[10] The loss is positively correlated with the degree of malignancy and invasiveness. Serum α3-Fuc-transferase levels as well as abnormally fucosylated haptoglobin increase in breast cancer and return to normal upon therapy.[11]

MUC1 has been intensively studied in normal and cancerous breast epithelium.[12] In carcinomas MUC1 is upregulated and its luminal distribution is often lost. MUC1 peptides are abnormally glycosylated in cancer with the consequence of a changed antigenic profile.[13-15] The MUC1 peptide epitopes in the O-glycosylated mucin tandem repeat region are recognized by several monoclonal antibodies, and appear in cancer mucins, probably due to a reduction in sugar chains normally masking this epitope. This change makes the molecule a potential target for immunotherapy (see chapter 27).[16-18]

Antibodies to the core protein of MUC1 have been identified in the serum of breast and ovarian cancer patients[19] and a cell-mediated immune response to MUC1 has been reported in patients with breast, ovarian, pancreatic and colorectal carcinoma.[20] Cytotoxic T-lymphocytes from breast cancer patients recognize and react with tumor cells presumably because they express mucin core protein epitopes lacking in normal cells. Antibody against one of these epitopes (SM3) is able to significantly inhibit cytotoxic lymphocyte-mediated tumor cell lysis.[21] Synthetic peptides derived from human MUC1 in ELISA assays demonstrate IgM-type immune responses in patients with breast cancer and other tumors.[22]

MUC2 mucin peptide epitopes of the tandem repeat region are also expressed in some breast carcinomas but rarely in normal breast epithelium, suggesting altered expression of MUC2 in breast cancer cells.[23]

The oligosaccharides of normal lactating mammary epithelial cells isolated from milk are relatively complex and have mostly disialylated and GlcNAcβ1-6Gal-elongated core 2 structures[24,25], and blood group I antigenic branches (Table 1). These may be extended at the GlcNAc moiety with poly-n-acetyl-lactosamine chains and may be terminated by sialylα3 residues linked to Gal and by Fucα3- or α4-linked to internal GlcNAc residues.[26] In contrast, BT20 breast cancer cells have simpler, shorter O-glycans of core 1, α3- and α6-sialylated core 1 structures and are lacking core 2.[27] MUC1 from the T47D breast cancer cells has less carbohydrate, reduced oligosaccharide chain length and strongly expresses Tn and sialyl-Tn, but low levels of Lewis antigens compared to MUC1 from normal breast cells.

Consistent with the decrease of GlcNAc and core 2 structures, and a high sialic acid content[28] in breast cancer cells, loss of core 2 β6-GlcNAc-transferase activity was observed in BT20 and T47D cells, due to the lack of mRNA expression (Fig. 11).[29] In the absence of O-glycan 1-6 branching, the carbohydrate chains occupy considerably less space and thus tend to expose more of the underlying peptide epitopes. However, another breast cancer cell line, MCF-7, also expresses MUC1 peptide epitopes but retains core 2 β6-GlcNAc-transferase activity, suggesting a different regulatory mechanism for peptide exposure in these cells. All of these breast cancer cells show an increase in α3-sialyltransferase activity acting on O-glycan core 1. The increased α3-sialylation may lead to premature termination of O-glycan chain synthesis. This assumes that the α3-sialyltransferase resides in relatively early Golgi compartments enabling a competition with enzymes involved in chain elongation. Such a process may be similar to that in leukemia cells and colon cancer where α3-sialyltransferase activity is increased and O-glycan chains are shorter and hypersialylated. Although the sialyl-Tn antigen has been reported in breast cancer, the α6-sialyltransferase acting on GalNAc-mucin is undetectable in these breast cancer cell lines.

Common to all breast cancer and normal mammary cell lines is the expression of enzymes to synthesize elongated, fucosylated and sialylated O-glycan chains. Polypeptide α-GalNAc-transferase is very active in normal and cancerous mammary cells towards MUC1 and MUC2 tandem repeat peptides but much less active toward those that contain Ser residues as the acceptor sites. The enzymes synthesizing core 1 and poly-n-acetyl-lactosamine structures are very active but α2- and α3-Fuc-transferase activities are relatively low in all cell types. α4-Fuc-transferase is high in normal and BT20 cells but very low in MCF-7 and undetectable in T47D cells. This suggests that some mammary cells may express Fuc-transferase III while others express a different type of α3-Fuc-transferase. All cells with the exception of T47D cells are capable of elongating core 1 by elongation β3-GlcNAc-transferase (Fig. 11). The enzymes synthesizing core 3 and 4 and the I antigen are absent from all mammary cells tested.[29]

Since T47D cells lack the enzymes synthesizing core 2, 3 and 4, as well as those elongating and branching core 1 structures, O-glycans lack poly N-acetyllactosamine chains. However, α2-

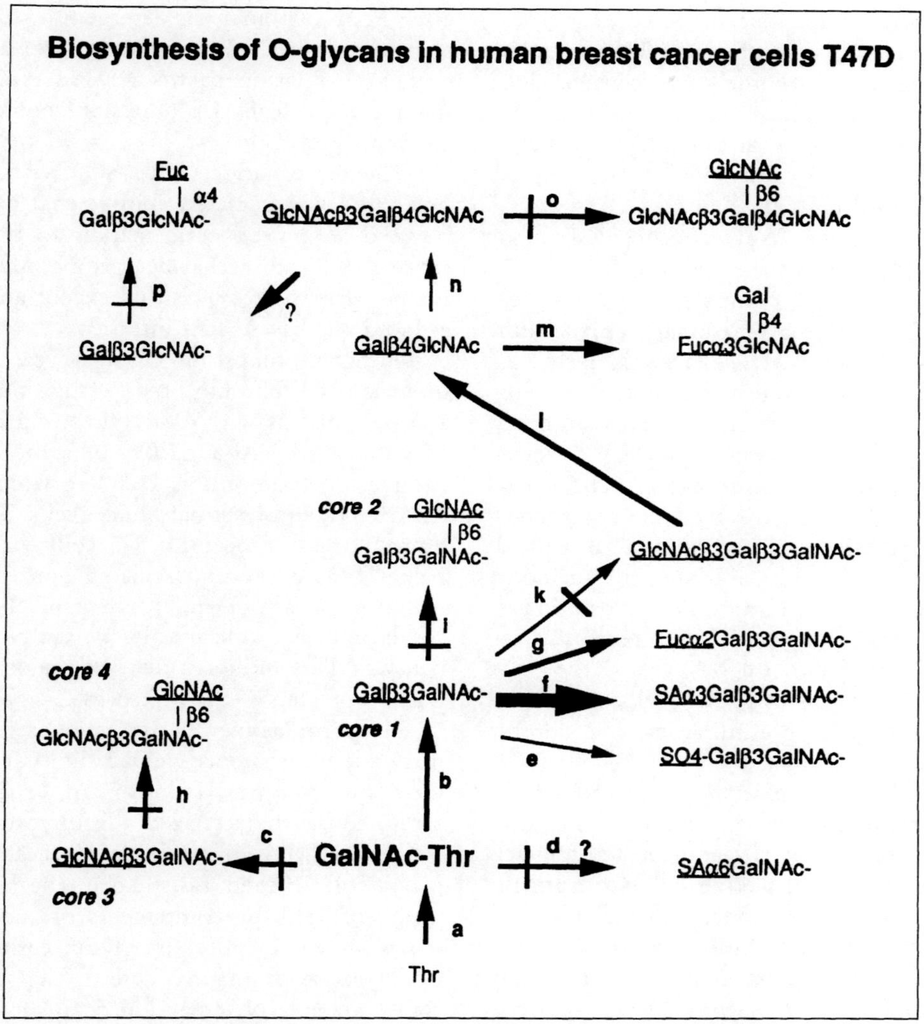

Fig. 11. O-glycan biosynthesis in human breast cancer T47D cells. The composite pathways show enzymatic steps involved in the synthesis of O-glycans in the human metastatic breast cancer T47D cell line.[29] The letters denote the transferases catalyzing the conversions: a, polypeptide α-GalNAc-transferase; b, core 1 β3-Gal-transferase; c, core 3 β3-GlcNAc-transferase; d, α6-sialyltransferase (O); e, core 1 sulfo-transferase; f, α3-sialyltransferase (O); g, α2-Fuc-transferase; h, core 4 β6-GlcNAc-transferase; i, core 2 β6-GlcNAc-transferase; k, elongation β3-GlcNAc-transferase; l, β4-Gal-transferase; m, α3-Fuc-transferase; n, iβ3-GlcNAc-transferase; o, Iβ6-GlcNAc-transferase; p, α4-Fuc-transferase. Blocked pathways are indicated by a bar across the arrow. Thick arrows indicate increased activity compared to mammary cells derived from normal human milk. In contrast to normal mammary cells, T47D cells cannot make core 2 structures due to lack of mRNA expression of core 2 β6-GlcNAc-transferase. There is also no elongation of core 1 (path k). None of the normal and cancerous mammary cell lines can synthesize core 3 or 4, the I antigen or sialyl-Tn in vitro and core 1 sulfation is decreased. T47D cells are therefore committed to synthesizing short O-glycan structures such as GalNAc- (Tn antigen) or fucosylated or sialylated core 1. This change in O-glycan structures in cancer cells leads to an exposure of underlying mucin peptide antigens. Human breast cancer cells BT20 show a similar alteration in O-glycan pathways. However, MCF-7 cells do express core 2 β6-GlcNAc-transferase and could therefore synthesize larger O-glycans. In these cells, however, the increase in α3-sialyltransferase (O) is thought to produce short, sialylated O-glycans leading to the exposure of peptide epitopes.

fucosylated and α3-sialylated core 1 structures can still be synthesized. These results are consistent with the sugar composition of immunoprecipitated mucins from T47D cells which indicate that GalNAc and sialic acid are present, but mucins are free of GlcNAc residues.[28]

Sulfation of O-glycan core 1 is also decreased in breast cancer BT20, MCF-7 and T47D cells compared to cells derived from normal mammary epithelium MTSV1-7.[29] Similarly, sulfation is decreased in colon cancer tissue and in polyposis cells differentiated to cancer cells. This suggests that core 1 3-sulfotransferase may be generally downregulated in cancer cells.

MCF-7 cells have been shown to contain glycoproteins carrying α2-8 linked sialic acid chains, using specific antibody and polysialic acid-specific endosialidase. While the polysialic acid residues on NCAM are found on N-glycans, they occur on O-glycans of MCF-7 cells since they are removed by β-elimination and not by N-glycanase.[30]

N-glycans released from normal breast and breast carcinoma tissues contain a higher amount of tri- and tetra-antennary chains with poly-N-acetyl-lactosamine structures in carcinomas. Primary mammary carcinomas exhibit higher N-glycan branching and contain more glycans that bind to *Datura stramonium* lectin (binding GlcNAc1-6 (GlcNAcβ1-2) Man-structures), i.e. more tri- and tetra-antennary branches with and without poly-n-acetyl-lactosamine chains compared to N-glycans from normal glands.[31] The degree of branching is increased in metastatic cancer as seen in a limited number of membrane glycoproteins. The enhanced expression of N-acetyllactosamine and GlcNAcβ1-6Man branches is correlated with progression of mammary carcinomas.[32] This is consistent with findings that a high proportion of human breast cancer tissue samples bind L-PHA (recognizing the GlcNAcβ1-6Man branch of N-glycans) in contrast to benign hyperplastic lesions. An elevation of L-PHA binding has been correlated with GlcNAc-transferase V activity synthesizing this GlcNAcβ1-6 branch in detergent extracts of breast tissues.[33] In another report, however, a correlation between L-PHA binding and breast cancer progression could not be confirmed.[34]

The disaccharide Galβ1-3GlcNAc is a breast tumor-associated epitope that could be used as a prognostic marker in breast cancer.[35] This disaccharide epitope appears in the early stages of breast cancer and is reduced in disease free intervals.

A high proportion of breast carcinomas has been found to produce sulfomucins and sialomucins with O-acetyl ester groups of sialic acid.[36] About 20% of sialic acid in breast cancer cells is Neu5Gc which is not detectable in normal human cells. Thus, human breast cancer MCF-7 cells express Neu5Gc as a cancer-associated epitope of MUC1 in a fashion independent of N-glycosylation but inhibitable by GalNAcα-benzyl.[37] This indicates that the major proportion of Neu5Gc is linked to O-glycans.

A spontaneously derived mouse mammary adenocarcinoma yielded the first surface mucin-type glycoprotein (epiglycanin) to be characterized.[38] Epiglycanin is thought to have a masking effect and to play a role in metastasis and in the recognition of cells by components of the immune system in mice. Its glycan composition comprises mainly core 1, sialylated core 1 and GalNAc-Ser/Thr chains as well as a high mannose and tetra-antennary N-glycans with poly-n-acetyl-lactosamine-type chains. Later studies of rat mammary adenocarcinoma sialoglycoproteins have shown a more complex profile of O-linked sialylated and non-sialylated oligosaccharides, as well as sulfated and sialylated core 2 derivatives on cell surface mucins.[3]

A cell surface transmembrane glycoprotein purified from ascites 13762 rat mammary cancer is overexpressed in some rat tumors. This molecule has two epidermal growth factor-like domains and could activate phosphorylation of the proto-oncogene receptor kinase p185 and in this way participate in tumor progression.[39,40] Cell surface sialomucins occur in many species and are believed to function as modulators of cell-cell recognition; for example, they may cover ligands for these interactions and confer resistance to cytotoxic immune responses.[40]

Tunicamycin treatment of rat mammary tumor cells increases their susceptibility to natural killer cell-mediated lysis and this is associated with a decrease in cell surface sialomucin. Since the major glycans on these cells are O-linked, it has been suggested that N-linked glycosylation is a prerequisite for sialomucin synthesis.[41] Two rat mammary cancer cell sublines expressing cell surface sialomucin differ in sialic acid content, xenotransplantability and other properties. Both cell lines appear to contain the α3-sialyltransferase (N) acting on N-glycans and other sialyltransferases acting on O-glyans although it was not established which sialyltransferase activity could account for the difference in sialic acid content.[42]

Altered sulfation has been observed on transformed mouse mammary epithelial cells, particularly the basal lamina where reduced N- and O-sulfate levels on heparan sulfate are present.[43] In the extracellular matrix of infiltrating human breast carcinomas, the density of sulfated proteoglycans is greatly reduced compared to the normal; the remaining sulfated proteoglycans are more elongated but are incapable of forming normal fibrillar structures, indicating that there are structural alterations of the extracellular matrix related to infiltrating cancer.[44]

References

1. Brockhausen I. Clinical aspects of glycoprotein synthesis. Crit Rev Clin Lab Sci 1993; 301; 65-151.
2. Lee AK, DeLellis RA, Rosen PP et al. ABH blood group isoantigen expression in breast carcinomas–an immunohistochemical evaluation using monoclonal antibodies. Am J Clin Path 1985; 83:308-319.
3. Hull S, Carraway K. Sulfation of the tumor cell surface sialomucin of the 13762 rat mammary adenocarcinoma. J Cell Biochem 1989; 40:67-81.
4. Springer GF. T and Tn pancarcinoma markers:autoantigenic adhesion molecules in pathogenesis, prebiopsy carcinoma-detection, and long-term breast carcinoma immunotherapy. Crit Rev Oncogenesis 1995; 6:57-85.
5. Springer G, Desai P, Wise W et al. Pancarcinoma T and Tn epitopes: autoimmunogens and diagnostic markers that reveal incipient carcimomas and help establish prognosis. In: Herberman R, Mercer D eds. Immunodiagnosis of Cancer 2nd ed. New York:Marcel Dekker, 1990:587-612.
6. Springer G, Desai P, Tegtmeyer H et al. Pancarcinoma T/Tn antigen detects human carcinoma long before biopsy does and its vaccine prevents breast carcinoma recurrence. Ann N Y Acad Sci 1993; 690:355-357.
7. Springer GF. Tn epitope (N-acetyl-D-galactosamine α-O-serine/ threonine) density in primary breast carcinoma: a functional predictor of aggressiveness. Molecular Immunol 1989; 26:1-5.
8. Brooks SA, Leathem AJ. Prediction of lymph node involvement in breast cancer by detection of altered glycosylation in the primary tumour. Lancet 1991; 338:71-74.
9. Zhuang D, Yousefi S, Dennis JW. Tn antigen and UDP-Gal:GalNAcα-R β1-3galactosyltransferase expression in human breast carcinoma. Cancer Biochem Biophys 1991; 12:185-198.
10. Idikio H, Manickavel V. Lewis blood group antigens (a and b) in human breast tissues. Cancer 1991; 68:1303-1308.
11. Thompson S, Cantwell BMJ, Matta KL et al. Parallel changes in the blood levels of abnormally-fucosylated haptoglobin and alpha 1,3 fucosyltransferase in relationship to tumour burden:more evidence for a disturbance of fucose metabolism in cancer. Cancer Letters 1992; 65:115-121.
12. Girling A, Bartkova J, Burchell J et al. A core protein epitope of the polymorphic epithelial mucin detected by the monoclonal antibody SM-3 is selectively exposed in a range of primary carcinomas. Int J Cancer 1989; 43:1072-1076.
13. Burchell J, Taylor-Papadimitriou J. Effect of modification of carbohydrate side chains

on the reactivity of antibodies with core-protein epitopes of the MUC1 gene product. Epith Cell Biol 1993; 2:155-162.
14. Taylor-Papadimitriou J, Stewart L, Burchell J et al. The polymorphic epithelial mucin as a target for immunotherapy. Ann NY Acad Sci 1993; 690:69-79.
15. Glebov OK, McKenzie KE, White CA et al. Frequent p53 gene mutations and novel alleles in familial breast cancer. Cancer Res 1994; 54:3703-3709.
16. McKenzie IFC and Xing PX. Mucins in breast cancer:recent immunological advances. Cancer Cells 1990; 2:75-78.
17. MacLean GD, Reddish M, Koganty RR et al. Immunization of breast cancer patients using a synthetic sialyl-Tn glycoconjugate plus Detox adjuvant. Cancer Imm Immunother 1993; 36:215-222.
18. Taylor-Papadimitriou J, D'Souza B, Burchell J et al. The role of tumor-associated antigens in the biology and immunotherapy of breast cancer. Ann NY Acad Sci 1993; 698:31-47.
19. Rughetti A, Turchi V, Ghetti CA et al. Human B-cell immune-response to the polymorphic epithelial mucin. Cancer Res 1993; 53:2457-2459.
20. Jerome K, Barndt D, Bendt K et al. Cytotoxic T-lymphocytes derived from patients with breast adenocarcinoma recognize an epitope present on the protein core of a mucin molecule preferentially expressed by malignant cells. Cancer Res 1991; 51:2908-2916.
21. Jerome KR, Domenech N, Finn OJ. Tumor-specific cytotoxic T cell clones from patients with breast and pancreatic adenocarcinoma recognize EBV-immortalized B cells transfected with polymorphic epithelial mucin complementary DNA. J Immun 1993; 151:1654-1662.
22. Kotera Y, Fontenot J, Pecher G et al. Humoral immunity against a tandem repeat epitope of human mucin MUC1 in sera from breast, pancreatic and colon cancer patients. Cancer Res 1994; 54:2856-2860.
23. Walsh MD, McGuckin MA, Devine PL et al. Expression of MUC2 epithelial mucin in breast carcinoma. J Clin Path 1993; 46:922-925.
24. Amano J, Straehl P, Berger EG et al. Structures of mucin-type sugar chains of the galactosyltransferase purified from human milk. Occurrence of the ABO and Lewis blood group determinants. J Biol Chem 1991; 266:11461-11477.
25. Hanisch FG, Uhlenbruck G, Peter-Katalinic J et al. Structures of neutral O-linked polylactosaminoglycans on human skim milk mucins. A novel type of linearly extended poly-N-acetyllactosamine backbones with Galβ(1-4) GlcNAcβ(1-6) repeating units. J Biol Chem 1989; 264:872-883.
26. Hanisch F-G, Peter-Katalinic J, Egge H et al. Structures of acidic O-linked polylactosaminoglycans on human skim milk mucins. Glycoconj J 1990; 7:525-543.
27. Hull SR, Bright A, Carraway KL et al. Oligosaccharide differences in the DF3 sialomucin antigen from normal human milk and the BT-20 human breast carcinoma cell line. Cancer Commun 1989; 1:261-267.
28. Lloyd K, Kudryashov V, Yin B et al. Analysis of carbohydrate composition and structure in radiolabeled immunoprecipitates: application to breast and ovarian cancer mucins. 3rd Internat. Workshop on carcinoma-assoc. mucins, Cambridge, UK 1994.
29. Brockhausen I, Yang J-M, Burchell J et al. Mechanisms underlying aberrant glycosylation of MUC1 mucin in breast cancer cells. Eur J Biochem 1995; 233:607-617.
30. Martersteck CM, Kedersha NL, Drapp DA et al. Unique α2,8-polysialylated glycoproteins in breast cancer and leukemia cells. Glycobiology 1996; 6:289-301.
31. Hiraizumi S, Takasaki S, Ohuchi N et al. Altered glycosylation of membrane glycoprotein associated with human mammary carcinoma. Jpn J Cancer Res 1992; 83:1063-1072.
32. Fernandes B, Sagman U, Auger M et al. β1-6 branched oligosaccharides as a marker of tumor progression in human breast and colon neoplasia. Cancer Res 1991; 51:718-723.
33. Dennis J, Laferté S. Oncodevelopmental expression of GlcNAc-β(1-6)Man-α(1-6) Man-β1- branched asparagine-linked oli-

gosaccharides in murine tissues and human breast carcinomas. Cancer Res 1989; 49:945-950.

34. Rak J, Miller FR. Expression of L-PHA binding oligosaccharides and progression of mammary neoplasia are unrelated. Internat J Cancer 1993; 54:524-526.
35. Rye PD, Bovin NV, Vlasova EV et al. Monoclonal antibody LU-BCRU-G7 against a breast tumour-associated glycoprotein recognizes the disaccharides Gal beta 1-3GlcNAc. Glycobiology 1995; 5:385-389.
36. Hanna WM, Corkill M. Mucins in breast carcinoma. Hum Pathol 1988; 19:11-14.
37. Devine P, Clark B, Birrell G et al. The breast tumor-associated epitope defined by monoclonal antibody 3E1.2 is an O-linked mucin carbohydrate containing N-glycolylneuraminic acid. Cancer Res 1991; 51:5826-5836.
38. Codington J, Haavik S. Epiglycanin-a carcinoma-specific mucin-type glycoprotein of the mouse TA3 tumor. Glycobiology 1992; 2:173-180.
39. Hull S, Laine R, Kaizu T et al. Structures of the O-linked oligosaccharides of the major cell surface sialoglycoprotein of MAT-B1 and MAT-C1 ascites sublines of the 13762 rat mammary adenocarcinoma. J Biol Chem 1984; 259:4866-4877.
40. Carraway KL, Fregien N, Carraway KL III et al. Tumor sialomucin complexes as tumor antigens and modulators of cellular interactions and proliferation. J Cell Sci 1992; 103:299-307.
41. Bharanthan S, Moriarty J, Moody CE et al. Effect of tunicamycin on sialomucin and natural killer cell susceptibility of rat mammary tumor ascites cells. Cancer Res 1990; 50:5250-5256.
42. Pratt T, Sherblom AP. Sialyltransferase of the 13762 rat mammary ascites tumor cells. Cancer Res 1984; 44:1148-1152.
43. David G, Van den Berghe H. Transformed mouse mammary epithelial cells synthesize undersulfated basement membrane proteoglycan. J Biol Chem 1983; 258:7338-7344.
44. Losa G, Alini M. Sulfated proteoglycans in the extracellular matrix of human breast tissues with infiltrating carcinoma. Int J Cancer 1993; 54:552-557.

CHAPTER 25

Glycoproteins in Other Cancers

25.A. SQUAMOUS CELL CARCINOMA

The expression of carbohydrate epitopes can be seen as a process of differentiation in stratified squamous cells. This form of epithelium is observed in the oral cavity, cervix and portions of the esophagus. The progeny of germinal cells in the basal layer changes their topographic position as they differentiate and these cells are eventually exfoliated. Markers such as ABO blood groups tend not to be expressed in basal epithelium but are present in overlying cell layers and are cell cycle-dependent (Table 6).[1] Thus, one usually observes a mixed population in stratified epithelium with typical topographic and structural features.

In carcinomas of the cervix the distinction disappears since ABO antigens are no longer expressed, in particular in metastatic carcinomas.[2,3] Loss of the H antigen in these tumors indicates a poor prognosis.[4,5] Similar losses of blood group antigens have been noted in buccal carcinomas where the normal cells adjacent to the tumor may still stain for blood group antigens. There is generally not a good correlation between differentiation and blood group expression but instead with invasive and metastatic properties. Some of the glycosylation changes in invasive squamous carcinomas are reminiscent of those observed during normal differentiation and in epithelial migration during wound healing, except that these normal changes are controlled and reversible.

It has been suggested that changes in blood group antigens in invasive cancers may facilitate metastasis by increasing cell motility. Reduced expression of Lewis antigens has also been typical in these cancers. In view of the interesting differentiation pattern shown by squamous cells, it is possible that blood group loss in cancer results from cell cycle arrest. In contrast to tumor losses of terminal sugar epitopes such as ABO and Lewis blood groups, sialyl-Tn is expressed in some squamous cell carcinomas of the esophagus but not in normal tissue. This was determined by immunohistochemical methods using monoclonal antibodies.[6]

No relationship has been found between sialyl-Tn and poor prognosis or lymph node metastasis.

25.B. MELANOMA

B16 mouse melanoma cells contain sialylated glycoconjugates including glycoproteins with N- and O-glycans.[7,8] Two variants of B16 melanoma cells selected by their resistance to toxic concentrations of wheat germ agglutinin show decreased tumorigenicity in mice compared to the parent B16 cells. The variants show less wheat germ agglutinin binding, probably due to the absence of sialic acid residues. A major glycoprotein is absent from the variants which may be responsible for their reduced lectin binding.

The presence of cell surface sialic acid may enhance the survival of cancer cells in the blood and therefore promote metastasis. Cell surface sialic acid of B16 melanoma cell variants accessible to cleavage by sialidase and to the action of cell surface sialyltransferase has been correlated with the ability of cells to form lung metastases.[9] In these studies whole cells were used as enzyme substrates.

The activity and mRNA expression of N-glycan branching enzyme GlcNAc-transferase V is induced in B16 melanoma cells by transforming growth factor β.[10] This increase is accompanied by morphological changes and increased L-PHA staining of a specific glycoprotein as observed by electrophoretic separation. Interestingly, an intron region of GlcNAc-transferase V is expressed in about 50% of human melanomas, resulting in the production of an antigenic peptide that is recognized by cytolytic T-lymphocytes. Normal tissues do not express the mRNA encoding this intron region in significant amounts.[11]

The growth of mouse melanoma cells may be inhibited with retinoic acid.[12] This is accompanied by changed glycosylation. A cell surface sialoglycoprotein containing sialylated complex N-glycans appears and there is increased binding to wheat germ agglutinin and Concanavalin A upon retinoic acid treatment. Retinoic acid treatment also results in increased sialyltransferase activity towards asialofetuin substrate[13,14] but the specific sialyltransferases involved remain to be identified.

Cultured human melanoma cells have been shown by combined chromatographic and immunostaining methods to contain small amounts of de-N-acetylated sialic acids (with a free amino group at the 5-position).[15] Synthetic de-N-acetyl sialic acid-containing gangliosides modulate cell proliferation and tyrosine phosphokinase reactions. The appearance of these sialic acid derivatives is induced by the tyrosine kinase inhibitor genistein.

25.C. TERATOCARCINOMA

Glycoproteins of murine teratocarcinoma cells are the main carriers of differentiation antigens Le^x, I and i (Table 1)[16] compared to glycolipids which contain a minor proportion of these carbohydrates. The expression of the I antigen occurs in early embryos and on undifferentiated mouse teratocarcinoma cells.[17]

Structural analysis of human PA1 embryonal carcinoma cells demonstrates the presence of tri- and tetra-antennary N-glycans with many repeating units of poly-n-acetyl-lactosamine chains carrying sialic acid in α2-3 and other linkages. For example, sialylα2-9 sialylα2-3 or 6 Gal-groups are found at the termini of poly-n-acetyl-lactosamine chains.[18]

Retinoic acid differentiation of mouse F9 teratocarcinoma cells results in the induction of multiple glycosyltransferases and an increase in fucosylated poly N-acetyl-lactosamine structures, and sensitivity to endo β-galactosidase of membrane glycoproteins such as lamp-1 glycoprotein.[19] Core 1 β3-Gal- and β4-Gal-transferases and in particular, core 2 β6-GlcNAc-transferase and GlcNAc-transferase V, are increased upon retinoic acid treatment. All of these activities may contribute to the enhanced synthesis of poly-n-acetyl-lactosamine chains in differentiated cells. Cummings and Mattox[20] reported that α3-Gal-transferase activity but not β3/4-Gal- or α6-Fuc-transferase activities increases upon retinoic acid differentiation.

25.D. TUMORS OF THE NERVOUS SYSTEM

Human neuroblastoma cell lines derived from neuroectoderm contain large amounts of Fuc compared to human fibroblasts.[21] Fuc was apparently predominantly linked in α1-3 linkages as shown by endoglycosidase and almond α-fucosidase (specific for Fucα3/4 linkages) digestion as well as by NMR (Table 6).[22]

The cell-cell interactions of tumor cells include those with platelets, leading to increased tumor cell aggregation and thrombus formation. Cell surface sialylation of human neuroblastoma cells SKNMC has been correlated with the ability to activate platelets.[23]

The expression of α2-Fuc-containing glycoconjugates changes dramatically during neural development in the rabbit.[24] The role of α2-fucosylation was studied by transfecting mouse neuroblastoma cells with the rabbit equivalent of the human H-dependent and the Se-dependent α2-Fuc-transferase genes. H-gene transfected cells exhibit α2-Fuc-transferase activity as well as the H antigen while transfection with the Se-type gene α2-Fuc-transferase resulted only in small phenotypic changes. The transfection of the H gene α2-Fuc-transferase significantly suppressed axonal outgrowth of neuroblastoma cells while multiple neurite sprouts were formed. This suggests that α2-fucosylation can modify the response of neural cells to growth stimuli.[25]

REFERENCES

1. Szulman A. The histological distribution of the blood group substances in man as disclosed by immunofluorescence. II. The H antigen and its relation to A and B antigens. J Exp Med 1962; 115:977-995.
2. Kuhns WJ, Bramson S. Variable behaviour of blood group H on HeLa cell populations synchronized with thymidine. Nature 1968; 219:938-939.
3. Davidsohn I, Kovarik S, Ni LY. Isoantigens A, B, and H in benign and malignant lesions of the cervix. Arch Pathol 1969; 87:306-314.
4. Dabelsteen E, Clausen H, Mandel U. Carbohydrate changes in squamous cell carcinomas. APMIS 1992; 100 Suppl 27:130-138.
5. To ACW, Soong S-J, Shingleton HM et al. Immunohistochemistry of the blood group A, B, H isoantigens and Oxford Ca antigen as prognostic markers for stage 1B squamous cell carcinoma of the cervix. Cancer 1986; 58:2435-2439.
6. Ikeda Y, Kuwano H, Baba K et al. Expression of sialyl-Tn antigens in normal squamous epithelium, dysplasia, and squamous cell carcinoma in the esophagus. Cancer Res 1993; 53:1706-1708.
7. Lin L-H, Stern JL, Davidson EA. Clones from cultured, B16 mouse-melanoma cells resistant to wheat-germ agglutinin and with altered production of mucin-type glycoproteins. Carbohydr Res 1983; 111:257-271.
8. Lotan R, Irimura T. Enhanced glycosylation of a melanoma cell surface glycoprotein by retinoic acid:carbohydrate chain analysis by lectin binding. Cancer Biochem Biophys 1987; 9:211-221.
9. Dobrossy L, Pavelic ZP, Bernacki RJ. A correlation between cell surface sialyltransferase, sialic acid, and glycosidase activities and the implantability of B16 murine melanoma. Cancer Res 1981; 41:2262-2266.
10. Miyoshi E, Nishikawa A, Ihara Y et al. Transforming growth factor β up-regulates expression of the N-acetylglucosaminyltransferase V gene in mouse melanoma cells. J Biol Chem 1995; 270:6216-6220.
11. Guilloux Y, Lucas S, Brichard VG et al. A peptide recognized by human cytolytic T lymphocytes on HLA-A2 melanomas is encoded by an intron sequence of the N-acetylglucosaminyltransferase V gene. J Exp Med 1996; 183:1173-1183.
12. Lotan R, Lotan D, Deutsch V. Growth inhibition of murine melanoma cells by antibodies to a cell surface glycoprotein implicated in retinoic acid action. Cancer Res 1987; 47:3152-3158.
13. Deutsch V, Lotan R. Stimulation of

sialyltransferase activity of melanoma cells by retinoic acid. Exp Cell Res 1983; 149:237-245.
14. Lotan R, Lotan D, Amos B. Enhancement of sialyltransferase in two melanoma cell lines that are growth-inhibited by retinoic acid results in increased sialylation of different cell-surface glycoproteins. Exp Cell Res 1988; 177:284-294.
15. Sjoberg ER, Chammas R, Ozawa H et al. Expression of De-N-acety-gangliosides in human melanoma cells is induced by Genistein or Nocodazole. J Biol Chem 1995; 270:2921-2930.
16. Childs RA, Pennington J, Uemura K-I et al. High-molecular-weight glycoproteins are the major carriers of the carbohydrate differentiation antigens I, i and SSEA-1 of mouse teratocarcinoma cells. Biochem J 1983; 215:491-503.
17. Feizi T, Kapadia A, Gooi HC et al. Human monoclonal auto-antibodies detect changes in expression and polarization of the Ii antigens during cell differentiation in early mouse embryos and teratocarcinomas. In:Muramatsu T, Gachelin G, Moscona AA, Ikawa Y, eds. Teratocarcinoma and Embryonic Cell Interactions. NY:Academic Press 1982:201-215.
18. Fukuda MN, Dell A, Oates JE et al. Embryonal Lactosamino-glycan–the structure of branched lactosaminoglycans with novel disialosyl, (sialyl α2-9 sialyl) terminals isolated from PA1 human embryonal carcinoma cells. J Biol Chem 1985; 260:6623-6631.
19. Heffernan M, Lotan R, Amos B et al. Branching β1-6-N-acetylglucosaminetransferases and polylactosamine expression in mouse F9 teratocarcinoma cells and differentiated couterparts. J Biol Chem 1993; 268:1242-1251.
20. Cummings RD, Mattox SA. Retinoic acid-induced differentiation of the mouse teratocarcinoma cell line F9 is accompanied by an increase in the activity of UDP-galactose: β-D-Galactosyl-α1,3-galactosyltransferase. J Biol Chem 1988; 263:511-519.
21. Santer UV, Glick MC. Presence of fucosyl residues on the oligosaccharide antennae of membrane glycopeptides of human neuroblastoma cells. Cancer Res 1983; 43:4159-4166.
22. Santer UV, Glick MC, van Halbeek H et al. Characterization of the neutral glycopeptides containing the structure of α-L-fucopyranosyl-(1-3)-2-acetamido-2-deoxy-D-glucose from human neuroblastoma cells. Carbohydr Res 1983; 120:197-213.
23. Bastida E, Almirall L, Jamieson GA et al. Cell surface sialylation of two human tumor cell lines and its correlation with their platelet-activating activity. Cancer Res 1987; 47:1767-1770.
24. Kusunoki S, Inoue K, Iwamori M et al. Developmental changes of fucosylated glycoconjugates in rabbit dorsal root ganglia. Neurosci Res 1992; 15:74-80.
25. Hitoshi S, Kojima N, Kusunoki S et al. Expression of the β-galactoside α1,2-fucosyltransferase gene suppresses axonal outgrowth of Neuro2a neuroblastoma cells. J Neurochem 1996; 66:1633-1640.

CHAPTER 26

Metastasis

The clinical stages of cancer are characterized by alterations in cancer cell survival and the progression to metastases. Metastatic spread poses a most formidable threat to cancer patients. Important stages in this complex cascade relate to tumor cell detachment from the primary sites, survival in the blood stream, adhesion to the endothelium and invasion of neighboring or remote tissues. These processes are likely to be facilitated by cell surface molecules such as cell adhesion glycoproteins.[1] Metastatic cells exhibit largely quantitative differences in carbohydrate structures that seem to be related to the primary tumor cell type.[2-4] Experimental models for metastasis have attempted to establish the functional role of cell surface carbohydrate, for example, within tumor-host cell interactions. The changes characteristic of metastatic variants may also be useful as prognostic indicators and may be exploited in designing therapy to specifically target metastatic cells.

The facilitating role of cell surface sialic acid in the metastatic process is demonstrated by the fact that hypersialylated cells are more metastatic than their less sialylated counterparts.[5-9] Sialic acid has been implicated in all stages of the metastatic process. The linkages and attachment points of sialic acid play a role in metastasis[8,10,11] but the change from N-acetyl to N-glycolyl-neuraminic acid in mouse lymphoma mutants apparently does not affect the metastatic phenotype.[12] Cells from human liver metastases demonstrate more sialic acid than the primary tumors, but the sialyltransferases responsible for sialylation remain to be identified.[13]

26.A. N-GLYCANS AND METASTASIS

Cell surface sialic acid[14] and polylactosaminoglycan chains[15] presenting sialic acid may be responsible for decreased adhesion and higher motility, and may therefore be involved in determining the malignant potential of cells. More asparagine-linked poly N-acetyl-lactosaminoglycan chains with increased sialylation and decreased fucosylation are present on highly metastatic human colon cancer cell lines compared with less metastatic variants.[16] The highly metastatic cell lines express more of lysosomal membrane proteins lamp-1 and lamp-2, and sialyl-Le^x antigens on the termini of their poly-n-acetyl-lactosamine chains. GlcNAc-transferase V, blood group i β3-GlcNAc-transferase and α6-sialyltransferase

(N) activities are unchanged, but α3-Fuc-transferase is decreased, and α3-sialyl-transferase acting on N-glycans is increased in the highly metastatic cells. These findings indicate that glycosyltransferases synthesizing the Le^x determinants on lamp proteins are probably not solely responsible for the increased antigen expression.

Liver metastases from colon cancer produced CEA carrying sialylated bi-, tri-, and tetra-antennary N-glycans with a high proportion of bisected chains and α6-fucosylation of the proximal GlcNAc residue. Type 2 chains with Le^x and Le^y and a small amount of type 1 chains with Le^a determinants are also found.[17] Chandrasekaran et al[18] also reported that CEA from colon and breast cancer metastases has a high proportion of fucosylated and sialylated tetra-antennary N-glycans. Sialyl-Le^x is increased in those colon cancers that have developed metastases,[19] and the increased expression is probably due to a higher number of Le^x -positive cells in such tumors.

In rat prostatic cancer, metastatic cells exhibit increases in both GlcNAc-transferase V and GlcNAc-transferase-III.[20] The GlcNAcβ1-6Manα1-6Man-R branch synthesized by GlcNAc-transferase V may be important by providing more attachment sites for the addition of sialic acid or other terminal structures. Metastatic cells may be protected by additional 1-6 branches due to their flexibility and sterically mask underlying epitopes. GlcNAc-transferase III, however, is expected to oppose these effects.

In a mouse lymphoma line, lectin-resistant mutants exhibit different cell surface oligosaccharides and degrees of sialylation associated with variations in their metastatic potential.[11] The formation of tumor metastases in murine tumor cells have been inhibited by inhibitors of N-glycan processing including swainsonine and castanospermine.[21-25] Mutant tumor cells deficient in N-glycan processing were found to have lower metastatic potential and reduced tumor growth.[12] This suggests that complex N-glycans may be involved in the metastatic process. Higher branching of N-glycans, detected by L-PHA binding, appears to be correlated with the metastatic potential of these tumor cells. Less metastatic cells are insensitive to L-PHA and show very low activity of branching GlcNAc-transferase V.[26] A cell surface glycoprotein (gp30) appears to be mainly affected by these changes. It appears that in several cell models, the metastatic potential correlates with β1-6 branching of N-glycans.

Human erythroleukemia cells K 562 are sensitive to NK cell cytotoxicity. The expression of GlcNAc-transferase III, which produces bisected N-glycans, promotes resistance to NK cytotoxicity in K 562 cells, resulting in spleen colonization in nude mice. Lectin flow cytometry shows that transfected cells exhibit more bisected structures, and a decrease of surface sialylation and tri- and tetra-antennary N-glycans.[27] Similarly, in B16 mouse melanoma cells, the transfection of GlcNAc-transferase III gene suppresses lung metastases in nude and syngeneic mice. The effect of GlcNAc-transferase III transfection is a shift of the balance of complex N-glycan structures to more bisecting (E-PHA binding) and less L-PHA binding structures. This indicates that GlcNAc-transferase V action has been reduced by the presence of GlcNAc-transferase III.[28] It is thus possible that bisecting structures have a beneficial effect in reducing metastases either directly or by preventing the formation of the metastasis-associated GlcNAcβ1-6 branching of N-glycans.

Sialylated tetra-antennary chains without repeated poly-n-acetyl-lactosamine units of B16 mouse melanoma cells were found to correlate with their lung metastasizing ability.[29] Several clones of ras-transformed rat fibroblasts were selected with *Sambucus nigra* lectin-recognizing sialylα2-6 linkages. A strong correlation was seen between invasive potential and high expression of α2-6 sialic acid linkages and α6-sialyltransferase activity towards asialo-orosomucoid substrate.[30]

26.B. O-GLYCANS AND METASTASIS

Mucin production appears to be important for the metastasis of human colonic cancer cells and is associated with poor prognosis in patients with mucin producing carcinomas.[31] O-glycan constituents of mucins are greatly modified in cancer and metastatic cells with a trend towards the synthesis of truncated chains. This often results in the expression of precursor structures and increased sialic acid.

The invasiveness of the human colon cancer cell line KM12 variants correlates with the expression of sialyl-dimeric Le^x attached to mucin-type chains.[32] Cells that express sialyl-Le^x excessively form liver metastases from intrasplenic tumors in nude mice.[33] In human tumors, sialyl-dimeric Le^x antigens are highly expressed in tissue from liver metastases, and these antigens increase with the progression of colorectal cancer.[19]

Highly invasive metastatic cells of murine origin display the Tn antigen which is normally cryptic, whereas a higher ratio of T to Tn antigen was observed in the less invasive counterpart.[34] The distribution of sialyl-Tn and Le^y antigen changes in metastases of cervical cancer.[35] In metastatic breast cancer cell lines there is also increased sialylation and a switch to shorter O-glycans. A loss of mucin O-glycan processing is seen in metastatic mammary cancer T47D cells which have been shown to lack core 2 structures due to the loss of core 2 β6-GlcNAc-transferase, but exhibit increased α3-sialylation.[36] It has not been established why sialyl-Tn expression is associated with a poor prognosis in humans, or how this is related to the E-selectin-Le^x binding mechanism.

Tumor cells with mucin-type O-glycans can bind readily to E-selectins on activated endothelium via sialyl-Le^x ligands. Sialyl-Le^x structures are present on O-glycans,[37] and may provide a mechanism for an early step in metastatic cell invasion. Many colon cancer cell lines exhibit cell surface O-glycan containing ligands for selectins.[16,33,38-40] Binding to E-selectin is abrogated in vitro by inhibition of O-glycosylation. This has been confirmed in human colonic cancer cells which exhibit sialyl-Le^x ligands whose attachment to E-selectin can be inhibited by GalNAcα-benzyl[37] as well as by antibody against E-selectin. In addition, in vivo experiments show that inhibition of O-glycan extension by GalNAcα-benzyl greatly reduces liver colonization of human cancer cell LS174T variants in nude mice. This suggests that the carbohydrate portion of mucins plays an important role in the metastatic process.[31] This is consistent with the finding that circulating mucins shed from cancer cells may not only interfere with binding to selectins but may also interfere with immune responsiveness.

Selectin-mediated adhesion as well as liver metastases formed in nude mice are also dependent on mucin O-glycan structures in human colon cancer cell lines LSB and LSC.[40] Other molecules facilitating adhesion include CEA and type IV collagen. Normal human colorectal cells bind to CEA and type IV collagen but not laminin on coated tissue culture wells.[41] There is evidence that in mice, as in humans, cancer cell adhesion to endothelium is facilitated by the presence of cell surface sialyl-Le^x, although the specificity of selectin interactions remains to be established in rodents.[33,42]

26.C. INHIBITION OF CELL ADHESION IN METASTASIS

Endogenous lectins have been implicated in the metastatic process because they promote cell adhesion. They may also be potentially useful for cancer therapy. Cell surface Gal may bind to Gal-specific lectins in the liver and thereby promote liver metastasis. Therefore, in subpopulations of mouse colon carcinoma cells, those expressing Gal would be most likely to express metastatic potential. Liver metastases in mice are prevented by galactose derivatives which are thought to inhibit the interaction of tumor cells with Gal-receptors on hepatocytes.[43-45]

E-selectins have been held responsible not only for the binding and integration of leukocytes but also for metastasis since human colon cancer cells bind to E-selectins even under flow conditions.[46] The role of E-selectin in metastasis has been confirmed in studies on transgenic mice where the tissue-specific expression of E-selectin has changed. As a consequence, metastases of B16F10 melanoma cells transfected with cDNA encoding α3-Fuc-transferase, the enzyme synthesizing the E-selectin ligand, have been rerouted to different tissues.[47] For example, in mice expressing E-selectins in all tissues or in the liver, metastases were formed in the liver instead of the lung, which is the site of metastasis in non-transgenic mice.

These findings have led to proposed cancer treatments as will be discussed more fully in chapter 27. The most prominent of these are the production of synthetic sialoglycoconjugates as selectin-binding agents,[48] the use of plant lectins for carbohydrate binding,[49] and the employment of inhibitors such as swainsonine to interfere with the processing of N-glycans to complex forms.[50]

When liver receptors are deficient, for example in liver disease, the incidence of liver metastases is reduced, presumably by the decreased ability of metastatic cells to invade through binding to liver receptors.[51,52] Inhibition of cell surface sialylation and galactosylation may also prove to be useful in the prevention of metastasis. Further studies are required to clearly define the functions of individual structures and the control of their biosynthesis in metastatic cells during various stages of the metastatic process.

References

1. Maemura M, Dickson RB. Are cellular adhesion molecules involved in the metastasis of breast cancer? Breast Cancer Research & Treatment 1994; 32:239-260.
2. Nicolson GL. Cell surface molecules and tumor metastasis. Exp Cell Res 1984; 150:3-22.
3. Jiang M, Passaniti A, Penno MB et al. Adrenal carcinoma tumor progression and penultimate cell surface oligosaccharides. Cancer Res 1992; 52:2222-2227.
4. Müthing J, Peter-Katalinic J, Hanisch F-G et al. The ganglioside GD1 alpha' $IV^3Neu5Ac$, $III^6Neu5Ac$-$GgOse_4Cer$, is a major disialoganglioside in the highly metastatic murine lymphoreticular tumor cell line MDAY-D2. Glycoconj J 1994; 11:153-162.
5. Dennis JW, Carver JP, Schachter H. Asparagine-linked oligosaccharides in murine tumor cells:comparison of a WGA-resistant non-metastatic mutant and a related WGA-sensitive metastatic cell line. J Cell Biol 1984; 99:1034-1044.
6. Finne J, Tao T, Burger MM. Carbohydrate changes in glycoproteins of a poorly metastatic wheat germ agglutinin-resistant melanoma clone. Cancer Res 1980; 40:2580-2587.
7. Yogeeswaran G, Salk PL. Metastatic potential is positively correlated with cell surface sialylation of cultured murine tumor cell lines. Science 1981; 212:1514-1516.
8. Passaniti A, Hart GW. Cell surface sialylation and tumor metastasis. J Biol Chem 1988; 263:7591-7603.
9. Bresalier RS, Rockwell RW, Dahiya R et al. Cell surface sialoprotein alterations in metastatic murine colon cancer cell lines selected in an animal model for colon cancer metastasis. Cancer Res 1990; 50:1299-1307.
10. Nabi IR, Raz A. Cell shape modulation alters glycosylation of a metastatic melanoma cell surface antigen. Int J Cancer 1987; 40:396-402.
11. Altevogt P, Fogel M, Cheingsong-Popov R et al. Different patterns of lectin binding and cell surface sialylation detected on related high- and low-metastatic tumor lines. Cancer Res 1983; 43:5138-5144.
12. Dennis JW. Different metastatic phenotypes in two genetic classes of wheat germ agglutinin-resistant tumor cell mutants. Cancer Res 1986; 46:4594-4600.
13. Aldaheff JA, Holzinger RT. Sialyltransferase, sialic acid and sialoglycoconjugates in metastatic tumors and human liver tissue. Int J Biochem 1982; 14:119-126.
14. Dennis JW, Waller C, Timpl R et al. Sialic acid on metastatic tumor cells reduces cell attachment to fibronectin and collagen type

IV. Nature 1982; 300:274-276.
15. Zhu BCR, Laine RA. Polylactosamine glycosylation on human fetal placental fibronectin weakens the binding affinity of fibronectin to gelatin. J Biol Chem 1985; 2060:4041-4045.
16. Saitoh O, Wang WC, Lotan R et al. Differential glycosylation and cell surface expression of lysosomal membrane glycoproteins in sublines of a human colon cancer exhibiting distinct metastatic potentials. J Biol Chem 1992; 267:5700-5711.
17. Yamashita K, Totani K, Kuroki M et al. Structural studies of the carbohydrate moieties of carcino-embryonic antigens. Cancer Res 1987; 47:3451-3459.
18. Chandrasekaran EV, Davila M, Nixon DW et al. Isolation and structures of the oligosaccharide units of carcinoembryonic antigen. J Biol Chem 1983; 258:7213-7222.
19. Hoff S, Matsushita Y, Qta D et al. Increased expression of sialyl-dimeric Le^x antigen in liver metastases of human colorectal carcinoma. Cancer Res 1989; 49:6883-6888.
20. Easton EW, Blokland I, Geldof AA et al. The metastatic potential of rat prostate tumor variant R3327-MatLyLu is correlated with an increased activity of N-acetylglucosaminyl transferase III and V. FEBS Lett 1992; 308:46-49.
21. Humphries MJ, Matsumoto K, White SL et al. Inhibition of experimental metastasis by castanospermine in mice:blockage of two distinct stages of tumor colonization by oligosaccharide processing inhibitors. Cancer Res 1986; 46:5215-5222.
22. Trimura T, Gonzalez R, Nicolson GL. Effects of tumicamysin on B16 metastatic melanoma cell surface glycoproteins and blood-borne arrest and survival properties. Cancer Res 1981; 41:3411-3416.
23. Dennis JW. Effects of swainsonine and polyinosinic:polycytidylic acid on murine tumor cell growth and metastasis. Cancer Res 1986; 46:5131-5136.
24. Pulverer G, Beuth J, Ko HL et al. Glycoprotein modifications of sarcoma L-1 tumor cells by tunicamycin, swainsonine, bromocondruritol or 1-desoxynojirimycin treatment inhibits their metastatic lung colonization in Balb/c-mice. J Cancer Res Clin Oncol 1988; 114:217-220.
25. Dennis JW, Laferté S. Tumor cell surface carbohydrate and the metastatic phenotype. Cancer Metastasis Rev 1987; 5:185-204.
26. Dennis JW, Laferté S, Waghorne C et al. β1-6 branching of Asn-linked oligosaccharides is directly associated with metastasis. Science 1987; 236:582-585.
27. Yoshimura M, Ihara Y, Ohnishi A et al. Bisecting N-actylglucosamine on K562 cells suppresses natural killer cytotoxicity and promotes spleen colonization. Cancer Res 1996; 56:412-418.
28. Yoshimura M, Nishikawa A, Ihara Y et al. Suppression of lung metastasis of B16 mouse melanoma by N-acetylglucosaminyl transferase III gene transfection. Proc Natl Acad Sci. USA 1995; 92:8754-8758.
29. Yamamura K, Takasaki S, Ichihashi M et al. Increase of sialylated tetra-antennary sugar chains in parallel to the higher lung-colonizing abilities of mouse melanoma clones. J Invest Dermatol 1991; 97:735-741.
30. Le Marer N, Stéhelin D. High alpha-2, 6-sialylation of N-acetyllactosamine sequences in ras-transformed rat fibroblasts correlates with high invasive potential. Glycobiology 1995; 5:219-226.
31. Bresalier RS, Niv Y, Byrd JC et al. Mucin production by human colonic carcinoma cells correlates with their metastatic potential in animal models of colon cancer metastasis. J Clin Invest 1991; 87:1037-1045.
32. Matsushita Y, Nakamori S, Seftor EA et al. Human colon carcinoma cells with increased invasive capacity obtained by selection for sialyl-dimeric Le^x antigen. Exp Cell Res 1991; 196:20-25.
33. Irimura T, Nakamori S, Matsushita Y et al. Colorectal cancer metastasis determined by carbohydrate-mediated cell adhesion:role of sialyl-Le^x antigens. Cancer Biol 1993; 4:319-324.
34. Springer G. T and Tn, general carcinoma antigens. Science 1984; 224:1198-1206.
35. Ogawa H, Inoue M, Tanizawz O et al. Altered expression of sialyl-Tn, Lewis antigens and carcinoembryonic antigens between primary and metastatic lesions of uterine cervical cancers. Histochemistry 1992;

97:311-317.
36. Brockhausen I, Yang J-M, Burchell J et al. Mechanisms underlying aberrant glycosylation of MUC1 mucin in breast cancer cells. Eur J Biochem 1995; 233:607-617.
37. Kojima N, Handa K, Newman W et al. Inhibition of selectin-dependent tumor cell adhesion to endothelial cells and platelets by blocking O-glycosylation of these cells. Biochem Biophys Res Comm 1992; 182:1288-1295.
38. Majuri M-L, Mattila P, Renkonen R. Recombinant E-selectin-protein mediates tumor cell adhesion via sialyl-Le a and sialyl-Le x. Biochem Biophys Res Comm 1992; 182:1376-1382.
39. Kitagawa H, Nakada H, Fukui S et al. Characterization of mucin-type oligosaccharides with the sialyl-Le a structure from human colorectal adenocarcinoma cells. Biochem Biophys Res Comm 1991; 178:1429-1436.
40. Bresalier RS, Byrd JC, Itzkowitz SH et al. Liver metastasis and adhesion to the sinusoidal endothelium by human colon cancer cells is related to mucin carbohydrate chain length. Proceed AACR 1995; abstract 421.
41. Ishii S, Steele G, Ford R et al. Normal colonic epithelium adheres to carcinoembryonic antigen and type IV collagen. Gastroenterology 1994; 106:1242-1250.
42. Berg EL, Magnani J, Warnock RA et al. Comparison of L-selectin ligand specificities: the L-selectin can bind the E-selectin ligands sialyl Le^x and sialyl Le^a. Biochem Biophys Res Comm 1992; 184:1048-1055.
43. Beuth J, Ko HL, Schirrmacher V. Inhibition of tumor cell colonization in two animal tumor models by lectin blocking with D-galactose or arabinogalactan. Clin Exp Metastasis 1988; 6:115-120.
44. Beuth J, Ko HL. Roszkowski W et al. Lectins:mediators of adhesion for bacteria in infectious diseases and for tumor cells in metastasis. Internat J Med Microbiol 1990; 274:350-358.
45. Yeatman T, Bland K, Copeland E et al. Tumor cell surface galactose correlates with the degree of colorectal liver metastasis. J Surg Res 1989; 46:567-571.
46. Goetz DJ, Brandley BK, Hammer DA. An E-selectin-IgG chimera supports sialylated moiety dependent adhesion of colon carcinoma cells under fluid flow. J Biomed Eng 1996; 24:87-98.
47. Biancone L, Araki M, Araki K et al. Redirection of tumor metastasis by expression of E-selectin in vivo. J Exp Med 1996; 183:581-587.
48. Hasegawa A, Fushimi K, Ishida H et al. Synthetic studies on sialoglycoconjugates 52:synthesis of sialyl Lewis x analogs containing azidoalkyl groups at the reducing end. J Carbo Chem 1993; 12:1203-1216.
49. Beuth J, Ko HL, Pulverer G et al. Importance of lectins for the prevention of bacterial infections and cancer metastases. Glycoconj J 1995; 12:1-6.
50. Goss PE, Baptiste J, Fernandes B et al. A phase I study of swainsonine in patients with advanced malignancies. Cancer Res 1994; 54:1450-1457.
51. Uhlenbruck G, Beuth J, Weidtman V. Liver lectins:mediators for metastases? Experientia 1983; 39:1314-1315.
52. Uetsuji S, Yamamura M, Yamamichi K et al. Absence of colorectal cancer metastasis to the cirrhotic liver. Am J Surgery 1992; 164:176-177.

CHAPTER 27

Diagnosis and Therapy

Recent developments in the structural analysis of carbohydrate chains have resulted in more sensitive techniques for defining carbohydrate structures in disease. This has enabled us to use the information for diagnosis or to monitor the progress of disease. There is an abundance of lectins and monoclonal anti-carbohydrate antibodies that are useful in rapid screening. In addition, the patterns of activities and mRNA expression of many biosynthetic enzymes may be measured in diseased tissues. Enzyme activities or the glycosylation patterns of glycoproteins in the serum have also been useful as diagnostic tools.

Carcinoembryonic antigen is a glycoprotein prevalent in the fetus, and is re-expressed in the adult in colon cancer and other malignancies.[1,2] In addition to tissue expression, CEA is present in the blood, which simplifies sample collection for patient followup. It has been of use in monitoring patients with cancers in remission for signs of recurrent disease, and increases may be detected before any clinical signs are evident; however, false negatives and positives do occur, and have fostered searches for additional serum glycoprotein markers which, combined with CEA, have improved diagnostic accuracy of colon cancer and gynecologic malignancy.[3-5] Changes in serum glycoprotein levels and sugar content may be indicative of tumor burden. Lymphocyte reactivity towards PHA correlates with levels of protein-bound sugars and acute phase proteins, e.g. haptoglobin and α_1-antitrypsin, but not always with CEA levels.[6-8]

The deletion of tissue blood group epitopes, and in some cases their aberrant expression, were early findings which delineate these changes as tumor-specific.[9-12] Later studies made use of lectins and monoclonal antibodies to determine that cryptic structures such as T and Tn antigens are carcinoma-related antigens,[13] and are of importance as potential immunogens in cancer therapy.[14-16] In breast cancer patients, the presence of sialyl-Tn antigen is predictive of a response to diminished adjuvant chemotherapy. Sialyl-T and sialyl-Tn as well as Lewis antigens are particularly prevalent in inflammation and metastases,[17-19] as are sulfated epitopes.[20] Their importance as ligands for selectin-mediated cell adhesion has spawned a new industry which is focused on the creation of oligosaccharide mimics or analogs for use in diagnosis and treatment.[21-24]

A systematic approach to treatment requires the identification of carbohydrate structures which are directly related to a biochemical anomaly characteristic of the disease. In addition, the location at which therapeutic interactions take place has to be defined, as well as the form of therapeutic agent delivery, so that only those events relevant to the disease will occur. Current treatment strategies are described below.

27.A. GLYCOCONJUGATES AS CARRIERS

Site-specific drug or gene delivery to diseased cells may be accomplished utilizing oligosaccharides which are recognized by membrane lectins. Examples include liver parenchyma asialo-receptors which recognize Gal residues, Kupffer and liver endothelial cells which recognize 4-sulfated GlcNAc residues, and a variety of cells which express Man-6-phosphate receptors. The drug carriers may be glycoproteins such as asialofetuin or lactosaminated albumin, or glycosylated polymers (Table 7). Drug-carrier linkages may be made via disulfide bridges or directly by amidation or esterification. These glycosylated drug carriers may differ in their immunogenicity, biodegradability and other properties. It is presumed that the ligand-receptor complex will be actively endocytosed and the drug made available at cellular target sites.[25]

Enzyme replacement therapy, for example, to replace glucocerebrosidase in Gaucher's disease, has been achieved by targeting macrophage receptors for mannose.[26] The glycosylation of the enzyme was engineered to form Man-terminating structures which are recognized by the macrophage lectins.

For gene targeting, partially glycosylated polylysine complexes can be made for selective transfer into cells expressing a lectin able to recognize the specific sugar moieties.[25] Plasmids complexed to sugars which bind to a specific membrane lectin may be superior to other agents such as calcium phosphate in producing long-term transfections.

Targeting to special locations such as the intestinal tract requires bioadhesive drug delivery systems to cope with problems such as movement of luminal contents, penetration of overlying mucus and specific attachment to mucosal cells. Specific adhesives include lectins, antibodies, purified fimbrial proteins and methacrylamide copolymers.[27]

27.B. INTERFERENCE WITH INFLAMMATORY OR OTHER CELL ADHESION PROCESSES

It has been shown that sialyl-Le^x oligosaccharides, which are ligands for endothelial P-selectin and neutrophil-mediated E-selectin, protect against acute lung injury induced by toxic proteins[28] or immune complexes in the rat.[29] Similarly beneficial effects of sialyl-Le^x have been demonstrated in cats protected from myocardial reperfusion injury, a condition which accompanies reperfusion of ischemic coronary vasculature. Thus, ligand analogs function as competitive ligands for selectin adhesion molecules that mediate neutrophil-endothelial interactions.[30] For example, cardioprotection was afforded by the intravenous injection of liposome-conjugated sialyl-Le^x prior to myocardial ischemia and reperfusion injury in a feline model.[31]

Certain forms of immune complex disease may respond to specific treatments with synthetic oligosaccharides. ABO hemolytic disease of the neonate caused by maternal-fetal ABO incompatibility respond favorably to treatment with A or B blood group trisaccharides which dissociates anti-A or anti-B antibodies from neonatal erythrocytes.[32] Xenograft rejection of pig tissues carrying Galα1-3 residues has been studied in the baboon model. α-Gal-containing carbohydrates such as blood group B disaccharide given intravenously reduces baboon serum cytotoxity to pig kidney cells and transplanted pig hearts.[33]

Derivatives of sialyl-Le^x and other sialylated Lewis antigens (Le^a, Le^y) as well as sialyl-T and sialyl-Tn antigens are all considered possible candidates as tumor metastasis inhibitors.[34] Chimeric molecules

Table 7.

Glycosylated macromolecular drug carriers

Starting material	*Serum albumin*	*Poly-L-lysine*
Available NH_2 groups	≃ 57	≃ 190
Substituted material	*Neoglycoprotein*	*Glycosylated polymer*
Structure	globular	α-helix
Form	spherical	cylindrical
Size (nm)	d = 6	d = 3; l = 30
M_r (Da)	≃ 68 000	≃ 80 000
Yield of linking (%)	≃ 50	≃ 95
Solubility (mg/ml)	≃ 100	> 200
Synthesis solvent	aqueous	organic
Sugar derivative	osidyl PTC	osidylphenylacetate
Sugars per molecule	≃ 25	≃ 50
Drugs per molecule	5–20	10–80
pI	≃ 4	neutral[a]
Biological efficiency		
Carried/free drug	up to 100	up to 100
Biodegradability	partial	total
Unspecific cell binding	±	−
Immunogenicity	+ + or ±[b]	−
Antigen presentation	+	−

[a]All remaining NH_2 groups are acylated.

[b]Heterologous serum albumin is highly immunogenic, i.e. glycosylated bovine serum albumin in mice, whereas homologous serum albumin is not immunogenic, i.e. glycosylated mouse serum albumin in mice.

d = diameter; l = length; PTC = phenylthiocarbamyl.

which combine a toxic protein with appropriate fragments of anti-Le^x are effective antitumor agents when tested in experimental systems.[35]

The metastases of murine tumor cells into the liver has been blocked by Gal and galactans.[36] Similarly, glycoprotein adhesion to liver cells containing asialoglycoprotein binding lectins was inhibited by Gal-terminating glycoproteins and thereby reduced metastasis in mice. Continuous infusion of Gal is now under trial in cancer patients in attempts to confine tumors and to avert liver metastases. Metastasis of murine sarcoma cells in lungs expressing Fuc-binding lectins could be blocked specifically by Fuc and fucoidans.[37]

27.C. IMMUNE CELL MODIFICATION OR ACTIVATION

Immune effector cells may be modified and activated by glycoconjugates; they may also be responsive to tumor oligosaccharide antigens. Thus macrophages and lymphocytes can be modified by glycoproteins to improve their efficacy as anti-tumor and anti-inflammatory agents. Immune stimulants may be neoglycoproteins which can activate macrophages to produce tumoricidal effects. The α-mannosidase inhibtor swainsonine[38] can improve immune functions mediated by lymphocyte-activated killer cells and natural killer cells as well as macrophages.

A safe human erythrocytic asialoglycophorin, given in repeated doses, has been utilized to produce anti-T humoral and cellular response with beneficial effect in women with advanced breast cancer.[39-41] Monoclonal antibodies prepared against the sialyl-Tn epitope have been capable of delineating tumor from non-tumor tissue and appear to be effective in diagnosis and treatment.[42] The immunization of several colorectal cancer patients with partially desialylated ovine submaxillary mucin containing sialyl-Tn and Tn antigens led to the production of anti sialyl-Tn antibodies without an adverse inflammatory response.[43]

Anti-idiotypic antibodies which carry on their antigen binding sites the surrogate image of a saccharide have promise as vaccines against carbohydrate antigens, especially in infants or immunologically compromised persons.[44] Antibody responses may also be mounted by the use of synthetic glycoconjugates or neoglycoconjugates as vaccines. The technique is based upon the conjugation or covalent coupling of polysaccharides to protein carriers[45] and could be applied to various bacterial polysaccharides or tumor-specific epitopes such as the sialyl-Lewis oligosaccharides. Mucin antigens have also been exploited for targeting. Interferon α conjugated to an antibody recognizing breast epithelial mucin was effective in improving the growth inhibitory action of interferon α on human breast cancer cells MCF-7 in nude mice.[46] This enhancement was attributed to the specific targeting mediated by the antibodies.

The synthesis of α–galactosyl epitopes on membranes of autologous tumor vaccines has a potential use in cancer treatment. Recent studies have shown that recombinant transferase activity could α3-Gal-convert cells pretreated with sialidase to expose N-acetyllactosamine acceptors to Galα3 Gal-expressing cells. Similarly treated cancer cells as vaccines would be opsonized in vivo by naturally occurring anti-Gal antibody and presented to immunocytes as an adjuvant which would hopefully mount effective anti-tumor responses in immunologically competent patients.[47,48]

27.D. INHIBITION OF GLYCOPROTEIN SYNTHESIS

Swainsonine is an α-mannosidase inhibitor which interferes with oligosaccharide processing in the Golgi and thus can inhibit complex N-glycan synthesis and the formation of tumor-type oligosaccharide chains. Minimal toxicity is observed when administered intravenously to cancer patients in doses that inhibit Golgi mannosidase II and lysosomal α-mannosidases, and some patients experience shrinkage of tumors.[38] Patients with liver dysfunction are assumed to be at a higher risk for the toxic effects of the drug. The possibility of hepatotoxicity and accumulation of oligomannosides may limit its use, although synthetic analogs of swainsonine may prove to be better in this respect.

Tunicamycin inhibits the first step of N-glycan synthesis; when injected into syngeneic mice it can inhibit pulmonary colonization of murine melanoma cells (Table 4). N-glycan processing inhibitors may also be of interest therapeutically because of their ability to reduce complex N-glycan formation.[49] The imino sugar N-butyldeoxynojirimycin inhibits the processing enzyme α–glucosidase I and this interferes with the N-glycan trimming process that precedes complex glycan synthesis. N-butyldeoxynojirimycin has been shown to suppress the secretion and increased intracellular retention of hepatitis B virus in HepG2 cells whose secretory function depends upon fully processed N-linked glycans.[50] A slightly different compound, N-butyl deoxygalactonojirimycin, inhibits glycolipid biosynthesis without affecting N-glycan processing, and could therefore be a potential selective therapeutic agent against glycolipid storage diseases such as Gaucher's disease.[51]

Castanospermine is another glucosidase I processing inhibitor which has shown promise in animal heart allograft recipients since it can downregulate the expression of cell adhesion glycoproteins such as leukocyte function-associated antigen-1 and intercellular adhesion molecule-1.[52] An analog of castanospermine, 6-O-butanoyl castanospermine, is effective in reducing virus load in the brain in a murine model of HSV type-1 infection, however, drug resistance develops over prolonged periods of usage.[53]

Inhibition of O-linked glycans can be demonstrated by aryl-N-acetyl-α-galactosaminides in human colon cancer cells; during exposure to this O-glycan extension inhibitor cells develop pleomorphic vacuoles and an increase in lysosomes. This compound non-specifically inhibits mucin glycosylation, and may be of interest in animal studies or for the treatment of mucin hypersecretory diseases or to expose the antigenic Tn antigen.

Sulfated sugars can inhibit the replication of HIV in vitro[54] but clinical trials in the therapy of AIDS have been disappointing because of side effects. Sulfated GAGs and sulfated polyanions appear to act by inhibition of HIV reverse transcriptase in H9 cells[55] and by interfering with virus-host cell adhesion.[56] Sulfated compounds have also been shown to attenuate the neurotoxic effects of β-amyloid by interacting with the β-sheet structure of amyloid.[57]

27.E. ENZYMATIC MODIFICATION OF TARGET CELLS

Sialidase treatment has been attempted to render tumors more antigenic and less tumorigenic.[58] Autologous bone marrow transplantation for acute myeloid leukemia has been carried out following in vitro treatment of cells with neuraminidase and monoclonal antibodies. The purpose was to unmask underlying sugars which would then be more accessible to elimination by anti-blast cell monoclonal antibodies plus complement, allowing for separations from non-AML cells.[59] The patient receiving treated bone marrow suffered no major complications and was independent of transfusion for three weeks.

Galactosidase removal of the xenogeneic Galα1-3Gal epitope prior to xenotransplantation has been tested comparatively against several tissues; removal was complete in rabbit erythrocytes, partial in pig endothelium, near complete in pig lymphocytes and absent in rabbit islets. The results have been assessed by lectin binding assay.[60]

27.F. GENETIC APPROACHES—INTERVENTION OF GLYCOSYLATION

Gene therapy approaches hold promise for the correction of carbohydrate-related gene defects, for example, for the engineering of oligosaccharides with tumor-modulating properties. Transgenic or gene knockout approaches may also be useful in the production of properly glycosylated recombinant glycoproteins or for the prevention of adverse reponses.

The pathogenesis of hyperacute rejection in xenotransplantation was studied by producing resistance to cell lysis in transgenic mice. Pigs have been inbred in order to express the human terminal complement inhibitor hCD59.[61] Both large vessel and capillary endothelium were found to express hCD59, and porcine cells were highly resistant to challenge with high titer anti-porcine antibody and human complement. A similar strategy may be useful to produce transgenic animals lacking Galα1-3Gal epitope that would not be reactive with naturally occurring antibodies. Knockout mice for this epitope have been developed (see chapter 18.D).

In breast cancer cell lines, the utilization of promoter DF3 in an adenoviral vector has been used to confer selective expression of heterologous genes such as β-galactosidase; this strategy may be of importance as an adjuvant in testing candidate therapeutic genes.[62] In human immortalized CF epithelial tracheal cells, adenovirus-mediated gene transfer of CFTR to CF and control cells, restored cAMP-secretion of chloride, as well as secretion of glycoconjugates.[63]

27.G. USE OF ANTIBODIES OR LECTINS

Certain cell surface carbohydrates of cancer cells do not occur elsewhere in the body in high concentrations and can therefore be exploited as immunogens. Different approaches have been taken in immunotherapy of breast cancer, exploiting either the characteristic exposure of peptide or carbohydrate epitopes on cancer cells. The appearance of anti-mucin antibodies and cytotoxic T cells in cancer patients[64] suggests that the increased mucin release from cancer cells led to an immune response. Anti-mucin peptide antibodies were capable of killing cancer cells in culture (see chapter 24).

Anti-mucin antibodies have been used in the diagnosis, prognosis and therapy for breast cancer, but the epitopes recognized by these antibodies are often not well defined. Most antibodies raised against cancer cells recognize a variety of mucin-type epitopes on cancer cell surfaces.[65] For diagnosis, cancer cells exhibiting these epitopes may be specifically detected with labeled antibodies.[65,66] In a therapeutic approach, cytotoxic drugs linked to antibodies could be targeted to cancer cells. The half life of therapeutic antibodies may be extended by administering humanized, chimeric antibodies.[67]

Clinical trials using mucin antigens have only been moderately successful and therefore model systems are being explored. In mice, preimmunization with MUC1 transfected cells was protective against tumor growth. Peptide epitopes of MUC1 are increasingly exposed in breast cancer; immunotherapy using a peptide epitope of MUC1 has been attempted to stimulate the production of antibodies directed against cancer cell surfaces.[68,69] The PDTR sequence in the tandem repeat region near a preferred O-glycosylation site with some adjacent amino acids appears to be a prominent epitope.[69-71]

Laurence et al[72] found that an anti-mucin antibody does not differentiate breast cancer cells in stages I, II and III. However, patients with metastases show significant levels of antigen, which fluctuate with the stage of the clinical status and tumor burden. This mucin antigen could therefore be useful as a marker for metastatic disease.

Anti-T antibody-based enzyme-linked immunosorbent assay has high efficacy in the detection of breast carcinoma since most breast cancer patients exhibit low levels of anti-T IgM.[65] A positive clinical response using T and Tn antigen immunotherapy has been reported by Springer et al.[16] Red cell extracts containing T and Tn epitopes elicit a strong humoral and cell mediated response in breast cancer patients although in a high proportion of patients, natural anti-T/Tn immune response is present before diagnosis.

A third of all breast cancers are sialyl-Tn positive associated with low patient survival and resistance to adjuvant chemotherapy. Synthetic sialyl-Tn glycoconjugates were therefore used as immunogens in patients with breast cancer. Keyhole limpet hemocyanin linked to sialyl-Tn with Detox adjuvant shows minimal toxicity.[73] Sialyl-Tn conjugated to human serum albumin elicits anti-sialyl-Tn IgM and IgG. All patients have developed antibodies against sialyl-Tn, but it is difficult to evaluate the clinical efficacy. A clustered sialyl-Tn antigen causes blastogenesis in a breast cancer patient's lymphocytes in culture in addition to significant γ-interferon production. Ovine submaxillary mucin carrying multiple sialyl-Tn epitopes used for immunization in mice stimulates an immune response and inhibits tumor growth.[74]

During the metastatic process, endogenous lectins present on cancer cells or on the endothelium contribute to adhesion and spreading of tumor cells. Lectins such as toxic mistletoe have been used to kill cancer cells in culture and have been used for cancer treatment.[75] At subtoxic doses lectins can also be used as vehicles to carry cytotoxic drugs that selectively kill cells recognized by the lectins.

Many of the bacterial and viral infections are mediated by binding of microbial proteins to cell surface glycoconjugate determinants which could be abrogated by blocking cell surface recognition sites with lectins. Therefore, lectins may be useful in the prevention of microbial infections or in the metastatic process.[36]

Bacterial carbohydrates may be recognized by cell surface lectins with activation of the immune system. The recombinant carbohydrate recognition domain of bovine conglutinin, a vertebrate C-type lectin, can enhance the clearance of bacteria or immune complexes from the circulation by promoting contacts between complement-coated bacteria and effector cells. Moreover, it recognizes high mannose oligosaccharides and by binding to influenza virus can reduce viral infectivity.[76]

Newer approaches to therapy have been based on refinements in enzyme purification and technologies to probe the binding sites of these proteins. Studies of interactions between functional groups and a biologically active molecule have been aided by the computer program GRID.[77] This program and the crystal structure of influenza virus sialidase enabled investigators to identify key regions of the active site necessary for the binding of natural substrate, including sialic acid. Binding appears to be enhanced by replacement of the carboxylate with a phosphonate group. These and similar studies may aid in the formulation of rational drug design techniques.[78]

References

1. Gold P, Freedman S. Specific carcinoembryonic antigens of the digestive system. J Exp Med 1965; 122:467-480.
2. Tamai M, Tanimura H, Yamaue H et al. Clinical significance of quantitative analysis of carcinoembryonic antigen assessed by flow cytometry in fresh human gastric cancer cells. Cancer Lett 1995; 90:111-117.
3. Schwartz P, Chambers S, Chambers J et al. Circulating tumor markers in the monitoring of gynecologic malignancies. Cancer 1987; 60:353-361.
4. Guadagni F, Roselli M, Cosimelli M et al. TAG-72 (CA 72-4 assay) as a complementary serum tumor antigen to carcinoembryonic antigen in monitoring patients with colorectal cancer. Cancer 1993; 72:2098-2106.
5. Huber K, Kirchheimer J, Sedlmayer A et al. Clinical value of determination of urokinase-type plasminogen activator antigen in plasma for detection of colorectal cancer: comparison with circulating tumor-associated antigens CA 19-9 and carcinoembryonic antigen. Cancer Res 1993; 53:1788-1793.
6. Bradley WP, Blasco AP, Weiss JF et al. Correlations among serum protein-bound carbohydrates, serum glycoproteins, lymphocyte reactivity, and tumor burden in cancer patients. Cancer 1977; 40:2264-2272.
7. Turner G, Skillen A, Buamah P et al. Relation between raised concentrations of fucose, sialic acid, and acute phase proteins in serum from patients with cancer, choosing suitable serum glycoprotein markers. J Clin Pathol 1985; 38:588-592.
8. Turner GA, Goodarzi MT, Thompson S. Glycosylation of alpha-1-proteinase inhibitor and haptoglobin in ovarian cancer: evidence for two different mechanisms. Glycoconj J 1995; 12:211-218.
9. Davidsohn I, Kovarik Ș, Ni LY. Isoantigens A, B, and H in benign and malignant lesions of the cervix. Arch Pathol 1969; 87:306-314.
10. Davidsohn I. Early immunologic diagnosis and prognosis of carcinoma. Am J Clin Path 1972; 57:715.
11. Levine P. Blood group and tissue genetic markers in familial adenocarcinoma: potential specific immunotherapy. Semin Oncol 1978; 5:25.
12. Levine P. Self-nonself concept for cancer and diseases previously known as "autoimmune" diseases. Proc Natl Acad Sci USA 1978; 75:5697-5701.
13. Springer G. T and Tn, general carcinoma antigens. Science 1984; 224:1198-1206.
14. Springer G. T and Tn pancarcinoma markers: autoantigenic adhesion molecules in pathogenesis, prebiopsy carcinoma-detection, and long-term breast carcinoma immunotherapy. Crit Rev Oncogenesis 1995; 6:57-85.
15. Springer G, Desai P, Wise W et al. Pancarcinoma T and Tn epitopes: autoimmunogens and diagnostic markers that reveal incipient carcimomas and help establish prognosis. In: Herberman R, Mercer D eds. Immunodiagnosis of Cancer 2nd ed. New York:Marcel Dekker, 1990; 587-612.
16. Springer G, Desai P, Tegtmeyer H et al. Pancarcinoma T/Tn antigen detects human carcinoma long before biopsy does and its vaccine prevents breast carcinoma recurrence. Ann N Y Acad Sci 1993; 690:355-357.
17. Miles D, Happerfield L, Smith P et al. Ex-

pression of sialyl-Tn predicts the effect of adjuvant chemotherapy in node positive breast cancer. Br J Cancer 1994; 70:1272-1275.
18. Ma XC, Terata N, Kodama N et al. Expression of sialyl-Tn antigen is correlated with survival time of patients with gastric carcinomas. Eur J Cancer 1993; 29A:1820-1823.
19. Hoff S, Matsushita Y, Qta D et al. Increased expression of sialyl-dimeric Le^x antigen in liver metastases of human colorectal carcinoma. Cancer Res 1989; 49:6883-6888.
20. Green P, Tamatani T, Watanabe T et al. High affinity binding of the leukocyte adhesion molecule L-selectin to 3' sulfated Le a and Le x oligosaccharides and the predominance of sulfate in this interaction demonstrated by binding studies with a series of lipid-linked oligosaccharides. Biochem Biophys Res Comm 1992; 188:244-251.
21. Giannis A. The sialyl-$Lewis^x$ group and its analogues as ligands for selectins: chemoenzymatic syntheses and biological functions. Angew Chem Int Ed 1994; 33:178-180.
22. Hotta K, Itoh K, Kameyama A et al. Synthetic studies on sialoglycoconjugates 65: stereocontrolled synthesis of positional isomers of tumor-associated ganglioside antigens, sialyl-$Lewis^x$ and sialyl paragloboside. J Carbohydr Chem 1995; 14:115-133.
23. Halcomb R, Wong C. Synthesis of oligosaccharides, glycopeptides and glycolipids. Curr Opin Struct Biol 1993; 3:694-700.
24. Rao BNN, Anderson MB, Musser JH et al. Sialyl Lewis X mimics derived from a pharmacophore search are selectin inhibitors with anti-inflammatory activity. J Biol Chem 1994; 269:19663-19666.
25. Monsigny M, Roche A, Midoux P et al. Glycoconjugates as carriers for specific delivery of therapeutic drugs and genes. Adv Drug Deliv Rev 1994; 14:1-25.
26. Barton N, Furbish F, Murray G et al. Therapeutic response to intravenous infusions of glucocerebrosidase in a patient with Gaucher disease. Proc Natl Acad Sci USA 1990; 87:913-916.
27. Helliwell M. The use of bioadhesives in targeted delivery within the gastrointestinal tract. Adv Drug Deliv Rev 1993; 11:221-253.
28. Mulligan M, Paulson J, DeFrees S et al. Protective effects of oligosaccharides in P-selectin-dependent lung injury. Nature 1993; 364:149-151.
29. Mulligan MS, Lowe JB, Larsen RD et al. Protected effects of sialylated oligosaccharides in immune complex-induced acute lung injury. J Exp Med 1993; 178:623-631.
30. Buerke M, Weyrich AS, Zheng Z et al. Sialyl-$Lewis^x$ -containing oligosaccharide attenuates myocardial reperfusion injury in cats. J Clin Invest 1994; 93:1140-1148.
31. Murohara T, Margiotta J, Phillips LM et al. Cardioprotection by liposome-conjugated sialyl $Lewis^x$-oligosaccharide in myocardial ischaemia and reperfusion injury. Cardiovasc Res 1995; 30:965-974.
32. Romano E, Soyano A, Montaño R et al. Treatment of ABO hemolytic disease with synthetic blood group trisaccharides. Vox Sang 1994; 66:194-199.
33. Ye Y, Neethling FA, Niekrasz M et al. Intravenous administration of α-galactosyl carbohydrates reduces in vivo baboon serum cytotoxicity to pig kidney cells and transplanted pig hearts. Transplant Proc 1994; 26:1399.
34. Hakomori S. New directions in cancer therapy based on aberrant expression of glycosphingolipids:anti-adhesion and ortho-signalling therapy. Cancer Cells 1991; 3:461-470.
35. Kuan C-T, Pastan I. Improved antitumor activity of a recombinant anti-$Lewis^y$ immunotoxin not requiring proteolytic activation. Proc Natl Acad Sci USA 1996; 93:974-978.
36. Beuth J, Ko HL, Pulverer G et al. Importance of lectins for the prevention of bacterial infections and cancer metastases. Glycoconj J 1995; 12:1-6.
37. Roszkowski W, Beuth J, Ko H et al. Blocking of lectin-like adhesion molecules on pulmonary cells inhibits lung sarcoma L-1 colonization in BALB/c mice. Experientia 1989; 45:584-588.
38. Goss PE, Baptiste J, Fernandes B et al. A phase I study of swainsonine in patients with advanced malignancies. Cancer Res 1994; 54:1450-1457.

39. Springer GF, Desai PR, Spencer BD et al. T/Tn antigen vaccine is effective and safe in preventing recurrence of advanced breast carcinoma. Cancer Detection and Prevention 1995; 19:374-380.
40. Springer GF, Chandrasekaran EV, Desai PR et al. Blood group Tn-active macromolecules from human carcinomas and erythrocytes: characterization of and specific reactivity with mono- and poly-clonal anti-Tn antibodies induced by various immunogens. Carbohydr Res 1988; 178:271-292.
41. Duk M, Steuden I, Dus D et al. Application of chemically desialylated and degalactosylated human glycophorin for induction and characterization of anti-Tn monoclonal antibodies. Glycoconj J 1992; 9:148-153.
42. Kjeldsen T, Clausen H, Hirohashi S et al. Preparation and characterization of monoclonal antibodies directed to the tumor-associated O-linked sialosyl-2-6 α-N-acetylgalactosaminyl (sialosyl-Tn) epitope. Cancer Res 1988; 48:2214-2220.
43. O'Boyle KP, Zamore R, Adluri S et al. Immunization of colorectal cancer patients with modified ovine submaxillary gland mucin and adjuvants induces IgM and IgG antibodies to sialylated Tn. Cancer Res 1992; 52:5663-5667.
44. Westerink M, Apicella M. Anti-idiotypic antibodies as vaccines against carbohydrate antigens. Springer Sem Immunopath 1993; 15:227-234.
45. Jennings H, Sood R. Synthetic glycoconjugates as human vaccines. In:Lee Y, Lee R, eds. Neoglycoconjugates: Preparation and Applications. San Diego CA: Academic Press, 1994:325-371.
46. Ozzello L, De Rosa CM, Blank EW et al. The use of natural interferon alpha conjugated to a monoclonal antibody anti mammary epithelial mucin (Mc5) for the treatment of human breast cancer xenografts. Breast Cancer Res Tr 1993; 25:265-276.
47. Galili U, Anaraki A. α-galactosyl (Galα1-3Galβ1-4GlcNAc-R) epitopes on human cells: synthesis of the epitope on human red cells by recombinant primate α1, 3galactosyltransferase expressed in *E. coli*. Glycobiology 1995; 5:775-782.
48. LaTemple DC, Henion TR, Anaraki F et al. Synthesis of α-galactosyltransferase for opsonization of human tumor cell vaccines by anti-galactose. Cancer Res 1996; 56:3069-3074.
49. Olden K. Adhesion molecules and inhibitors of glycosylation in cancer. Cancer Biol 1993; 4:269-276.
50. Block TM, Lu X, Platt FM et al. Secretion of human hepatitis B virus is inhibited by the imino sugar N-butyldeoxynojirimycin. Proc Natl Acad Sci USA 1994; 91:2235-2239.
51. Platt F, Neises G, Karlsson G et al. N-butyldeoxygalactonojirimycin inhibits glycolipid synthesis but does not affect N-linked oligosaccharide processing. J Biol Chem 1994; 269:27108-27114.
52. Hibberd A, Grochowicz P, Smart Y et al. Castanospermine downregulates membrane expression of adhesion molecules in heart allograft recipients. Transplant Proc 1995; 27:448-449.
53. Bridges CG, Ahmed SP, Kang MS et al. The effect of oral treatment with 6-O-butanoyl castanospermine (MDL 28,574) in the murine zosteriform model of HSV-1 infection. Glycobiology 1995; 5:249-253.
54. Okutani K, Shigeta S. Inhibitory effect of sulfated derivatives of a marine bacterial polysaccharide on replication of human immunodeficiency virus in vitro. Nippon Suisan Gakkaishi 1993; 3:1433.
55. Chang R, Tabba H, He Y et al. Dextran sulfate as an inhibitor against the human immunodeficiency virus (42811). PSEBM 1988; 189:304-309.
56. Baba M, Pauwels R, Balzarini J et al. Mechanism of inhibitory effect of dextran sulfate and heparin on replication of human immunodeficiency virus in vitro. Proc Natl Acad Sci USA 1988; 85:6132-6136.
57. Sadler I, Hawtin S, Tailor V et al. Glycosaminoglycans and sulfated polyanions attenuate the neurotoxic effects of β-amyloid. Biochem Soc Trans 1995; 23:106S.
58. Schauer R. Chemistry, metabolism and biological functions of sialic acids. Adv Carbohydr Chem Biochem 1982; 40:131-234.
59. Ball ED, Vredenburgh JJ, Mills LE et al. Autologous bone marrow transplantation for

acute myeloid leukemia following in vitro treatment with neuraminidase and monoclonal antibodies. Bone Marrow Transplant 1990; 6:277-280.

60. Cairns T, Hammelmann W, Gray D et al. Enzymatic removal from various tissues of the Galactose α1-3 Galactose target antigens of human antispecies antibodies. Transplant Proc 1994; 26:1279-1280.
61. Fodor W, Williams B, Matis L et al. Expression of a functional human complement inhibitor in a transgenic pig as a model for the prevention of xenogeneic hyperacute organ rejection. Proc Natl Acad Sci USA 1994; 91:11153-11157.
62. Chen L, Chen D, Manome Y et al. Breast cancer selective gene expression and therapy mediated by recombinant adenoviruses containing the DF3/MUC1 promoter. J Clin Invest 1995; 96:2775-2782.
63. Mergey M, Lemnaouar M, Veissiere D et al. CFTR gene transfer corrects defective glycoconjugate secretion in human CF epithelial tracheal cells. Am J Physiol 1995; 269:L855-L864.
64. Taylor-Papadimitriou J, Stewart L, Burchell J et al. The polymorphic epithelial mucin as a target for immunotherapy. Ann NY Acad Sci 1993; 690:69-79.
65. Desai PR, Ujjainwala LH, Carlstedt SC et al. Anti-Thomsen-Friedenreich (T) antibody-based ELISA and its application to human breast carcinoma detection. J Immunol Methods 1995; 188:175-185.
66. McKenzie IFC and Xing PX. Mucins in breast cancer:recent immunological advances. Cancer Cells 1990; 2:75-78.
67. Co MS, Baker J, Janzek BE et al. Humanized anti-Lewis Y antibodies:In vitro properties and pharmacokinetics in rhesus monkeys. Cancer Res 1996; 56:1118-1125.
68. Gendler S, Taylor-Papadimitriou J, Duhig T et al. A highly immunogenic region of a human polymorphic epithelial mucin expressed by carcinomas is made up of tandem repeats. J Biol Chem 1988; 263:12820-12823.
69. Taylor-Papadimitriou J, D'Souza B, Burchell J et al.. The role of tumor-associated antigens in the biology and immunotherapy of breast cancer. Ann NY Acad Sci 1993; 698:31-47.
70. Burchell J, Taylor-Papadimitriou J, Boshell M et al. A short sequence, within the amino acid tandem repeat of a cancer-associated mucin, contains immunodominant epitopes. Int. J. Cancer 1989; 44:691-696.
71. Burchell J, Taylor-Papadimitriou J. Effect of modification of carbohydrate side chains on the reactivity of antibodies with core-protein epitopes of the MUC1 gene product. Epith Cell Biol 1993; 2:155-162.
72. Laurence V, Forbes MA, Cooper EH. Use of mucin like cancer associated antigen (MCA) in the management of breast cancer. Br J Cancer 1991; 63:1000-1004.
73. MacLean GD, Reddish M, Rao Koganty R et al. Immunization of breast cancer patients using a synthetic sialyl-Tn glycoconjugate plus Detox adjuvant. Cancer Imm Immunother 1993; 36:215-222.
74. Singhal A, Fohn M, Hakomori S. Induction of α-N-acetylgalactosamine-O-serine/threonine Tn) antigen-mediated cellular immune response for active immunotherapy in mice. Cancer Res 1991; 51:1406-1411.
75. Schumacher U, Stamouli A, Adam E et al. Biochemical, histochemical and cell biological investigations on the actions of mistletoe lectins I, II and III with human breast cancer cell lines. Glycoconj J 1995; 12:250-257.
76. Lim BL, Holmskov U. Expression of the carbohydrate recognition domain of bovine conglutinin and demonstration of its binding to iC3b and yeast mannan. Biochem Biophys Res Com 1996; 218:260-266.
77. Goodford P. A computational procedure for determining energetically favorable binding sites on biologically important macromolecules. J Med Chem 1985; 28:849-857.
78. von Itzstein M, Dyason JC, Oliver SW et al. A study of the active site of influenza virus sialidase:an approach to the rational design of novel anti-influenza drugs. J Med Chem 1996; 39:388-391.

CHAPTER 28

CONCLUSIONS

As a result of many different approaches to examine the glycobiology of disease, a number of predictable patterns of glycosylation changes in disease have emerged. These should be considered in context with the biological functions of oligosaccharides. Some of the notable examples are hypersialylation of cancer and leukemic cells and the expression of Le^x type antigens in cancer cells which can be related to their biological behavior.

Structural information may be biased since only a fraction of the total cellular carbohydrate has been isolated and analyzed. Similarly, glycosyltransferase activity measurements suffer from in vitro artifacts and the mixing of cell populations when tissues have been homogenized.

Cellular glycosylation is extremely sensitive to changing conditions of growth and differentiation and to the extracellular environment. It is therefore not surprising that changes in oligosaccharide structures occur during disease processes where there is abnormal growth or differentiation due to cancer, infection, inflammation or other conditions. Thus changes may be secondary to the disease-triggering factors, yet they are major contributors to cell surface properties and the phenotype of a cell. Structural differences between healthy and diseased tissues are usually quantitative rather than qualitative in nature. These carbohydrate changes may contribute to the disease process by virtue of presenting themselves at cell surfaces, or they may serve a protective function. The primary structures, conformations, flexibility of oligosaccharides and their interactions with other cell surface components are the basis of their functions that remain to be explored further.

Carbohydrate alterations in different conditions may be similar but can be triggered by different mechanisms. Alterations of glycosyltransferase activities are commonly observed in disease, but the underlying mechanisms have not been sufficiently studied and may be very complex. It is the aim of future research to examine in further detail the mechanisms leading to glycosylation changes as well as their functional implications.

The knowledge of the structures and biosynthesis of glycoproteins in disease can be exploited in designing new diagnosis and therapy. Research of new carbohydrate-based therapy should be aimed at the following two goals. We need to explore ways to reprogram altered glycoprotein synthesis in order to convert cells back to a non-pathological phenotype. This concept is based on the knowledge of mechanisms leading to the disease phenotype and the three-dimensional interactions between carbohydrates and carbohydrate-binding molecules. The second goal is to use abnormal glycosylation as a means for diagnosis and as a target to specifically select out diseased cells and treat these directly or elicit immune reponses against them.

INDEX

A

B

C

P

R

S

T

U

V

W

X

Z

SPRINGER NATURE

GPSR Compliance

The European Union's (EU) General Product Safety Regulation (GPSR) is a set of rules that requires consumer products to be safe and our obligations to ensure this.

If you have any concerns about our products, you can contact us on ProductSafety@springernature.com

In case Publisher is established outside the EU, the EU authorized representative is:

Springer Nature Customer Service Center GmbH
Europaplatz 3
69115 Heidelberg, Germany

FSC
www.fsc.org
MIX
Papier | Fördert
gute Waldnutzung
FSC® C083411

Kolibri 360 GmbH
Ferdinand-Jühlke-Straße 7
99095 Erfurt, Deutschland
produktsicherheit@kolibri360.de